Lecture Notes in Civil Engineering

Volume 26

Lecture Notes in Civil Engineering (LNCE) publishes the latest developments in Civil Engineering - quickly, informally and in top quality. Though original research reported in proceedings and post-proceedings represents the core of LNCE, edited volumes of exceptionally high quality and interest may also be considered for publication. Volumes published in LNCE embrace all aspects and subfields of, as well as new challenges in, Civil Engineering. Topics in the series include:

- Construction and Structural Mechanics
- Building Materials
- Concrete, Steel and Timber Structures
- Geotechnical Engineering
- Earthquake Engineering
- Coastal Engineering
- Hydraulics, Hydrology and Water Resources Engineering
- Environmental Engineering and Sustainability
- Structural Health and Monitoring
- Surveying and Geographical Information Systems
- Heating, Ventilation and Air Conditioning (HVAC)
- Transportation and Traffic
- Risk Analysis
- Safety and Security

To submit a proposal or request further information, please contact the appropriate Springer Editor:

- Mr. Pierpaolo Riva at pierpaolo.riva@springer.com (Europe and Americas);
- Ms. Swati Meherishi at swati.meherishi@springer.com (India);
- Ms. Li Shen at li.shen@springer.com (China);
- Dr. Loyola D'Silva at loyola.dsilva@springer.com (Southeast Asia and Australia/NZ).

More information about this series at http://www.springer.com/series/15087

Giuseppe Amoruso · Rossella Salerno
Editors

Cultural Landscape in Practice

Conservation vs. Emergencies

Springer

Editors
Giuseppe Amoruso
Dipartimento di Design
Politecnico di Milano
Milan, Italy

Rossella Salerno
Dipartimento di Architettura e Studi Urbani
Politecnico di Milano
Milan, Italy

ISSN 2366-2557 ISSN 2366-2565 (electronic)
Lecture Notes in Civil Engineering
ISBN 978-3-030-11421-3 ISBN 978-3-030-11422-0 (eBook)
https://doi.org/10.1007/978-3-030-11422-0

Library of Congress Control Number: 2018967435

This Springer imprint is published by the registered company Springer Nature Switzerland AG
The registered company address is: Gewerbestrasse 11, 6330 Cham, Switzerland

Preface

Landscape Cultures, A Contemporary and Global Societal Challenge

The book addresses *Cultural Landscape* understanding as driver for societal challenges, economic development, social inclusion, place assessment and conservation of heritage. The book disseminates issues growing from the relation between conservation and emergencies and identifies analytical and descriptive tools with the aim of sharing knowledge already available and with the aim of generating new knowledge so that it can be transformed into skills, seismic culture and social resilience.

In 1972, it was approved the *UNESCO Convention concerning the Protection of the World Cultural and Natural Heritage* that aimed at the identification, protection, conservation, presentation and transmission to future generations of cultural and natural heritage of outstanding universal value.

For the purpose of the *Convention*, they are considered as "cultural heritage" monuments, groups of buildings and sites. Particularly important for the research included in this book are the groups of buildings: «*groups of separate or connected buildings which, because of their architecture, their homogeneity or their place in the landscape, are of outstanding universal value from the point of view of history, art or science*»; and sites: «*works of man or the combined works of nature and man, and areas including archaeological sites which are of outstanding universal value from the historical, aesthetic, ethnological or anthropological point of view*» [1].

Urban or natural landscape has to be admired as an artistic form?, is it only an aesthetic concept? is it only to be lived or consumed rather than an ethical and ethological concept linked to regional habitat?

Landscape in not only a mere physical extension to be understood in a technical or economic sense, but also a place where the "cultural mediation" is developing and giving form to space. The awareness that culture "informs" landscapes (even from a concrete point of view) derives from this as well as the legitimate question if each landscape is really a cultural landscape, as the definition given by the World

Heritage Convention seems to recognise: «*landscape has different characteristic in each region depending on the cultural background and geographic condition*». Nevertheless, the interpretation of UNESCO and of ICOMOS considers a cultural landscape as an "outstanding universal value", introducing a hierarchy of priorities, even truly different values; the idea of "cultural landscape" is however almost recent—the first cultural landscape classified by UNESCO, Tongariro National Park in New Zealand, dates back to 1993—and today it is right for the scientific community and all the communities to rethink this concept. The interpretation «*cultural landscape represents the combined works of nature and man*» considers some typological evaluation of landscapes, used today in preserving world cultural and natural heritage: landscapes designed and created intentionally by man such as parks, recreational gardens, plazas, squares, cemeteries, promenades, yards; gardens related to monumental buildings and/or ensembles; organically evolved landscapes; continuing evolving landscape; associative cultural landscapes connected with religious/cultural natural elements. The recognition of cultural landscape is, however, focused on the rarity of type, regarded as a natural monument, although a debate is still running in ICOMOS to find useful interpretations to widen the field of preservation [2].

If buildings and cities are a portrait of the human condition, institutional operators and citizens, according to their skills and possibilities, have to invest resources so that this heritage could remain alive, in the uses and forms of daily life but also in memory, in rites, in traditions [3].

According to the 2003 *Convention for the Safeguarding of Intangible Cultural Heritage*, intangible cultural heritage (ICH)—or living heritage—is the mainspring of humanity's cultural diversity and its maintenance is a guarantee for continuing creativity. It is defined as follows:

Intangible Cultural Heritage means the practices, representations, expressions, knowledge, skills—as well as the instruments, objects, artefacts and cultural spaces associated therewith—that communities, groups and, in some cases, individuals recognize as part of their cultural heritage. This intangible cultural heritage, transmitted from generation to generation, is constantly recreated by communities and groups in response to their environment, their interaction with nature and their history, and provides them with a sense of identity and continuity, thus promoting respect for cultural diversity and human creativity [4].

Our landscape is also the material expression of meanings that have value, hierarchy and relations according to the different systems from which they derive (economic, social and cultural) and that have the role of making evident the quality of democracy that is rooted in a community.

Cultural heritage consists of products and processes of a culture that are stored and transmitted through the regeneration of memory and the processes related to it.

Natural disasters and degradation phenomena represent the setback of the homogeneous and constant "evolution" of a historically relevant landscape or area; this phenomenon invests, therefore, a vulnerable and weakened system in the settlement infrastructure, which inevitably becomes exposed to risk, in a territorial context already characterised by high seismicity or reduced resilience.

This book shares examples that can be proposed for restoring and reconstructing the physical and, through it, the human condition, investigating the representation and enhancement of cultural heritage and historical urban landscapes, as defined by UNESCO, as well as placemaking methodologies and strategic design that support the traditions of inherent to a place [5].

Knowledge of values and meanings that belongs to the territory, appropriately documented and shared within each community, is the milestone for informing every decision-making process that intends to transform it or build it. Applying this principle, for example, the American physiologist Ancel Keys unveiled essential concepts for the peoples of the Mediterranean countries and almost invisible to the common perception revealing the values and millennial qualities of a diffuse heritage, which were transmitted without interruption from generation to generation. Thanks to his studies on the epidemiology of cardiovascular diseases, he formulated, starting from the 1950s, the hypotheses on the influence of diet on these pathologies and on the benefits brought about by the adoption of the so-called *Mediterranean Diet*. This intuition led, in 2010, to the V° Intergovernmental Committee for the Intangible Cultural Heritage of Humanity, UNESCO, to proclaim "*The Mediterranean Diet*", Intangible Cultural Heritage of Humanity.

The Mediterranean Diet is therefore a broader concept of cultural landscape, understood as a paradigmatic set of knowledges, knowledge and practices closely intertwined in a conceptual continuum, from the urban and agricultural landscapes, to the ways of social aggregation, trades and the table. The element of identity of Mediterranean communities characterises the personal experience of each individual who lives there.

But how is it possible to investigate, map, describe and visualise the domains that represent this largely intangible cultural heritage and which can be highlighted in the various territorial contexts?

The cultural landscape can be translated into a "diet"—from the Greek word *díaita*, lifestyle—it is a social practice based on a series of skills, knowledge, practices and traditions that vary from landscape to food chain, which in the basin of Mediterranean concerns collection, cultivation, fishing, conservation, management, preparation and, in particular, consumption. In other words, it is the main infrastructure for cultural landscapes that was developed in different European regions.

A few months after the seismic events in Italy or after the destruction of the triumphal arch of Palmira, the researches here presented promote an adequate response to the demands of the people: "to interpret and represent a collective need, to transform memories, testimonies and signs into knowledge". Recalling past catastrophes, experts have noted that it is necessary to "have access to all the available knowledge, which will in turn generate new knowledge, which ultimately results in increased competence, professionalism and awareness on seismic risk reduction." [6] In this framework, the book introduces and promotes the role of digital technologies and the effective solutions to foster the engagement with heritage for its preservation and, consequently, for social innovation participation, risk of loss reduction and the development of identity.

The *EU Framework Programme for Research and Innovation* recognises several societal challenges the Horizon 2020 reflects the policy priorities of the Europe 2020 strategy.

The *Societal Challenge 6 Work Programme 2014–2015* focused, among others, on "Reflective societies" through the transmission of European cultural heritage, uses of the past and 3D modelling for accessing EU cultural assets; and also on the innovation in the public sector, open government, business model innovation, social innovation community, ICT for learning and inclusion.

According to these issues, the book opens to multiple fields of investigation related to design for places and smart growth: digital heritage, documentation and representation of places, agrarian urbanism, environmental design, architecture, sustainable design, smart cities.

A relevant application concerns (re)building for the future and how to reconstruct after a disaster. The book presents examples of best practices for the reconstruction of settlements affected by disasters, focusing on recent earthquakes in Italy. How can an area's built heritage support regeneration and reconstruction efforts? How can buildings be made safer and more resilient? How can communities be involved in rebuilding? And, how can the process of reconstruction protect local identity?

The book addresses also the role of education and learning and how to develop our heritage and knowledge of the complexity of places by using tools and techniques for representation, documentation and communication. Another focus is on the application of the most appropriate design tools to respect and develop local identity, to take advantage of the cultural context that the territorial system expresses and to suggest how place identity can inform new design that will build community.

What are the methods for place mapping that are recognised by UNESCO as signifiers of environmental, economic and cultural values? How can technologies provide tools and methodologies to support documentation and representation of a cultural landscape?

In this volume, authors present a selection of decision-making and participatory processes that are dealing with complexity and valorisation of cultural landscapes, in order to share good practices for post-disaster construction and rehabilitation, but also to deliver placemaking methodologies and strategic actions for the preservation of territories traditions and community social welfare.

The research, through the documentation of meanings and semantic characteristics of territories affected by seismic events, wants to promote the practical application of tools and processes for the configuration and implementation of cultural landscape: an information system that highlights the tangible and diffuse value of a community through its constructive characterisation according to environmental and natural issues.

In the characterisation and representation of cultural landscape, stability and continuity issues are highlighted, and organicity and difference are also recognisable parameters. The concept of place is inextricably linked to the concept of limit and boundary, its spatial relationship and connotation, which intertwine with a physical-perceptive delimitation and give a representation of its soul. The place is a set of identities with boundaries, in which there is always a link between subjects and the space. It is therefore something specific, with its own character, which identifies it and, at the same time, makes it unique.

The scientific challenge is to integrate different tools for settlements description and visualisation in their complex nature: it is necessary to make this knowledge available through technological platforms and participatory and inclusive models at the service of economic, tourist and government agencies that makes available the resources available, the potential for development and growth of new entrepreneurship and which makes the opportunities of the territory accessible to a wide international public in all the meanings linked to cultural landscapes.

Visual technologies and tools are studied in their specific fields of application and represent a rapidly growing sector that involves a variety of users in the contemporary society, addressing the societal challenges.

Organisation of the Book

The book represents an exchange of researches and best practices for placemaking and cultural landscape studies.

With contributions from leading experts, including university researchers, professionals and policy-makers, the book promotes transmission and dissemination of universal principles and contemporary applications to address the emerging societal challenges.

Critical to the presentation of case studies and current practice is the use of emerging tools for representation and documentation and to promote research on new core knowledge in the field, and on applications that are contributing to the evolution of such expertise and skills.

The book is the result of extensive researches developed by Rossella Salerno and Giuseppe Amoruso for national and international research programs. The overall organization of the book was designed and developed by both authors, collecting the 22 chapters in two macro areas: (I) Landscape and Territorial Practices and (II) Historic Settlements and Environmental Design. All the essays were preliminary reviewed by the scientific committee and then subjected to a final evaluation process by two anonymous referees.

The first part is presented and edited by Rosella Salerno is more focused on landscape and territorial issues according to natural conditions of land and places while the second part, is presented and edited by Giuseppe Amoruso, is focusing more on historic settlements, built environment and preservation issues.

Chapters will also ultimately help people's efforts to overcome the emergency phase of reconstruction after natural disasters and, introducing references and relevant issues on recent studies, describe emerging tools to understand such paradigmatic knowledge continuously inspiring practices that affect the agrarian, natural and urban landscapes. The work presents also a collaborative framework encouraging international cooperation and exchange of best practices and fosters the work of charities in different countries.

According to the holistic and multidisciplinary issues related to cultural landscape, the target audience could be heterogeneous and wider, including: scholars of cultural heritage, architecture, urbanism, humanities, landscape/regional and urban planning, urban geography, environmental and sustainability education, building construction, maintenance and design but also scholars of digital media, ICT, 3D modelling and computer graphics. The book also benefits scholars, academics and practitioners that are involved in the process of understanding, designing and transforming places and aims to foster an international exchange of research, case studies and best practices to confront the practical challenges of maintaining cultural landscapes alive and enabling community to improve cultural networks and promote education.

A brief description of each of the chapters follows:

Part I—Landscape and Territorial Practices

In Chapter "Fragile Cultural Landscapes: A Regenerating Case Study in East Veneto", Rossella Salerno (*Politecnico di Milano, Italy*) reflects on the case study of a reclamation landscape in the East Veneto Region (Italy), framing it in the European strategies and policy about landscape and heritage: it will be taken into account what has till know been done by local government, first of all by the *Osservatorio del paesaggio del Veneto Orientale* respecting the *European Landscape Convention* guidelines, then the paper will point out both sustainable innovation and potentialities of this fragile territory.

Further, the paper will try to outline the digital infrastructures' communication abilities of in sharing knowledge, connecting them to main questions about preserving and promoting landscape and scattered heritage, again keeping in the background the European policies, mostly about digitalisation.

The research illustrates also the recent best practice of "Alpinescapes" web platform, implemented to collect and share information about the cultural landscape between Lario and Ceresio lakes and lastly to map and merge Digital Cultural Heritage data from Italian and Swiss territories.

In Chapter "Rural Landscape in Sardinia. Historical Settlement in the West Coast of Sardinia: The "Ager Bosanus"", Andrea Pirinu (*Università di Cagliari, Italy*) is describing the rural architecture that still strongly characterises the landscape of Sardinia and composes a complex mosaic of constructive cultures. The Sardinia Regional Office of Planning promoted the publication of seven manuals and an Atlas of building cultures focused on knowledge, documentation and diffusion of historical memory and aimed to strengthen the capacities of protection and valorisation of traditional building. The survey has interested different geographical regions and during the definition of PPR (Regional Landscape Plan) determined the selection of landscape patterns, identified through complex analysis of the interrelationships between environmental, historical and cultural framework and settlement model. It follows, therefore, starting from the indications of the Regional Plan, the identification of terraced hills and rural villages bordering the edge of the plateau such as landscape matrix centres and key points of the project.

In Chapter "Riverscapes and Watersheds: Cultural Heritage Layers Along the River Guadalbullón (Jaén, Spain)", Pilar Chías and Tomás Abad (*University of Alcalá, Spain*) give an insight on the river landscape; from the perspective of the natural processes, the river flows across a narrow valley showing varied land forms and topographical features. From the cultural point of view, the river was the borderline between the Muslim territories and the Christian kingdoms throughout the Middle Ages. They were linked by means of an ancient royal road that was recently transformed into a highway from Madrid to Málaga, putting pressure on the landscape. But the narrow valley still keeps old archaeological sites dated back to the Bronze Age. The old route is still in use, holding lodgings, bridges and watermills as described by the travellers since the seventeenth century. It still keeps the castles that controlled the passage, and the *atalayas* where visual signals alerted to the dangers. All of them live together with ancient crops as oil groves, with elements of vernacular architecture such as *alquerías*, and with ancient opencast mines. The research aims to prevent disappearance of all these structures, to avoid the extinction of species and to preserve the memory of territory and landscape.

In Chapter "Indian Villas in the Valencian Landscape (Spain): *Casino del Americano*", Pablo Rodríguez-Navarro and Sergio Estruch González (*Universitat Politècnica de València, Spain*) introduce the architectural style appeared in the Canary Islands and the northern provinces of the Peninsula after the phenomenon of emigration from Ultramar. La Quinta de Nuestra Señora de las Mercedes, or also known as Casino del Americano, is a recreational villa that since 1869 is part of the neighbourhood of Benicalap, and whose architectural and construction characteristics make it a property with great heritage value, and one of the last Indian palaces in the city of Valencia. Due to the lack of bibliography, the information will be mostly obtained through the study of the building. The graphic surveying is the main working method, which has an intrinsic value because of the danger of collapse the building. The research presents mainly the historical, stylistic and constructive study, as well as a definition of the main characteristics of Indian architecture, through the most representative case of Indian house in the city of Valencia.

In Chapter "Traditional Identity and the Progressive Loss of Local Character in La Sagra Region (Toledo, Spain)", Alejandro García Hermida (*Universidad Alfonso X el Sabio de Madrid, Spain*) presents the basic aspects of a research dealing with the identity features of the architectural tradition of La Sagra region, in the Spanish province of Toledo, and the needed review of the criteria applicable to its conservation in order to avoid its disappearance. The problems presented in this regional analysis are very similar to those in other places and, therefore, similar studies may be valid for them.

For this research, the architectural tradition had to be studied first, as the existing diversity and complexity in the local traditions of this area were unknown before undertaking this study, all previous studies considering it as a homogeneous entity.

Finally, towns have been studied building by building, recording the transformations operated in each one of them. This is, the evolution of the building elements which define local urban landscape. To this aim, the composition, volumes and various constructive solutions used both in the new buildings and in the interventions made in the last decades on existing buildings have been analysed, then contrasting them with the traditional local identity municipal regulations of each town seek for preserving. Finally, the results obtained have been compared with that stipulated by the regulations in question, trying in this way to identify their strengths and capacities.

In Chapter "The Liberty Network in Varese Province: Strategies for Its Knowledge and Enhancement", Anna Anzani and Claudia Caramel (*Politecnico di Milano, Italy*) propose a study on the phenomenon that, between the end of nineteenth and the beginning of twentieth centuries, produced various architectural experiments in Europe including the development of Liberty, according to the research of a national style. In the province of Varese as well as in the adjacent Canton Ticino, the diffusion of Art Nouveau was fostered by the creation of new rail links that became a source of tourism development. The purpose of this research work is to suggest possible perspectives to improve the people daily experience by enhancing the identity characters of a specific urban environment, inevitably including the preservation of the ecological conditions.

Solutions based on a "space outfitting" approach will be proposed, which can be a response to current social changes, but also a transdisciplinary perspective to address the relationship between present and memory, relying upon reuse instead of new construction, low environmental impact, no waste, energy sustainability, social network and shared economy, participatory design, attention to the details, comfort and wellness.

In Chapter "Survey and Digital Representation of an Architecture in a Landscape Between Karst and the Sea", Paola Cochelli and Veronica Riavis (*Università di Trieste, Italy*) focus on the reconstruction of the Casa Alberi by Romano Boico obtained by integrating different types of survey to the cadastral documentation. As described by Boico the House in the Rock is like "a large rock shattered here and there and reassembled" that is camouflaged among the karst rocks of the promontory, characterised in plan and elevation by jagged shapes and lines. The villa is also characterised by the consistent use of raw materials for the structural elements and details such as the autochthonous stone of Aurisina and larch wood.

The study has previously analysed the historic evolution of project and construction phases of the villa, and the artistic context through historical and photographic sources. The next phase involved the digital reconstruction of the building, in which have been illustrating its architectural peculiarities. The traditional survey has been integrated with the photogrammetric one, that allowed to effectively reconstruct the orography and the territorial character in which the building is inserted, with particular attention to the passage in the rock that leads to the pier and the natural cave.

In Chapter "Creating a Map of the Underground Heritage in the Mediterranean Area: A Visual Representation for a Comprehensive Research", Beniamino Polimeni (*De Montfort University, UK*), Roberto Bixio, Carla Galeazzi, Carlo Germani, Mario Parise, Stefano Saj and Mariangela Sammarco (*Italian Speleological Society, Italy*) present a study in human-made (or artificial) cavities in the Mediterranean Area which has led the Italian Speleological Commission to study and catalogue some of the most common troglodyte types of the region. From 2000 onwards, the Commission has drawn up a study of the geographical distribution of the rock-cut structures by a project initially developed by Mario Mainetti and Erica Besana in 1994. This geographical catalogue has been realised integrating the information available in the international bibliography with the scientific research carried out by the Commission and the research groups affiliated. The result of this work is a list of 1948 rupestrian sites distributed over 31 countries. In this chapter, a general overview of the project is presented, along with a description of some case studies from different countries, including Tunisia, Libya, Turkey and Italy (Apulia and Lazio).

In Chapter "The Challenges of the "Divided" Heritage of Cyprus", Kokan Grchev and Ozgur Dincyurek (*Eastern Mediterranean University, Famagusta, TRN Cyprus*) intend to discuss different continuities as well as internationally established consensuses on cultural heritage and its identity by opening the issues of "divided" heritage of Cyprus. Heritage is discussed as representing and evolving contemporary culture, problematising human existence and guiding the future of the societies and cultures towards valuable identities. Perceived heritage of the island as a specific resource is questioned theoretically and by following the activities of the "Technical Committee on Cultural Heritage in Cyprus", in their efforts to bring together some practices aimed for heritage positioning in the focus of contemporary culture. Specific discourse is open towards the evident need of monuments on one side, and the social integration of heritage in cultural and development activities on the other. Unique emergencies are recognised as challenging the existence of both.

In Chapter "Conservation Issues on UNESCO World Heritage Sites in Russia. From the Roerich Pact to Contemporary Challenges", Polina Mironenko (Politecnico di Milano, Italy) presents an overview on the heritage protection within the Russian Federation. The idea to settle an organised preservation of the artistic and scientific world treasures belonged to the famous artist and Russian public character Nikolaj Konstantinovich Roerich. On 15 April 1935, in Washington, representative members of the USA and other twenty nations of the American continent signed an agreement "On the protection of the artistic and scientific institutions and the historical monuments" that later became famous in the international legal practice as Pact Roerich. From 1999, this day has been considered the *Universal Day of Culture under the Banner of Peace*. On 12 October 1988, Russian Federation accepted the *World Heritage Convention* from UNESCO for the conservation and promotion of the heritage of humanity. Currently in Russian Federation, there are 26 properties that are inscribed on the *World Heritage List* and considered as a universal outstanding value: 16 cultural sites and 10 natural sites.

Nowadays, the protection of the cultural heritage in the modern Russian society is becoming more complex and intense and, for this purpose, it requires a constant attention and the development of specific heritage skills.

In Chapter "Placemaking Workshops: Application of the PPS Method", Tomasz Jeleński (*Cracow University of Technology, Poland*) describes three cases of placemaking workshops conducted by the author in three different places: Ukraine, Tunisia and Poland, and against different cultural and political backgrounds. Public space is a priceless environment for human communication, interaction and local economies. It is a space for culture and a medium for symbolic content. A good public space is fundamental for the sense of community and thus an indispensable component of sustainable urban structures. In each case, application of placemaking methods encouraged public participation, showed the potential to facilitate the decision-making process and helped resolve potential or existing conflicts while building confidence in democratic procedures and institutions. The methodology is based mostly on the abundant experience of the organisation *Project for Public Spaces* (PPS), a non-profit established in 1975, which set itself a goal to assist local communities in reclaiming spaces that were socially degraded or car-dominated, and to create or strengthen more traditional characteristics of public places.

Part II—Historic Settlements and Environmental Design

In Chapter "Place Identity Graphic Assessment and Post-disaster Reconstruction", Giuseppe Amoruso (*Politecnico di Milano, Italy*) identifies the tools to understand and assess values and meanings of a territory that have to be appropriately documented and communicated in order to inform every decision-making process that intends to transform it or build it.

Natural disasters and degradation phenomena interrupt the organic and natural evolution of a place or an area rich in heritage; this phenomenon invests, therefore, a vulnerable and weakened system in the settlement infrastructure, which inevitably becomes exposed to risk, in a territorial context already characterised by high seismicity or reduced resilience.

The landscape documentation and the different identities and the graphic transcription of their semantic expressions, also according to new technology applications, provide a cognitive framework but also an operative vision to regenerate places and building according to local traditions; with the aim of generating new knowledge so that it can be transformed into skills, seismic culture and social resilience.

The research proposes the integration, in decision-making processes, of models, representations and visualisations based on repertoires, high-iconic databases and predictive simulations. The promotion of local identity and psychological and environmental well-being requires the definition of the tools for collecting and documenting local characters: analysis of urban patterns, construction techniques and tonal analysis of the urban environment, classification of architectural and landscape vocabulary.

In Chapter "L'Aquila Model. Strategies and Restoration Processes for Historic Centre", Mario Centofanti, Stefano Brusaporci and Pamela Maiezza (*Università dell'Aquila, Italy*), nine years after the earthquake of 6 April 2009, analyse strategies and current processes the reconstruction of L'Aquila and the centres of its territory, highlighting positivity and criticality, in order to identify the possible optimisation of methodologies and procedures, useful for present and future emergencies.

A first criticality is the "non-synchronic reconstruction". Both in terms of financial resources and in terms of validation and implementation procedures, the planning of the interventions and the execution management are structured on parallel and independent lines, which cause even considerably different process speeds.

A second problem is the "parcellisation of interventions". The "Plan of reconstruction" is as a mosaic of the individual parcelled proposals for intervention, caused by the procedural chain, without an overall strategic vision aimed at re-identifying urban identity and spatial and figurative qualities.

A third problem is denotable as "divided knowledge". The process of knowledge is discretized, according to the individual building or block project's validation procedure. In the case of the historic centre of L'Aquila, the paradox is that we have an extraordinary knowledge, unique and without precedents, with thousands of surveys, tests and analysis on materials. However, this knowledge is dispersed among the archives of the various institutions responsible for project validation and procedures control.

Purpose is to recognise the identity values, for the foundation of shared architectural/urban restoration projects and of reconstruction processes of the historic centre of L'Aquila that for the historical centres of the territory. It aims to a urban and territorial re-composition, coherently with the history of constitution and transformation of L'Aquila as territorial city.

In Chapter "Vulnerability of Architectural Heritage in Seismic Areas: Constructive Aspects and Effect of Interventions", Maria Rosa Valluzzi and Luca Sbrogiò (*Università degli Studi di Padova, Italy*) document the vulnerability of architectural heritage that is particularly prone to damage in seismic area, due to possible constructive defects or limitations, e.g. the low masonry quality, the scarce connection among components (walls, floors and roof), the structural irregularities, which combine together with the effects of lack of maintenance and deterioration exposure of materials. Nevertheless, the analysis of historic centres struck by earthquakes in the last 30 years in Italy revealed severe damages and collapses attributable to additional vulnerabilities provoked by heavy retrofitting techniques, ideally conceived to improve the mutual collaboration of the structures, but actually entailing a more complex and hybrid behaviours. It is the case of use of reinforced concrete to substitute floors and roof (aimed at increasing the in-plane stiffness), to strengthen vaults, or adopted as ring beams at floor and roof levels (to connect walls against overturning). This practice was commonly adopted in masonry buildings from the ‘80s onwards, according to the knowledge and recommendations available at those times.

In this work, the effects of past interventions applied to historical centres struck by a scale of earthquakes over time are analysed in terms of local and overall damage. Three villages of central Italy, namely Castelluccio di Norcia (PG), Campi Alto di Norcia (PG) and Castelsantangelo sul Nera (MC), all summing up to about 150 buildings, are examined. A classification of damage based on the *European Macroseismic Scale* is proposed. The study provides vulnerability maps able to take into account the influence of intervention on buildings in historical centres, which can contribute to evaluate possible damage scenario and maintenance plans.

In Chapter “Reducing the Loss of Built Heritage in Areas of Tourist Interest”, Giuliana Cardani (*Politecnico di Milano, Italy*) introduces issues on the potential of tourism as driver to sustaining the maintenance of built heritage. Tourism is an essential resource for the promotion of cultural sites and for keeping them alive. Traditional and modern technology should be employed whether for a single building or a whole village or even a small island. On the other hand, increasing exploitation of such sites may reveal itself to be a double edged sword. By attracting an uncontrolled number of visitors, this could result in a more rapid decline of the resource. Some places have been radically transformed to accommodate ever increasing numbers of tourists, with detrimental results. Rather than being enriched by the authenticity of the site, mass tourism is liable to damage the authenticity that it seeks. Sometimes, the final result is less authentic and the traditional spirit of the place destroyed, sacrificed to a need to comply to standard “hit-and-run” tourist destinations. The paper wants to present one of the many Italian cases where the need to deal with an ever-growing tourist pressure has reached such a high level that intervention now becomes urgent: the small Isola Superiore of Stresa, named Fishermens’ Island located in the centre of Lago Maggiore (Italy). Some suggestions are here reported in an attempt to reduce the loss of cultural heritage to the small island as well as in other similar situations.

In Chapter “Small Historic City Centers of Mediterranean Europe: Critical Points and Potentialities for Environmental Sustainability”, Valentina Pica (*Marie Sklodowska-Curie Fellow, University of Roma Tre, Italy*) addresses the importance of the conservation of historic urban environments with a multidisciplinary and operative approach. This is currently one of the most universally urgent and challenging cultural heritage conservation issues that also deals with environmental sustainability. More specifically, southern Europe small towns are progressively being abandoned and are far from being accurately requalified. The current conditions and historic resources of the Albaicín quarter of Granada (Spain), as well as of other small historic city centres in Italy, are being pointed out. A method for a comprehensive recovery plan approach of these centres is proposed, that should start from a territorial analysis, throughout different steps, such as: identifying existing assets; historical and critical studies of their diachronic evolution; mapping and general classification of the built heritage and existing infrastructures; study of the accessibility of the historical centre’s sites; a deep analysis of their critical points and opportunities and a critical valuation of the existing planning regulations. The chapter also argues in favour of numerous international projects, aimed at the renewal and resiliency of small historic Mediterranean centres, in order to foster good practices in their conservation worldwide.

In Chapter "Ichnography, Orthography and Scenography as Forms of Tracing the Past: A Reconstitution of the Roman *Forum* of *Ebora Liberalitas Julia*", Mariana Martins de Carvalho (*Universidade do Porto, Portugal*) presents an attempt to better understand the central core of a roman city—the forum—from three different forms of representation: ichnography (plan), orthography (elevation) and scenography (perspective).

Our laboratory is the monumental centre of Evora, located in Alentejo, Portugal, where the urban ruins of the roman temple are located. Here, we are attempting to trace the roman forum of *Ebora Liberalitas Julia.*

In order to comprehend the whole remaining structure, a dissection of each fragment was carried out. We analysed and drew all the archaeological vestiges over the current plans of the city; and then ichnographic, orthographic, scenographic hypothetical projections started rising from this ground. These three forms of expression are the result of thought and invention, such as described by Vitruvio, in his chapter on the fundamental principles of architecture (I, II, 2). In these representations, we can see how each one is increased with a dose of interpretation and hypotheses analysis.

The developed representations are supported not only by archaeological sources, but also iconographic, literary and historic sources. To study and to draw roman architecture in an urban context, it is fundamental to recognise the urban history and the phases by which each building has passed, in order to provide data that will add knowledge to the remaining architecture. Although the purpose of this study is to recognise roman architecture, it is impossible, in an urban scenario as this one, not to consider other times as well.

Considering the importance of the sources and its semantic expression, this essay tries to fill the gap between different types of representation in order to increase the possibilities of a more accurate reconstitution.

The ruins, and the archaeological remains that were exhumed, will be our field of experiment to try to trace the presence of the past and the different configurations of this monumental centre, reasoned by the evidence, knowledge and imagination.

In Chapter "Shortsighted Solutions Versus Long Term Planning", *Effects of Rapid Infrastructure Developments in the World Heritage Site of Gjirokastra*, Kreshnik Merxhani (*Polytechnic University of Tirana, Albania*) and Valmira Bozgo (*Environmental planner, Albania*) document the case study of an approved infrastructure project for the construction of a new road known as the "Bypass", which endangers the historic centre of Gjirokastra, a World Heritage Site since 2005. The project's initial aim was the diversion of automobile traffic from the city's Old Bazaar, for a more tourist-friendly pedestrian area. After few procedural problems were noted during public consultations, a more in-depth look at the project revealed a diversion form the original aim, and technical solutions that posed a threat to the *Outstanding Universal Values* for which the city enjoys its UNESCO *World Heritage* status. This generated a wide and largely publicised discourse among professionals opposing the development, and the government representatives that put through the proposal. Several technical and legal arguments, showing the perceived damages of this project to the structural integrity and the urban historical landscape of Gjirokastra, were presented to responsible national and international authorities. Currently, the project is in an "limbo" kind of state, having been suspended by means of several media declarations from national authorities, while still being depicted in the local development plan. The case study will be discussed for its technical and professional problems, disagreements with the priorities of the historical centre and the threat on the *Outstanding Universal Values* of the city. The bypass case, will be further utilised to illustrate the effects that shortcut solutions with no sound bases on strategic studies, can have on an urban historical landscape. Possible resolutions will be discussed for breaking the trend of rushed "politically impressive" developmental projects that present negative ramifications on historical landscapes.

In Chapter "Functional Heritage. Reconnecting with the Iron Web", William Howse and Renata Jadresin-Milic (*Unitec Institute of Technology Auckland, New Zealand*) present the case of historic buildings in New Zealand that have been underutilised and redundant in the face of contemporary development, although the potential for adaptation for reuse is increasingly being considered. The purpose of this project has been to develop an architectural solution for the adaptive reuse of an historic building, with the intention for the building to become functionally relevant in the twenty-first-century Dunedin in response to a current urban situation. The site in Dunedin is the currently neglected historic former Mornington Cable-Car 17 Depot. As a general methodology of the site condition valorisation, a systematic study of architectural, historical and urban analysis was conducted to provide insight into the appropriate approach to adaptive reuse. Further, field trips, examination of relevant literature and precedents, and the examining of different designs concepts through drawings and 3D models were methods applied in this project. The final design proposes "alternative representation" as a new and "experimental" approach to heritage developed for the purposes of the project. As a result, the building maintains its integrity while simultaneously each historical period is acknowledged for its contribution to the building itself and to the history of Dunedin. Hence, the building acts as a reminder of what has gone before, while fulfilling its original function as a cable-car depot, and ultimately stimulates a momentum for change in the local urban landscape.

In Chapter "The Ottoman Landscape of Büyükçekmece. A Case of Oversight or Misinterpretation of the Past?", Luca Orlandi (*Istanbul Technical University, Turkey*) focuses on the importance of the Ottoman site at Büyükçekmece Lake in Thrace, incorporated today in the outskirts of the fifteen millions inhabitant's megalopolis of Istanbul, and its relations to the surrounding environment. The Ottoman site analysed as case study was designed in the second half of the sixteenth century by the master-builder Sinan for the will of Sultan Suleiman and Grand Vizier Sokollu Mehmet Pasha. The Büyükçekmece complex, a post station on the road, includes a monumental stone bridge for crossing the lake, a capacious caravanserai, a small mosque and a fountain, and for centuries was an important centre along the Imperial road connecting the capital of the Ottoman Empire with the West. A recent project consisting of a tourist-sports centre and an entertainment park done by the Great Municipality of Istanbul in order to revitalise the area and attract more visitors and residents, had instead compromised the historical buildings and the original site, making impossible to read the traces of the Ottoman past in that region. The impact on the historical buildings has been completely ignored, and although it is still perceivable, the set of homogeneous buildings and the redevelopment process did not take into account the strong cultural values of these architectural elements, their legacy and their relation with the landscape. This paper aims to describe how that process of regenerating the Ottoman heritage in Büyükçekmece, hastily done in an inappropriate way, had definitely compromised the whole area, giving no more possibilities to read neither the buildings in the proper way nor the surrounding landscape that once characterised the area.

In Chapter "Ottoman Heritage in Southern Balkans: The Multicultural Port Town of Kavala", Velika Ivkovska (*Bahçeşehir University, Istanbul, Turkey*) introduces issues on multiculturalism as one of the keystones in understanding the complexity of the Ottoman civilisation. Especially in the provincial areas of the vast empire established by the Ottomans in almost six hundred years, it is possible to observe how multiculturalism and multi-ethnical components were a fundamental tool to glue together such diverse of populations and religions, in which each community had the possibility to maintain their distinctive collective identities, culture, rituals and practices. In the case of the town of Kavala in the Southern Balkans, that passed from Byzantine into the hands of the Ottoman rule after its conquest in 1391, it is possible to track how the town—through the centuries—became more and more a multicultural centre, due to the economic and social dynamics in the society itself, up till the dramatic events of the Balkan Wars in 1912. The Ottoman town developed rapidly around its harbour and the trade activities, and this paper wants to outline the urban environment of Kavala and its growth in the Ottoman time, from the conquest until its annexation to Greece. In a contemporary perspective, today it seems significant to widen the perception of how a vast tangible heritage left by the Ottoman civilisation, crossing the borders of different nations, religions, customs and cultures, especially in the Balkan context, could be analysed and incorporated in a bigger system to redefine the importance of such multicultural asset in a global manner.

In Chapter "A Cultural Reading of the Territory. Practices and Interdisciplinary Approach for the Making of Beauty in Apulia", Valentina Battista (*Euro Mediterranean Youth Academy, Italy*) presents an overview on beauty that, according to the ancient Greek culture, was inherently visual; beauty represented a well-suited harmony between finite and infinite, an overarching force to grasp those preternatural laws shaping human life, our earthly life. A measure, which if observed carefully, in architecture as well as in the other arts, is able to produce Eudaimonia, happiness. And here's how "a fragment of a temple, wrote Simon Weil, is still beautiful because we recognise the symbol of the temple in its entirety, as everything in the universe is beautiful because it symbolises the universe".

Since any cultural reading of the territory and its own beauty should start from the sources of law/the legal instruments available to ensure its preservation, the chapter considers the pivotal example of Apulia and the regional law on beauty, an ambitious project of interdisciplinary bottom-up co-drafting/programming, which ties together seemingly antithetic criteria like quality and well-being, focusing on the citizen, on his needs and dignity. This law provides tools, methods and actions to experience the beauty of our territory and to appreciate the vast diversity of Apulian identities' mosaic. The need to protect and preserve cultural heritage is a human rights issue and that cultural heritage is relevant not only in itself but also in relation to its human dimension, in particular in its meaning for individuals and community as well as in their identification and development processes.

Milan, Italy

Giuseppe Amoruso
Rossella Salerno

References

1. Convention concerning the Protection of the World Cultural and Natural Heritage (1972) https://whc.unesco.org/en/conventiontext/
2. Salerno R, Casonato C (2009) Paesaggi culturali—Cultural landscapes. Rappresentazioni, esperienze, prospettive. Gangemi, Roma, pp 8–9
3. Eco U (1998) Fondo Young sulla memoria e la mnemotecnica. Presentazione del Fondo. http://web.unirsm.sm/young/eco.htm
4. Convention for the Safeguarding of the Intangible Cultural Heritage (2003) https://ich.unesco.org/en/convention
5. Amoruso G (2015) The image of historic urban landscapes. Representation codes. In: Brusaporci S (ed) Handbook of research on emerging digital tools for architectural surveying, modeling, and representation. IGI Global, Hershey, pp 566–595
6. Nuzzolillo B (2002) Cerreto Sannita: un modello di ricostruzione post-sismica. Edizioni Sanniopress, Benevento, p 7

Contents

Foreword

> *Socrates: The good cannot be beautiful, nor the beautiful be good, if the beautiful and the good are two different things.*
> Plato, *Hipias Mayor*, 304 a

Presentation

This book project, edited by Giuseppe Amoruso and Rossella Salerno, is an extraordinary success and a great opportunity to deeply think about one of the key themes of heritage, the cultural landscape with its multiple faces. A landscape contextualised in the framework of an apparent fragmentation: *Conservation vs. Emergencies*, an invite to the dialectic idea that belongs to the pure Platonic tradition, since as Socrates, the Plato *alter ego*, compared these two concepts, so we could say that they are different without a doubt, but we could never assure that they are opposites. For sure we are talking about complementary words because they both need each other to complete or perfect themselves since there would be no use in an action aimed at restoring the ashlars of the facades of a Gothic cathedral if the buttresses that sustain them are not checked first.

Therefore, this statement clearly expresses the need to address the problems and challenges of architectural conservation from a multidisciplinary perspective that considers the perspective of natural disasters as an integral part of any intervention, so that heritage conservation organisations and society, as a whole, can face them adequately and with the lowest possible material and vital costs.

To achieve this goal, the editors, who in addition to sharing their own contributions, have done a magnificent and important work of documentation, recruitment and selection, and have managed to gather an interesting and exhaustive series of articles on the subject from some of the greatest specialists in the international field, who have contributed with their own experiences.

In this way and thanks to the various themes that are addressed in the production, ranging from the loss of identity of the landscape or the challenges of public spaces to post-disaster reconstruction, through sustainable innovation strategies, such as those pointed out by Rossella Salerno in her article, Chapter "Fragile Cultural Landscapes: A Regenerating Case Study in East Veneto", they offer us an inter-cultural vision in which, through the dialogue between past and future, between tradition and vanguard, building up a complete and innovative academic corpus which in my opinion, constitutes a unique tool and indispensable for all who, like me, take part in the process of understanding, designing and transforming the places in which we live in.

Therefore, this work is the expression of the vocation of its editors to contribute to social welfare through the optimal preservation of what identifies us and makes us unique: our culture, both in its intangible manifestation -uses, representations, expressions, knowledge, and techniques [1]—as in the material one-goods and natural and cultural spaces that are inherent to them-, since from the combination of both arises the substrate that build what it is known as "cultural landscape".

Approach to the Notion of "Cultural Landscape"

Although it is generally thought that the expression that gives its name to this publication is relatively recent, its origin dates back to 1925, when Carl O. Sauer, the father of North American cultural geography and founder of the Berkeley school, published the article entitled "The morphology of landscape", which contained its already famous definition: "The cultural landscape is fashioned out of a natural landscape by a culture group. Culture is the agent, the natural area is the medium, the cultural landscape the result. Under the influence of a given culture, itself changing through time, the landscape undergoes development, passing through phases, and probably reaching ultimately the end of its cycle of development" [2].

However, its use was not generalised until 1992, the year in which UNESCO approved the incorporation of the cultural landscape as a special type of site. Since its creation in 1972, the *Convention on the Protection of the World Cultural and Natural Heritage* established two major categories of protected sites: cultural and natural, regulated respectively in articles 1 and 2 of the same. The inclusion of the cultural landscape in the first category was not without controversy and after long and arduous discussions, the *World Heritage Committee* ruled that "Cultural landscapes are cultural properties and represent the 'combined works of nature and of man' designated in Article 1 of the Convention. They are illustrative of the evolution of human society and settlement over time, under the influence of the physical constraints and/or opportunities presented by their natural environment and of successive social, economic and cultural forces, both external and internal" [3]. Regarding the requirements that should be fulfilled for their declaration as such, the entity stated that "They should be selected on the basis both of their outstanding universal value and of their representativity in terms of a clearly defined geo-cultural region and also for their capacity to illustrate the essential and distinct cultural elements of such regions" [4].

As we see, this definition is far from shedding light on the subject, instead obscures it, and the reason for this decision remains a mystery, since, as Adrian Phillips points out: "In a sense, all landscape is 'cultural', since no part of the earth is immune to human influence of some kind" [5]. And at the same time, every landscape is or has been natural, since all human intervention takes place on a physical medium. To which we should add the following questions: Isn't it a monumental construction, the combined work of nature and man? does not show the evolution of a given society over time?

UNESCO itself maintains that "Cultural heritage means monuments, groups of buildings and sites that have historical, aesthetic, archaeological, scientific, ethnological or anthropological value". While "Natural heritage means exceptional physical, biological and geological formations, habitats of threatened animal and plant species, and areas that have scientific, conservation or aesthetic value".

Is it then the criterion of the minimum human intervention that should prevail or is the exclusion of the municipal or urban term what determines its classification as cultural landscape? What happens with the construction that is erected in the forest nearby the city?

Phillips, in drawing the guiding lines for the identification of the protected landscapes of the International Union for Conservation of Nature (IUCN), is aware of the difficulty involved in defining the concept, when he points out that "'Landscape' is a difficult word. It has many meanings and is interpreted differently by different people. Some societies have no word for it" [6]. Regarding the previous comment, it would be possible to add that if landscape is difficult, of what is the culture nor we speak, because it is a term for which the challenge of achieving a unanimous and universal conceptualisation is equal or similar to that of the *beauty* or *art*, that is to say, impossible.

Not much more enlightening is the definition contained in the *European Landscape Convention* (CEP) [7], approved in Florence in 2000. It is the first international agreement exclusively dedicated to the subject, which describes it as "any part of the territory as perceived by the population, whose character is the result of the action and the interaction of natural and/or human factors" [8]. As can be seen in the previous description, as well as in a large part of the recent literature on the landscape, the idea that landscape arises from the interaction of society with its environment over time, highlighted again as a shared defining element. In other words, the definition given by Sauer is still fully in force and it is the one that has inspired most of the international agreements and texts on the subject.

However, the text of the CEP contains an essential difference with the approaches of UNESCO and it is the fact of considering "that the landscape is a quality of the territory, of each territory and, therefore, not only of the territories with exceptional landscape values" [9]. This is so because it is based on the premise that what must be protected is the product of our interaction with the territory, which will be defined by what differentiates us and makes us unique: our culture. Therefore, it is not so much that the cultural landscape is something particularly beautiful or exceptional, but simply different and unique. In addition, the opposite would imply affirming that there are better or worse cultures than others within the Union, which would undermine the essential principles of the Union.

Now, the main problem lies in identifying those intangible attributes that make a landscape unique, since, as Hernández points out, "Society as a whole, for example, is capable of perceiving and, above all, assigning a historical and patrimonial transcendence to a certain architectural construction, for example, a cathedral, although it lacks training in artistic matters. On the other hand, contemplating a landscape, it will assign a positive or negative value according to the perception that it provides (nice, pleasant, etc.), but with greater difficulty, it will be able to recognise a historical meaning related to its long configuration process. Therefore, it is necessary to sensitise the society, but also to educate it about the value of the landscape as a patrimonial element" [10].

It is, therefore an intangible heritage whose value transcends the ownership of the land and the goods it contains, or in other words, a cultural landscape is a special type of heritage asset considered "probably, in its reference, more open to be perceived as a morphological and functional expression and symbolic of historical and current relations between society and nature" [11].

This is because the CEP falls within the *European Territorial Strategy* (ETS), adopted by the ministers responsible for planning the territory of the EU a year earlier in Potsdam, which points out the need to "use a creative approach in order to reverse the trend of abandonment, deterioration and destruction in many regions" of "the Europe's cultural heritage which extends from cultural landscapes in rural areas to the historic centres of cities' and 'it is the expression of his identity" (§133). The ETS does not clarify what is the border that separates the first from the second, but it does expose a series of problems and examples that can help us to better approach the object of the study.

Thus, we find some great types of causes that affect the conservation of our landscapes: the lack of continuity in human intervention (such as the abandonment of the fields), the inadequate or excessive exploitation (such as the installation of industrial or residential areas and the uncontrolled extraction of raw materials that can destroy entire landscapes) and also the standardisation or loss of identity due to constant evolution.

To avoid the disappearance of this rich heritage, the ETS recommends the conservation, development and creative recovery of the landscapes that have a particular cultural, historical, aesthetic and ecological significance. Among the examples worthy of conservation is the *bocage*, the *open fields* systems or the historical roads that cross different countries.

And it proposes the use of traditional methods of landscape management such as agricultural exploitation, the development of rural tourism or reforestation, which could represent an alternative to the total abandonment of a place avoiding problems such as soil erosion and the abandonment of the undergrowth and silviculture, which have fostered terrible forest fires in recent years in southern Europe.

Now, what conclusion can be drawn from all of the above? That the cultural landscape must be rural? This seems to be deducted from the text of the ETE, especially when referring later to "urban landscapes", assuring that they are often the product of a random evolution, just like in rural areas.

There are many theorists who have addressed this issue with a similar approach, such as the aforementioned María Hernández, who argues that "The rural landscape is the result of the interaction between society and the environment in which it is based", and that after a long adaptation, gives place to the cultural landscapes [12].

If we look for the meaning of rural in the Spanish dictionary, we will find something like: "Belonging to or relating to the life of the field and its work". While if we look for landscape, we will read something like: "Part of a territory that can be observed from a certain place". Where it is inferred that in its current meaning there is nothing that limits it to a country view. Now, if we go a little deeper into the origin of the word, we will see that in most Latin languages comes from French, such is the case of Spanish and Italian, and more specifically of the term *pays*, which literally means field, and that in turn comes from the Latin *pagus*, which during the Republic was the Roman demarcation that identified the village or rural district, which unlike *civitas* or *urbs* [13], was used to name sown or cultivated land, mainly with vines and olive groves.

As we can see, this correlation between cultural landscape and rural environment, which probably comes from a more or less conscious association of landscape with its etymological meaning, has given a restrictive or limited interpretation of the concept that concerns us and presents several problems. Among others, it establishes a division, by the opposition, between rural and urban landscapes, rejecting the identification of a territory as a cultural landscape by the mere fact of being in the middle of an urban environment. Thus, it would happen that the beautiful landscape that can be observed when approaching the bridge of Rande in Vigo (Galicia, Spain), with the rafts where the mussels are grown floating on the estuary of the city, would be excluded from such classification. *A contrario sensu*, if a territory is located in a non-urban environment could not be cultivated or sown, as, for example, in desert or rocky environments, a literal interpretation would exclude certain places of extraordinary cultural richness as those characterised by the small dwellings that the Galician sailors built on the rocks by the sea, from which they went directly to fish as in the small fishing port of Redes (Galicia, Spain).

On the other hand, it could be objected that a rural landscape does not have to produce a cultural one by itself, since even over the centuries there has been a long adaptation, from man to his environment, in fact, the current situation may have been reached without particularly worthy contributions in the cultural order, such as architecture, music or dance, among others. But, what really characterises a folk and defines it are not the greatest events, and their monumental commemorative constructions, which however relevant they may be, are generally punctual. What really identifies a territory and its people is their day to day, which will always revolve around their own homes and small public and religious buildings, but above all and fundamentally, around the humble anonymous architectures on which they articulate their activity. For example, we can find small accessory architectures such as village fountains, laundries or granaries to store the grain outside the houses, usually on pillars to preserve the grain of corn and other foods, with peculiar characteristics that personalise it, often defining broad regions.

The Polish anthropologist Eugeniusz Frankowski, the author of *Horreos y palafitos de la Península Ibérica*, drew attention to these constructions by studying them in detail. His work was published in 1918 by the *Commission of Paleontological and Prehistoric Research of the Board for the Extension of Spanish Studies and Scientific Research*. In it he also paid attention to the evolution of these types of popular buildings in other countries, although he focused specifically on Spain and especially in Galicia, citing as a very similar example the grain deposits in Bosnia called *kukuruzniak* (name of corn in Bosnian), some baskets interweaved with reeds and supported on six pillars, illustrated with an engraving from a photograph of the ethnologist Paul Labbé.

It is important to highlight the importance of these architectures that we call anonymous and the need to study it, as an important cultural element. In this sense, Frankowski himself quotes an especially illustrative anecdote: "It is said that a foreign writer who travelled through Galicia, in a work he wrote, said that the Galicians were so religious that almost all of them had a chapel next to their house, because confused with chapels the granaries finished with a cross or a church facade". And he interprets: "Who knows the importance of this granary in the life of the Galician villager, who holds the whole harvest of the year, and has seen the facades of some poor churches with the bell tower on the facade, will recognise at once that these ornaments are the imitation of the bell tower, the church tower, etc" [14]. I would add, knowingly, that it is certain that the Galician considers the *hórreo* as so sacred or more, since "*orandum est ut sit mens sana in corpore sano*" [15].

It is therefore not strange that the small fishing village of Combarro, 7 km from Pontevedra, essentially characterised by a small cluster of these popular stone buildings facing the sea, has been declared a Historic Site. Therefore, it is an asset of Spanish Cultural Interest and taking into account its landscaping characteristics, it could be an example of "small cultural landscape", although it is located in the urban core.

From the above, we can deduce that for a landscape to be considered cultural, it should meet several conditions. First, have crops or pastures, among others and/or plantations of native and/or exotic species of special botanical interest. But also infrastructures and real estates had to be related to previous crops and plantations that have arisen and developed specific cultural elements under them. That is, it has to gather assets of agricultural and/or forestry interest, architectural assets and cultural traditions in a distinctive or characteristic set. So, we can consider that a cultural landscape must meet both natural heritage and cultural heritage, within which we must consider not only the agricultural or forest heritage but also the architectural, including archaeological sites.

This is the position assumed by all the member countries of the EU and, therefore, the Government of Spain, which in its *National Plan for Cultural Landscape* affirms that it is a dynamic concept and difficult to manage, given the complexity of its own nature, in which "natural and cultural, material and immaterial, tangible and intangible components" [16] intervene and defines it, for the purpose of applying the Plan, as the "result of the interaction over time of people and the natural environment, whose expression is a territory perceived and valued for its cultural qualities, product of a process and support of the identity of a community".

However, Giuseppe Amoruso and Rossella Salerno have taken a step further in the conception of this work of extraordinary importance, since it is the first publication on the subject that assumes such a broad and open notion of the cultural landscape, able to cover all the approaches and expositions, emphasising practical examples that can help to better understand the difficult approach to this concept.

Overcoming Terminological Differences

Although today it is recognised that the landscape is, as conceived by the CEP, an integrating reality of nature and culture, and therefore it is not possible to face both perspectives in order to influence the historical and cultural dimension of the values that characterise certain landscapes, it is relatively frequent in the academic field and, especially, in certain national and international heritage standards, the use of the notion of the cultural landscape in opposition to the one of the traditional or natural landscape.

Another interpretation is the one provided by UNESCO, which argues that with this consideration of the heritage in its dual cultural and natural aspect, "the *Convención* reminds us of the ways in which man interacts with nature and, at the same time the fundamental need to preserve the balance between the two" [17]. ETS also refers to this necessary balance when it states that "The conservation of these landscapes is important, but it should not hinder, or even make impossible their economic exploitation".

In any case and beyond the terminological differences, what does seem to exist is the need to generate creative and innovative strategies for the achievement of that noble objective that is the preservation of our cultural landscapes. With this aim, the publication that I have the honour and the pleasure of introducing is born, in which starting from an idea as open and as extensive as possible, and from an inclusive perspective, the multiple and diverse facets of the cultural landscape are analysed, whether this is rural or urban, through a selection of theoretical approaches, case studies, pilot projects and practical experiences of such diverse nature that cover the entire range of possibilities, overcoming the now obsolete distinction between natural and cultural landscapes.

The Cultural Landscape, an Endangered Species

In this way and assuming that a cultural landscape is the result of the intervention of man in such a way that it reflects the history of his interaction with the environment, no kind of landscape reflects that evolution more than the one derived from agricultural exploitation, whose origin is traced back to the Neolithic. In this sense, it seems clear why the EU ensures that the best way to avoid or prevent the destruction of these cultural landscapes is to continue and promote agricultural use or something similar that has defined them.

Thus, we find the *Landscape of the Vineyard Culture on the Pico Island* in Portugal and the *Colombian Coffee Cultural Landscape*, both declared *World Heritage Sites* in 2004 and 2011, respectively. In the first, UNESCO has highlighted the harmonious conjunction of vine crops in rectangular shape delimited by stone walls, homes, warehouses, churches and ports conditioned by an agricultural practice of several centuries. The second constitutes the *Coffee Axis* or *Coffee Route* that extends throughout four departments, in which this crop is combined with rural housing, production, and marketing facilities, infrastructures, etc., but also with other original manifestations of popular culture such as music, dance or gastronomy.

Although this last example may seem exotic, this work collects similar and possibly even more exotic landscapes for a western European mentality, but which nonetheless are found in Europe. I am referring for example to "The Ottoman landscape of Büyükçekmece", in Thrace. Luca Orlandi (Chapter "The Ottoman Landscape of Büyükçekmece. A Case of Oversight or Misinterpretation of the Past?") analyses this complex on the outskirts of Istanbul, designed in the sixteenth century by the master-builder Sinan by order of Suleiman the Magnificent. The author denounces that the uncontrolled expansion of the megalopolis has endangered a way of life that preserved the culture and popular architecture of centuries ago until very recent times.

A conceptually similar case in terms of the negative consequences of "politically impressive" developments, as described by the authors of Chapter "Shortsighted Solutions Versus Long Term Planning", Kreshnik Merxhani and Valmira Bozgo, is the "Effects of the Rapid Creation of Infrastructures on the World Heritage Site of Gjirokastra", in Albania. Merxhani and Bozgo raise the case of a new highway known as "Bypass", currently under discussion among technicians and professionals critical of the project and the government that proposes it, in order to save the Old Bazaar of the city.

Not only the uncontrolled expansion of cities or inappropriate actions can destroy heritage, but also tourism itself. Giuliana Cardani (Chapter "Reducing the Loss of Built Heritage in Areas of Tourist Interest") raises the case that in principle, a laudable intention, such as attracting this resource to be able to make a better conservation of heritage, can achieve exactly the opposite, its rapid deterioration. From the case of the small island in the centre of Lake Maggiore (Italy), the Isola Superiore of Stresa poses a series of suggestions to reduce the loss of its cultural heritage.

The case of the rural landscape of Sardinia, with a human implantation, totally integrated into the territory of the Planargia and with its own fluvial port in the city of Bosa, in the mountains facing the sea, supposes the opposite example. It is a model of peaceful coexistence and perfect harmony of tourism with the cultural landscapes. Andrea Pirinu (Chapter "Rural Landscape in Sardinia. Historical Settlement in the West Coast of Sardinia: The "Ager Bosanus"") vividly and minutely describes this surprising place, physically close to the island's saturated tourist centres, but far enough away to maintain its idiosyncrasy between olive groves and vineyards, until today.

Another of the great dangers that threaten our landscapes is abandonment, a theme for which Valentina Pica (Chapter "Small Historic City Centers of Mediterranean Europe: Critical Points and Potentialities for Environmental Sustainability") proposes solutions, especially for the errors that are currently being made in the conservation of small historic centres of Mediterranean Europe, like the Albaicín neighbourhood in Granada and other small Italian cities, bearing in mind the environmental sustainability.

More difficult to perceive is the progressive loss of local character, as in the example of the Sagra district, in Toledo (Spain), analysed by Alejandro García Hermida (Chapter "Traditional Identity and the Progressive Loss of Local Character in La Sagra Region (Toledo, Spain)"). The author honestly recognises that before carrying out this study, he did not know the diversity and complexity of local traditions, which previous studies considered homogeneous. This gave rise to an exhaustive investigation: village by village and house by house in order to obtain a grounded vision of the local urban landscape.

Also, in relation to the loss of identity, Kokan Grchev and Ozgur Dincyurek (Chapter "The Challenges of the "Divided" Heritage of Cyprus") bravely face "The challenges of heritage 'divided' from Cyprus", by questioning the internationally established consensus on cultural heritage and his identity. The authors question, theoretically, the consideration of the perceived heritage of the island as a specific resource aimed at generating a "valuable" identity, in relation to the efforts of the *Technical Committee on Cultural Heritage* in Cyprus to position such heritage in the centre of contemporary culture. The specific speech is open to the evident need of monuments, on the one hand, and to the social integration of heritage in cultural and development activities, on the other hand.

The Singular Constructions as a Symbol of Cultural Identity

But it also can be other types of human intervention not derived from agricultural exploitation that characterise a territory, such as the *Cultural Landscape of Sintra*, inscribed in 1995 as a *World Heritage Site*, which includes part of the Sintra Mountain Range and the historic centre of the villa, with its romantic architecture of the nineteenth century and with parks and gardens that harmoniously combine native and exotic species. However, it must be recognised that Sintra is, above all, known for its National Palace of Sintra, of Arab and Manueline style, and for the National Palace of Pena, of the nineteenth century, in romantic style, located on the top of a hill. That is, for its architectures and its immediate surroundings. But it does not need to be especially a spectacular or surprising architecture to characterise a site or its people. Thus, William Howse and Renata Jadresin-Milic (Chapter "Functional Heritage. Reconnecting with the Iron Web") have decided to develop an architectural solution for the reuse of a single building in Dunedin, New Zealand. It is the historic Mornington Cable-Car 17 Depot, a deposit of funiculars that stopped working in 1957, of which a detailed architectural, historical and urban analysis was carried out through numerous drawings and 3D models. The final idea proposes an "alternative representation" with a new and "experimental" approach to the heritage.

Another research is essentially focused on a singular building and on the ways of representing it, as a key to understanding it, is that of Mariana Martins de Carvalho (Chapter "Ichnography, Orthography and Scenography as Forms of Tracing the Past: A Reconstitution of the Roman *Forum* of *Ebora Liberalitas Julia*") for the Roman forum of Evora Liberalitas Julia, in the Portuguese city of Alentejo. Martins challenge with the interpretation of the *ichnografia*, which he interprets as planes, the *orthografia* as elevation and the *scaenographia* as perspective, from Vitruvius. His essay attempts to fill the gap between the different types of representation in search of a more precise reconstruction. For this, she used "evidence, knowledge and imagination".

Similarly, Pablo Rodríguez-Navarro and Sergio Estruch González (Chapter "Indian Villas in the Valencian Landscape (Spain): *Casino del Americano*") also focus on an emblematic building from 1869, Quinta de Nuestra Senora de las Mercedes, also known as Casino del Americano, in the Benicalap district, which is one of the last palaces of Indians of Valencia. At the moment it is in a ruinous state, although the idea is to turn it into a public place for the district. Given that, they are facing the risk that its ruin is going to be irreversible, with it the main and of course one of the few buildings of this type would disappear, a symbol of an era and a migratory movement that shaped the character of the city. The aim is to present a historical, stylistic and constructive study of this building, with the graphic topography as the main method of work.

Paola Cochelli and Veronica Riavis (Chapter "Survey and Digital Representation of an Architecture in a Landscape Between Karst and the Sea") also focus on a building, the Casa Alberi from Romano Boico, built near Trieste, Italy, in a rugged coastal landscape, reason why is known as the House on the Rock. The authors have focused their study on the history of the development of the project since its commission in 1957, its artistic context and later its digital reconstruction with a traditional survey integrated with the photogrammetric. In this way, I consider that they adequately reflect that the architectural expression of the house corresponds to a contained and rested assimilation of its natural environment.

If the previous chapter deals with the rocks and the architecture integrated into them, Beniamino Polimeni, leading a large team of the Italian Speleological Society

(Chapter "Creating a Map of the Underground Heritage in the Mediterranean Area: A Visual Representation for a Comprehensive Research") has created a map of the underground heritage of the Mediterranean area, which considers a visual representation for a thorough investigation. The study, a 1994 project by Mario Mainetti and Erica Besana, has catalogued 1948 troglodyte deposits from 31 countries and also presents some cases among them, including Tunisia, Libya, Turkey and Italy (Apulia and Lazio). It should be noted that this is an original research on the integration of human work with nature.

Public Spaces and Itineraries: Places of the Confluence of Cultures

But we can also focus on "constructions" that are even less visually tangible. For Tomasz Jeleński (Chapter "Placemaking Workshops: Application of the PPS Method"), public space is an invaluable environment for human communication, interaction and local economies. The article describes three interventions made by the author in Ukraine, Tunisia and Poland, that is, with different cultural and political contexts, helping the three counties to solve potential or existing conflicts. The methodology was one of the non-profit organisation *Public Spaces Projects* (PSS) dedicated to help the recovery of degraded spaces and to create or reinforce their traditional characteristics.

It is that way if there is a typology of public space that serves for human interconnection, especially when it accumulates centuries of history and multiple cultures have left their traces in it. Hence, the ETS proposes the protection of historic roads such as the pilgrimage routes of the Camino de Santiago de Compostela, which crosses several countries and the Via Francigena in Italy, both classified as *European Cultural Routes* in 1987 and 1994 respectively. In 2004, both roads were qualified as *Great European Cultural Itinerary*.

The French Way of Santiago in Spain and the routes of the same in France were declared World Heritage in 1993 and 1998, respectively. Since 1970, the number of pilgrims has gone from only 68 to 301,006 in 2017, with peaks coinciding with the Holy Years (those in which the day of the Apostle, on July 25th, falls on Sunday). The Via Francigena originates from the English city of Canterbury and ends in St. Peter's Square in the Vatican after crossing France from the north to the south and Switzerland.

In my particular case, I had the opportunity to study the different roads of Santiago in Galicia, as well as to see how the main religious monuments of the country were related to them. As I once described in a book about the five cathedrals of Galicia, these are: "the Cathedral of Mondoñedo on the Northern Way; the Cathedral of Lugo on the Primitive Way; the Cathedral of Ourense on the Sur Way; the Cathedral of Tui on the Portuguese Way; and at the end of all the roads, the Cathedral of Santiago".

But the roads of Santiago not only have great cathedrals, convents, churches and monasteries, but also multiple population centres and small popular constructions, anonymous architectures such as the popular housing, fountains and granaries, which ultimately characterise much better the deep idiosyncrasy of the people, in this case Galician. As the *General Director of Cultural Heritage of the Government of Galicia* and art historian Ángel Sicart pointed out another book that we published in the year 2000 about the French Camino de Santiago: "The effort involved in the graphic representation of such an abundant cultural heritage, is pleasantly rewarded from the possibilities it offers, not only to professionals and experts but for the understanding on the part of a public less used to the graphics support of the spatial reality in which it moves daily. The information that is handled gives a clear idea of the cultural load of the Camino as a tangible support of a civilisation that transcends epochs and borders".

In this sense, it is important to highlight the importance of the works collected in this book that I am pleased to present, which point out the value of the recovery and enhancement of old itineraries. So Pilar Chías and Tomás Abad (Chapter "Riverscapes and Watersheds: Cultural Heritage Layers Along the River Guadalbullón (Jaén, Spain)") analyse what they call, with an apt expression, "layers of cultural heritage" along the Andalusian river Guadalbullón, with military castles, bridges, farmhouses, water mills, adjective constructions, anonymous architectures among, of course, the ubiquitous olive groves. The authors emphasise that, from a cultural point of view, during the Middle Ages, the river was the border between the Muslim territories and the Christian kingdoms of the north and they amusingly describe the old route followed by the travellers.

Finally, Velika Ivkovska (Chapter "Ottoman Heritage in Southern Balkans: The Multicultural Port Town of Kavala") also presents us with an itinerary. A journey through the urban environment and the history of the Kavala, the ancient Macedonian Neapolis, which belonged from 1371 to the Ottoman Empire until its annexation to Greece, thanks to which it is possible to follow how the city became more and more a multicultural centre, through the centuries. The author proposes to analyse all the vast tangible heritage left by the Ottoman civilisation, which crosses the borders of different nations, religions, customs and cultures, especially in the Balkan context, to incorporate it into a broader system that allows redefining the importance of this multicultural resource in a global mode.

Conservation Versus Emergencies

This work is remarkable in many aspects, but mainly because in addition to studying some landscapes damaged by the lack of intervention or the excess of it, it goes a step further and also analyses the consequences of the human errors, either by the action or the omission, in the case of natural catastrophes. Indeed, many of the disasters have been bolded by the inappropriate architectural or engineering interventions, with poor-quality materials or inadequate constructive solutions.

In this sense, the Managua earthquake of 1972 is unfortunately paradigmatic, which ironically took place just a month after UNESCO approved the *Convention for the Protection of the World and Cultural Heritage*. With 6.2 degrees on the Richter scale, it destroyed the capital of Nicaragua in just 30 seconds, causing 20,000 deaths and 300,000 victims as estimated. This tragedy, which was compared with the destruction of Hiroshima and Nagasaki, could have been much less serious if, after the previous earthquake of 1931, the destroyed popular dwellings, constructed with the traditional construction system in Nicaragua of the *taquezal* or *bahareque*, constituted essentially by a latticework of reeds lined with mud and stones, would have been adequately restored (some modern buildings with anti-seismic or old systems with more noble building systems, however, endured much better). The fires and subsequent looting ended the work of nature.

Unfortunately, it was decided to demolish the city centre, located between several faults, and leave the ruins as a witness to a war that was still to come. And what was rebuilt or made new, it was monopolised by the Somoza family, which ultimately facilitated the Sandinista revolution.

The L'Aquila earthquake of 2009 took place in Europe, with an intensity of 6.3 degrees according to the United States Geological Survey, with more than 300 deaths and significant material damage. The wreckage, that to a greater or lesser degree, destroyed private houses, affecting 50,000 people, as well as numerous historical monuments. As highlighted in the book, some examples of good practices, for the reconstruction of the areas affected by the latest earthquakes, are presented and the emphasis is placed on how the most resistant buildings can be made without distorting the identity forged over the centuries.

To solve these situations, the application of a universal model such as "Place identity graphic assessment and post-disaster reconstruction" is proposed by Giuseppe Amoruso (Chapter "Place Identity Graphic Assessment and Post-disaster Reconstruction") in this work. The author proposes a method of integrating decision-making processes, models and all kinds of representations, that includes predictive simulations and an exhaustive use of new technologies, as well as other convergent strategies in collaboration with citizens, always taking into account the traditional construction systems of the area. This should have been the case of the last earthquake in Managua already mentioned.

Going from the general to the particular, Mario Centofanti, Stefano Brusaporci and Pamela Maiezza (Chapter "L'Aquila Model. Strategies and Restoration Processes for Historic Centre") present the "Model of L'Aquila. Strategies and reconstruction processes for the rehabilitation of the historic centre", nine years after the tragedy occurred. That is because it is about making a "virtual pause", not really, to analyse the process that is followed in the reconstruction, seeking to optimise the methodologies used for the future. They raise interesting doubts, such as "non-synchronous reconstruction", "subdivision of interventions" and the problem of the "divided knowledge". It is logical the need to thank the authors for such a contribution, where they offer concrete solutions.

To close this group of contributions to the book, Maria Rosa Valluzzi and Luca Sbrogiò (Chapter "Vulnerability of Architectural Heritage in Seismic Areas:

Constructive Aspects and Effect of Interventions") also focus on areas with seismic damages, but specifically on the effects, unfortunately bad, of the interventions carried out on monuments or buildings with architectural interest that were restored after previous earthquakes: specifically Castelluccio di Norcia (PG), Campi Alto di Norcia (PG) and Castelsantangelo sul Nera (MC), with a total of 150 buildings. Obviously, seeing what has not worked before us can help to learn how to improve the future.

Our Vision Always Pursues Beauty [18]

As it could not be less, in a work promoted by Italy, the closing part is written by Valentina Battista (Chapter "A Cultural Reading of the Territory. Practices and Interdisciplinary Approach for the Making of Beauty in Apulia"), with "A cultural reading of the territory" based on the beauty of Apulia, the surprisingly forgotten heel of the Italian boot. From Bari, to Taranto, passing through Brindisi and Lecce, the most extensive coast of an Italian region is an impressive succession of spectacular landscapes, especially seafaring, and the Greek, the Roman, the Norman, the Turkish, the Aragonese architecture, but above all the baroque (*il barocco leccese*) of Spanish origin in many cases. It is not surprising that she dedicates her chapter to the "*Legge sulla bellezza della Puglia*", since for Battista "Beauty is the only idea that reveals the real world" and that "it is capable of producing *Eudaimonia*, happiness".

On 31 July 2018, the president of the Region of Puglia presented in Bari the draft of the *Legge sulla Bellezza*, which seeks to enhance and protect the territory and its cultural identity. It is a law that I consider of great interest, because, free of charge, it has been prepared by a technical–scientific committee that seeks to valorise the territory and cultural diversity, searching for the beauty through the memory of the community, the work of the man and nature, but also the elimination of the detractors of the beauty of the "*Mosaico Pugliese*"(*Pugliese mosaic*, TN) that is, the diversity of the territory. The law raises "*un vero e proprio diritto alla bellezza delle città*" (*a real and proper right to the beauty of cities*, TN).

With the same idea, Anna Anzani and Claudia Caramel (Chapter "The Liberty Network in Varese Province: Strategies for Its Knowledge and Enhancement"), in their study on "the network of freedom in the province of Varese" in Italy, ask for an awareness towards the importance of beauty as a fundamental right of the human being, in the belief that the aesthetic experience is a normal attribute of our life. They consider that certain architectural styles such as Liberty and Art Nouveau, spread through the province of Varese and the canton of Ticino, thanks to the new railway communications of the time, are now extremely vulnerable as they are not subject to the protection of the law. Hence, its holistic approach to its conservation must be accompanied by an improvement of the life quality.

I want to finish wishing all readers of this work to learn as much as I have done with it, in order to get a better conservation of landscapes and monuments, but also of agrarian lands and rural constructions, which can often teach us much more about the people who gave and/or give the true meaning to a cultural landscape since, after all, each territory is what it is for the culture that inhabits it and gives it shape.

From these pages, you can enter other worlds contemplating them through the eyes of those who know them best. Because, as Vitruvio affirmed, our vision always pursues beauty; here you will find it in every new path and approach to acts and places, often unjustly unknown, that will expand the vision of the world enabling to apply all this knowledge to the future.

A Coruña, Spain

José Antonio Franco Taboada
Universidade da Coruña

References

1. Convention for the Safeguarding of the Intangible Cultural Heritage. UNESCO 2003, Art. 2. https://ich.unesco.org/en/convention
2. Sauer CO (1925) The morphology of landscape. University of California Publications in Geography 2(2):19–53, p 46
3. Operational Guidelines for the Implementation of the World Heritage Convention. UNESCO, 2008, p 14, §47
4. Operational Guidelines for the Implementation of the World Heritage Convention. UNESCO, 2008, p 86, §47
5. Phillips A (2002) Management guidelines for IUCN category V protected areas: protected landscapes/seascapes. IUCN Gland, Switzerland and Cambridge, UK, p 6
6. Phillips A (2002) Management guidelines for IUCN category V protected areas: protected landscapes/seascapes. IUCN Gland, Switzerland and Cambridge, UK, p 5
7. "The absence of cultural adjectives in the statement of the concept does not imply in any way denial of its deep cultural meaning, repeatedly recognized in the previous Report, in the Preamble and in the articles of the Convention". Carrión A (2015) Plan Nacional de Paisaje Cultural. Ministerio de Educación, Cultura y Deporte, Madrid, p 6
8. Translation of the Instrument of Ratification of the *European Landscape Convention*, published in the Official Gazette of the Spanish State on February 5, 2008 (BOE Nº31)
9. Carrión A (2015). Plan Nacional de Paisaje Cultural. Ministerio de Educación, Cultura y Deporte, Madrid, p 25
10. Hernández M (2009). El paisaje como seña de identidad territorial: valorización social y factor de desarrollo, ¿utopía o realidad?. Boletín de la A.G.E, 49:181
11. Carrión A (2015). Plan Nacional de Paisaje Cultural. Ministerio de Educación, Cultura y Deporte, Madrid, p 4
12. Hernández M (2009). El paisaje como seña de identidad territorial: valorización social y factor de desarrollo, ¿utopía o realidad?. Boletín de la A.G.E, 49:170
13. Berger A (2002) Encyclopedic Dictionary of Roman Law. The Lawbook Exchange, Clark (New Jersey), p 616
14. Frankowski E (1918) Hórreos y palafitos de la Península Ibérica. Museo Nacional de Ciencias Naturales, Madrid, p 26

15. Décimo Junio Juvenal, Sátira X (línea 356)
16. Carrión A (2015) Plan Nacional de Paisaje Cultural. Ministerio de Educación, Cultura y Deporte, Madrid, p 26
17. World Heritage Center (2008) Information folder about world heritage. UNESCO, Paris, p 7
18. Vitruvio (1995) De Architectura. Alianza Editorial, Madrid, p 88 (III, 3, §13)

Part I
Landscape and Territorial Practices

Fragile Cultural Landscapes: A Regenerating Case Study in East Veneto

Rossella Salerno

Abstract The paper reflects on the case study of a reclamation landscape in the East Veneto region (Italy), framing it in the European strategies and policy about landscape and heritage: it will be taken into account what has till now been done by local government, first of all by the "Osservatorio del paesaggio del Veneto Orientale" respecting the *European Landscape Convention* guidelines, then the paper will point out both sustainable innovation and potentialities of this fragile territory. Further the paper will try to outline the digital infrastructures' communication abilities in sharing knowledge, connecting them to main questions about preserving and promoting landscape and scattered heritage, again keeping in the background the European policies, mostly about digitalization. As regards the relationship between landscapes and heritage, it will be paid specific attention both on the called "Top-down" actions and on knowing and giving values from below, from "Bottom-up" practices. In fact, the two strategies are complementary, because the first—*Presentation*—regards the careful goal of informing and making heritage available in a way normally carried out by scholars, professionals and more in general experts in the heritage field. The second strategy, better known as *Interpretation*, refers on the other hand, to research and creativity activities and results generated in a heritage site. In such perspective, involving visitors as well local people and communities is crucial for interpreting and changing sites as cultural landscapes, into places where legacies of the past can turn to a resource for the future. Referring in a specific way to those last processes, as they are more innovative, the paper will illustrate, like an example useful to be applied, the recent experience of "*Alpinescapes*" web-platform, implemented to collect and share information about the cultural landscape between Lario and Ceresio lakes, and lastly, to map and merge Digital Cultural Heritage data from Italian and Swiss territories.

Keywords Cultural landscape · East Veneto · Alpinescapes

R. Salerno (✉)
Department of Architecture and Urban Studies, Politecnico di Milano, Milan, Italy
e-mail: rossella.salerno@polimi.it

G. Amoruso and R. Salerno (eds.), *Cultural Landscape in Practice*, Lecture Notes in Civil Engineering 26,
https://doi.org/10.1007/978-3-030-11422-0_1

1 A Point of View on Cultural Landscapes

Today it's possible looking at cultural landscape as a combination of tangible and intangible goods, rooted in a territory, in other words it's feasible considering the leading idea of cultural landscape as a *unicum* able to collect all the different and intertwined legacy dimensions, characterizing a specific site: this point of view is based on a rich and well-structured cultural and regulatory framework.

Such position is an outcome of different and combined perspectives: the *European Landscape Convention* guidelines [6], but also of the cutting-edge debate which in Italy produced, in 2004, the *Codice dei beni culturali e del paesaggio* [3], and last but not least, on a global scale, of the Unesco approach to cultural landscapes [9].

So the point of view here adopted, considers as inseparable the relationship between territory, inhabitants, tangible and intangible culture; the most evident synthesis as concerns such a complex connection finds its more appropriate expression just in the idea of cultural landscape: that is an idea able to join nature and changes made on the nature itself by human work.

It seems evident that such interpretation realizes a wide "cover" able to include the outstanding landscapes considered by the Unesco as *World Heritage*, and also the so-called everyday landscapes, "normal" landscapes which are far from mass tourism interests, better intended as non-outstanding, in the same time rich in memories for people living them and potentially open to a new sustainable tourism.

2 A Case Study: A Fragile Cultural Landscape in East Veneto

From this point of view, the reclamation landscape in the East Veneto region has, as matter of fact, the specific characteristic of a cultural landscape; how we can read on the *Manifesto per la tutela e la valorizzazione del paesaggio della bonifica del Veneto Orientale*, the reclamation landscape is an outcome of a long time work devoted to change the water system existing between lagoons and rivers, so regenerating a land to be employed for agriculture, human settlements, industry, tourism [4].

The reclamation landscape is a key-element characterizing for the East Veneto geographic-environmental infrastructure, together with the river Piave basin, next to the Cadore mountain area, the Treviso hills' piedmont system and further the valleys obtained from the reclamation, just next the Venetian lagoon.

As the VeGAL (Venetian Local Actions Group) affirms, the East Veneto territory is a clear example of "a new land" arisen in the XX Century; today the landscape maintained most of signs and characteristics in respect the ones emerged from the swampland during the '50s of the past century [10].

We are dealing with a new landscape, almost completely artificial, marked since it appeared, by a series of land and water infrastructure and very specific buildings (canals, banks, roads between small farms, turning bridges, basins, water pumps) which permitted and still today safeguard both a stable emergence and a civilized life. So there, we can find land and water geometries, colored by plantations, where it's possible to look at a kind of 'minor' architecture made of rural houses, farms and landowners' villas, all perfectly respecting the landscape context [1].

The landscape, arisen between the XIX and XX Centuries, finds out actually, since few time, a proper touristic potentiality, based on historic, cultural and environmental enhancing.

We don't consider such so long and complex changing process ended, it's in fact up to administrators and inhabitants trying to identify how the landscape evolves in respect of nature, rivers, valleys and coasts settings.

Anyway, as wrote Federica Letizia Cavallo, even if the reclamation landscape represented in different ways a shared reference for most Italian people, today it seems instead not so visible and known both for its history and functions [2].

The reclamation territory in East Veneto outlines a way of interpreting the landscape as a clear symbol of the relationship among a cultural, anthropic, and natural processes, where innovative approaches are strictly intertwined to new use practices by tourists, residents and metropolitan citizen.

The large number of restored projects both in historic centers and single buildings, rural too, the museums started, the green and wood areas regenerated, a special care for facades on canal or rivers, have been thought as taking part of an itineraries system.

Almost two hundreds of structural recycling interventions have been made on valuable manufactured goods as palaces, religious buildings, archeological settings, rural and reclamation houses, mills, etc., the most remarkable among them have been classified and mapped; at the same time, in some pilot areas, also each artwork belonging to buildings, palaces and so on, has been classified.

Considering the entire process, a turning point happened in 2012 with the *Osservatorio sperimentale per la tutela del paesaggio della Bonifica del Veneto Orientale* set-up [7]; it has been established by a regional law (L.R. 10/2011) and respecting art. 133 of the *Codice del paesaggio* (D.Lgs. 42/2004) [3], both aim to support the inhabitants in actions and practices devoted to recognizing local identity through heritage enhancing.

The East Veneto's administrators have been able to start a rural regeneration course, applying a new rurality strategy and investing in several and converging directions leading to setting-up itineraries, to re-discovering land products, to restoring unique buildings and small centers or historic villages, last but not least to organizing a structured research and studies plan about the territory.

The outcome is a strategic vision on the North East Italian Region, intended as a mix of coast and farmed environments which aspire to integrate each other; so the backcountry is not a land able to be crossed only or to provide human resources, but intends to be attractive for a new kind of tourism, slow tourism in another seasons; so travelers can find something different in respect of the coast where the rural defines the territory and, in the end, redefines the built-up area too. It dealt with the work's results which can today produce, if conveniently supported, a new kind of economy. It needs also underlining that such policy is first of all a cultural effort. Intervening on historic centers, valuable buildings, regenerating and enhancing local products, designing itineraries, etc., so that everything can contribute to a social and economic growth, would be impossible without a new way to find out the "soul", identity and vocations of the territory [10].

3 Feasible Innovative Regeneration Strategies for Fragile Cultural Landscapes

What has till now has been done to enhance the reclamation cultural landscape in east Veneto region is an interesting example for the fragile areas in Italy: a re-emergence of a valuable identity in a place, also for the relationships with a new strategy for the economic growth, shouldn't turn out to be constraining for other, diverse and innovative ways able to connect people and heritage.

Conditions exist to go beyond, to experience innovative methodologies allowing local communities to realize networks for protecting and running spread heritage, also employing sustainable management plans to foster economic and work-related effects.

The goal of enhancing the fragile territories can also aim to make Europeanization stronger and shared, using new catalyst to support and re-think the connections between people and their heritage.

Experiencing innovative communication forms can turn into opportunities to start positive processes affecting *non-outstanding* landscapes characterized by interesting spread heritage, aiming at the same time at setting-up shared good practices for sustainable tourism, signified by technological innovation and rural smart hub planning to promote culinary excellence.

Knowledge connection concerns essentially a deep understanding about geographical and historic contexts, where heritage becomes embedded and then approaches, technologies, innovative tools can ease both sharing and increasing actions for managing tangible and intangible heritage.

The heritage digitization issue opens for sure new scenarios: this represents a crucial aspect in the *European Agenda for Culture*, setting first of all to increase the public accessibility to different forms of cultural and linguistic expressions, but also to boost opportunities for people to be involved in protecting and enhancing heritage.

4 Digital Infrastructure and Strategies for Communicating Heritage

A European Commission Communication, *Towards an integrated approach to Cultural Heritage for Europe* [5], supports in a specific way a heritage vision aimed to share values and promote social cohesion, in other words it fosters heritage in a framework oriented to enhance cultural diversities and intercultural dialogue, considering culture as a catalyst for creativity, in the end to interpret culture as an active element for the Union's international dimension.

Following such mainstream, the crucial question this paper wants to focus on is the relevance today is assuming to know and share heritage through the digital, employing in particular, the digital technologies for communicating a cultural landscape.

Among the numerous infrastructure already operative in this field, a relevant position is represented by the *Europeana* (www.europeana.eu) web platform which interprets in a different way how approaching heritage, through the triad: Digitizing, Communicating, Sharing.

The *Europeana* web platform makes access free to more than fifty millions of digitized items, among them there are books, music, artworks and much more, in order to share heritage with recreational, educational and research purposes.

As well *Europeana*, other on-line platforms adopt a similar strategy; it's important to recall at least http://www.digitalmeetsculture.net, Europeana Space, RICHES, Civic Epistemologies, Photoconsortium.

Unfolding the multiple local identities to a European dimension, going out each regional reference point to make the cultural resources in a territory accessible in a wider scale: all this represents a new challenge because, from that point of view, the digital issue becomes crucial as it is intertwined to the heritage communication one. The Icomos (International Council of Monuments and Sites) identifies two possible approaches to the "heritage communication" subject: on one hand a "top-down" strategy, better known as *Presentation* strategy, on the other hand we can find the "bottom-up" practices, defined instead as *Interpretation* [8].

The first strategy concerns a careful information and access (also physical) plan to heritage places, normally it is a approaching way for scholars, professionals, specialists in the field of heritage: so it's mostly a one-directional communicating system. The second manner is instead centered on the activities, results, research and creativity, which a heritage site can implement. In such perspective, involving visitors and inhabitants is crucial in order both to interpret and change the site and the cultural landscapes where past legacies can evolve in a resource for a future growth. The communication strategy *Presentation* requires experts both in the research field and in representing through digital technologies the outcomes, employing 3D models and GIS, in case of archeological sites, architectural heritage and more in general in tourist context of great acclaim.

A innovative and effective digital technologies' development and thus the possibility to realize rendered digital models in real time, joins together a feasibility in producing photorealistic images of tridimensional objects and at the same time a possibility in making available information in a visual way: further architecture and complex sites representation can be improved by texts, documents, images, iconography, and by locating buildings in their proper context through geo-referred 3D models.

Today the most updated tools in this specialized field are however represented by the H-BIM (Heritage-Building Information Model).

Digital models for heritage anyway play a crucial role in communicating and sharing knowledge too; from this point of view, that is more social and cultural, the combined employ of ICT and web allows to outline new interesting scenarios for heritage, as for example the multimedia employ for cultural heritage, the use of data in documenting the memory of places, the image and information technologies to share the places' experience, last but not least the employ of Open data and web based tools.

Lately also AR/VR technologies, applied in a touristic field, have been implemented to boost knowledge in fact, of complex buildings or sites transformations.

5 A Possible Way to Follow

To paying back a tangible idea about a possible use of digital and technological infrastructures able to convey heritage communication on a wide scale, that is in a European horizon, we intend here suggesting for the case study of the reclamation cultural landscape in East Veneto, an available way to be pursued to direct the regenerating process towards a wider communication about heritage; so, as example, we would describe one of the on-line platforms—*Alpinescapes*—based on data collecting and visual representation of a cultural landscape; *Alpinescapes* deals with the territory between the Lario and Ceresio lakes, allowing to match and map heterogeneous data coming from different sources about Digital Cultural Heritage in the Italian and Swiss territories.

The web platform Alpinescapes (http://www.paesaggiculturali.polimi.it/) is an outcome of a research project developed under an Italy/Swiss research program, named "Il paesaggio culturale alpino su Wikipedia: Valorizzare il paesaggio culturale attraverso dati aperti, Wikipedia, Sit e allestimenti analogico digitali".

So *Alpinescapes*—as we can see on Wikipedia—is a project which focuses on open data and the Creative Commons attribution-share alike license to provide access to existing content on the Alps (cultural and natural heritage, geography, anthropology…); it links data from the Geographic information system to Wikipedia; it contributes to Wiki Loves Monuments 2014 and produces the design of two museums in Italy which take into account also the link between digital and analog (https://meta.wikimedia.org/wiki/The_Alps_on_Wikipedia).

The platform user can overlap different information levels and choose to manage a 500 m area around his location or contributing directly acting on Openstreetmap (OSM) o Wikipedia, using a dynamic map.

Further it has been provided by a new web based map, careful for alpine territories characteristics, and equipped with a web-app allowing to visualize and put in relationship heterogeneous data, as mentioned above, with local projects in fact of enhancing the cultural alpine landscape: all the sources are open and can be increased and the users can add items on Wikipedia, Wikivoyage or Openstreetmap.

Alpinescapes respects the European Union guidelines as regards *Digital Cultural Heritage*, facilitating a wider possibility to access data, boosting a sustainable tourism experience, promoting inclusive and participative processes by inhabitants.

It deals with an experience which represents a change of point of view towards aware, participative actions to protect and regenerate heritage to be opened not only to local people but to those digital communities involved by touristic or cultural reasons.

6 Conclusions

Heritage digitization can so enclose a very wide issues' range able to answer a boosting request of social inclusion, participation and accessibility to information sources; digitization doesn't consist in fact in a trivial "translation" from analogic to digital in order to get digital copies of text, drawings, maps, videos and so on.

Rather the utility of digitization should be addressed to share knowledge more and more and finally can represent an effective answer to an increasing request of social inclusion.

So it seems very suitable considering all those and knew scenarios able to link together knowledge from experts to a heritage vision from the inhabitants and the tourists visiting cultural landscape sites.

To conclude, digital technologies should so allow to build "a common ground" to protect and preserve heritage, at the same time to promote actions where several and different actors can "interpret" heritage and decide the future.

References

1. Baldo G (2015) Il paesaggio ritrovato. Vocazioni della Venezia Orientale, On-line Resource
2. Cavallo FL (2014) Ri-conoscere il paesaggio di bonifica. In: Rapporto Attività, cit., pp 20–22
3. Codice dei beni culturali e del paesaggio, ai sensi dell'articolo 10 Legge 6 luglio 2002, n. 137. Decreto Legislativo 22 gennaio 2004, n. 42
4. Comitato Scientifico dell'Osservatorio del Paesaggio (2014) Linee guida per un Manifesto per la tutela e la valorizzazione del paesaggio della bonifica del Veneto Orientale. In: Rapporto Attività, cit., pp 8–9
5. Communication from the Commission to the European Parliament, The Council, The European Economic and Social Committee and the Committee of the Regions (2014) Towards an integrated approach to cultural heritage for Europe
6. Council of Europe (2000) European landscape convention. English Official Text. Council of Europe, Strasbourg, France

7. Rapporto Attività dell'Osservatorio del Paesaggio della Bonifica del Veneto Orientale (2014) On-line Resource
8. The ICOMOS charter for the interpretation and presentation of cultural heritage sites. Reviewed and revised under the auspices of the ICOMOS International Scientific Committee on Interpretation and Presentation. In: Proceedings of the 16th General Assembly of ICOMOS, Quebec, QC, Canada, 4 Oct 2008
9. UNESCO (2013) Managing world cultural heritage; world heritage resources manual. World Heritage Centre, UNESCO, Paris, France
10. VeGAL (2016) Rural-Urban_Coast Lab. Incontri sulla Venezia Orientale Futura, On-line Resource

Rural Landscape in Sardinia. Historical Settlement in the West Coast of Sardinia: The "Ager Bosanus"

Andrea Pirinu

Abstract The rural landscape object of this research is located in Planargia, a geographic region of north-west Sardinia. This area is characterized by a territorial system constituted by the town of Bosa, a network of very small settlements placed on the edges of the basaltic plateau and rustic villas dispersed in the country, strongly linked among them by a tight net of connections. Historic settlement and technical-productive model comes directly from the morphology of the territory, where in a limited area, volcanic plateaus and depression of Miocene marls are interlaced, where the deep canyon of the Temo river and of several tributary streams creates a landscape characterized by heavily steep slopes with mixed soil, ideal to grow olives and grapes (Le Lannou in Pâtres et paysans de la Sardaigne, Tours Arrault, Traduzione italiana (a cura di Manlio Brigaglia): Pastori e contadini di Sardegna, Cagliari, Edizioni della Torre [1]). The disperse settlement is still present in the organization of the territory, but the time of living are solely those related to agriculture activity and summer season; starting from the early Middle Ages, as confirmed by archeological surveys, a gradual renunciation of the permanent residence in the country has produced the development of the villages located on the edge of the plateau. A further testimony of the existence of this historic residential model, is the presence of isolated churches (today ruined), that confirms the existence of *villas*, aggregates of few buildings that in the Byzantine period were the centers of the country (Day in Villaggi abbandonati in Sardegna dal Trecento al Settecento, Parigi [2]).

Keywords Survey · Historical buildings · Rural landscape · Bosa · Sardinia

A. Pirinu (✉)
DICAAR, Faculty of Engineering and Architecture,
University of Cagliari, Cagliari, Italy
e-mail: apirinu@unica.it

G. Amoruso and R. Salerno (eds.), *Cultural Landscape in Practice*,
Lecture Notes in Civil Engineering 26,
https://doi.org/10.1007/978-3-030-11422-0_2

1 Rural Landscape in Sardinia

The rural architecture still strongly characterizes the landscape of Sardinia and compose a complex mosaic of constructive cultures. The existence of an historical heritage *result of immense deposit of work and planning ability of local communities gives to the territory of the Island an unique character and offers to the rural landscape a new centrality in which technologies and local materials and their re-use as design appears in consistent with an eco-design and a contemporary art of building and living* [3]. The attention and the cultural sensitivity shown by the Sardinia Regional Office of Planning has made possible the creation of seven Manuals and an Atlas of building cultures focused on the knowledge, cataloging and diffusion of historical memory and aimed to strengthen the capacities of protection and valorization of traditional building. The survey coordinated by the Department of Architecture of the University of Cagliari has interested different geographic regions and during the definition of PPR (Regional Landscape Plan) determined the selection of landscape sectors, identified through complex analysis of the interrelationships between environmental, historical and cultural framework and settlement model. In particular, the indication of PPR concerning the rural countryside of the coastal context of Planargia (Fig. 1) is: *keep the ratio of the widespread settlement system of Magomadas in the southern part of landscape patterns with his own countryside, and identify, the terraced hills, the shore line and the deep valleys as landscape matrix centres and rural centers as internal territorial emergencies, in order to ensure the identity of place.* It follows, therefore, starting from the indications of the Regional Plan, the identification of terraced hills and rural villages bordering the edge of the plateau such as landscape matrix centers and key points of the design.

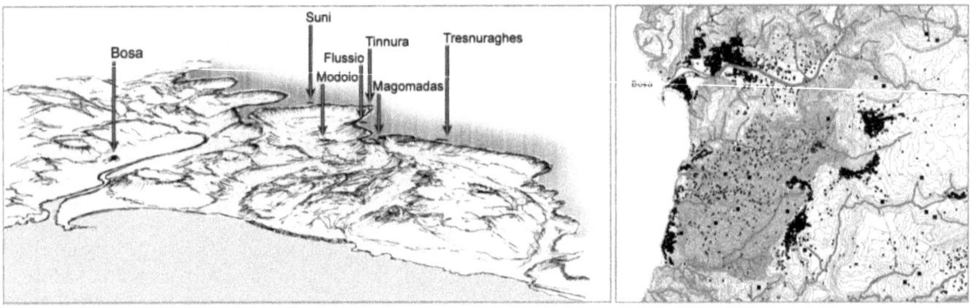

Fig. 1 The coastal context of Planargia and identification of the area (C.T.R. 1998)

1.1 The "Ager Bosanus" in the West Coast of Sardinia: Historical and Geographical Overview

The *ager bosanus*, characterized by sparse settlements located between the edge of the Planargia plateau and the trachytic ridge overlooking the sea, includes the municipalities of Bosa, Magomadas, Tresnuraghes, Flussio, Tinnura, Suni and Modolo that lies in the homonymous valley.

Since the Neolithic period human activity is documented in the entire area when a number of sites are located along the routes of obsidian that from *Monte Arci* was exported along the coasts of northern and central Italy and France. The *Domus de Janas* (Neolithic period tombs), the high number of *nuraghe* placed along the coastline and the plateau, the archaeological finds and place-names that bring us back to a Phoenician-Punic presence in the whole region, allows to define an almost uninterrupted human presence. Phoenician domination is followed by the Roman presence that, in the Time of Augustus, defines an organization of a territory articulated on the center of Bosa, small settlements and rural rustic villas, located on the slopes of hills and in few narrow valleys.

The area, in Roman period, was crossed by the coastal road "*a Tibulas Sulcos*" [4] that connected *Karales* (Cagliari) to *Turris Libissonys* (Porto Torres) and characterized by a dense network of roads between the centers of the area and the hinterland; the confirmation of the commercial activity that took place is suggested by the place name of Magomadas that derives from the Punic word *mqm hds* that means *the place where market, or trade, is held.*

In Planargia, as in general in Sardinia (with the possible exception in the plain of Campidano) in the Roman period, there isn't a real rural landscape based on the system of the farm, as it happens in the Italian peninsula, and that model is directly adapted on the regional reality based on the Nuraghic period settlement.

To confirm this trend, there is a reoccupation of nuragical sites left in the Punic period, mainly related to the agricultural exploitation of the surrounding area.

Recent archaeological investigations [5] in the whole territory of Magomadas led to the identification of a medieval settlement in the area of *San Giovanni/Santu Maltine*, and showed an almost uninterrupted frequentation in the Republican period (I–II century BC) until the beginning of the fifteenth century.

In fact the area, between the edge of the plateau and the coast line, corresponds with the agro of ancient Roman town of *Bosa Vetus* that was located on the right bank of the Temo river and not far from the current modern settlement; since the beginning of the twelfth century, the family of Malaspina (arrived in Sardinia to give help to the Judges involved in the fight against the Moors) began the construction of a castle on the hill of Serravalle and the territory of the Planargia becomes his fief.

In this time the *curatoria* of *Frussia*, that included in the Judicial period the settlements of the district, become the hinterland of the Malaspina's castle, and therefore his name becomes *curatoria* of *Serravalle*. As we know, in Sardinia, since the thirteenth century are reported some example of "organized city" that will interest a limited number of settlements; for most of cases there is a system of villages included in *curatorie*, which achieves a sort of widespread settlement in the territory. This organization is disrupted in the second half of the fifteenth century at the end of the war between the Judge of Arborea and the Crown of Aragon and as a result of the subsequent epidemics that cause a population decline of one-third of the population and a general impoverishment of the centers that partially disappear.

The birth or rebirth of urban centers take place after the middle of the fifteenth century and their development continues throughout the sixteenth century reaching the shape and dimension that retain, for most of them, until the early twentieth century, with the exception of those that assumed a certain economic and political significance in the seventeenth century or in the nineteenth century. Some of these will rise again with the same name in other sites than the previous, the rural life will be centralized in more compact nucleus, emphasizing the rural depopulation and originating, in the studies area, the development of centers on the edge of the plateau.

Today, with the exception of the city of Bosa, who has had a considerable expansion, the other centers have maintained approximately the extension shown in the Cadastral maps of the nineteenth century, preserving a land use strongly related to rural activity [6] (Figs. 2 and 3).

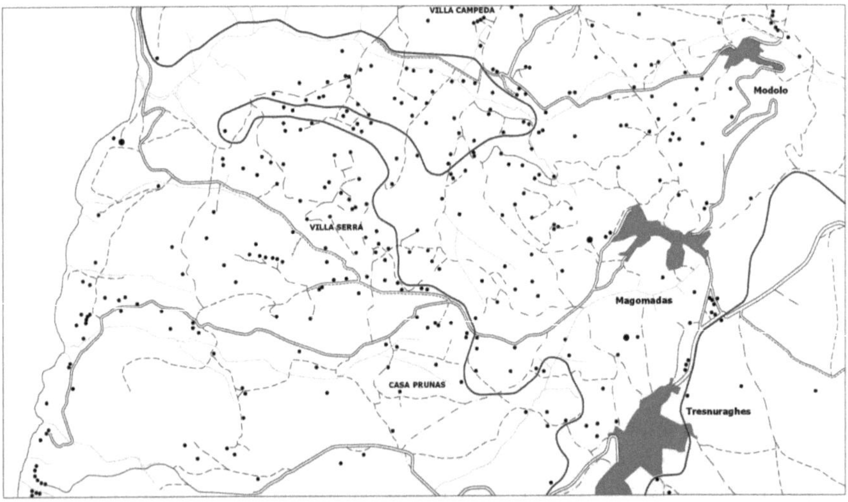

Fig. 2 The route of the nineteenth century's railway and some historical "villas" on a cartographic representation (IGM 1958)

Fig. 3 Rural landscape in Modolo's valley

1.2 The Sparse Settlement. Historic Building Heritage

The historical forms of settlement in *Planargia* are characterized by a production model set on a dense rural road network that supports a landscape of steep slopes redesigned through the use of terraces (Fig. 4) and offers a case study representative

Fig. 4 Terraced hills in the territory of Magomadas

of a constructive model realized with the use of historic building techniques and local materials. The rural house of *Planargia*—an intermediate model between the ones of Northen *Campidano*'s centers and of mountainous ones of *Meilogu* [7]—is designed in the service of agricultural activity and generates customs and practices that transform the workplace into a temporary place of residence and define a broad repertoire of building models.

The different examples of rural houses, abandoned, partially recovered and often flanked by recent building (that represent the result of the changing needs and habits of today's society) are testimony of a landscape characterized by the presence of vineyards and olive groves and sometimes orchards placed along the perimeter of the fund, inside of which, in some cases there was (until recent times) a space dedicated to the culture of grain.

The observation of the survey's data and the analysis of historical maps has shown a system of routes survived almost unchanged consisting of main roads that collect all the centers of the district and a dense network (Figs. 5 and 6) represented by mule tracks and paths bordered by dry-stone walls largely preserved. Some sections of the nineteenth century's main communication network, particularly in the section that connected Tresnuraghes to Bosa, have conserved the stone-paved and the only changes observed are an increased section of the road and a replacing of the unpaved road with a mantle of asphalt. The survey's analysis also offers the

Fig. 5 Ancient road in the territory of Magomadas (1), Magomadas/Modolo (2), Magomadas/Tresnuraghes (3) and the so called *Su caminu osincu* (the street of Bosa)

Fig. 6 Structure of territory in locality *Santa Lughia* (Magomadas)

opportunity to acquire a great number of information on traditional historical buildings and allows to identify the "minimal cell", several variants that come from it (through depth or lateral juxtaposition) and the identification of several stable residences located near the railway line realized in the second half of the nineteenth-century [8]. This "module" have generally a net size equal to (4.00 × 4.00) meters, consists in a single living space with a single access, often windowless and built with local stone, wooden roof (with inclined pitches—mostly double) covered with a mantle of tiles. Inside the building we can find furnishing accessories, including benches, shelves built-in cupboards and often a system to collect, convey and keep rainwater in underground tanks (also made of local stone). The data acquired do not allow to hypothesize an increase of minimum side produced through the repetition of the module and leads to the definition of basic variants only where the dimension of rural buildings is mainly due to the budget of the owners; we can find single-cell unit and a number of variants that, regarding the size, result in some cases the enlargement of the "minimal cell" but more frequently are, in origin, double, triple with depth or lateral development. Within this typological catalog we report some variations that shows a more refined design, dating back to the late nineteenth century, characterized by larger rooms, symmetrical main façade with central door and windows realized with decorative elements such as molded cornices and jambs realized with worked trachyte. The smallest module that we can find is represented by the wine cellar (4.00 × 4.00 m), which is the

Fig. 7 Basic variant 1. Base model (1) and some variations of the basic module (2) in locality *Santa Lughia, Pischinas* (Magomadas) *and Sorrighes* (Bosa). We can observe a reuse of "ancient" materials, in this case, employed in the corner brickwork

room mainly used for grape processing and custody of working tools, where sometimes are present niches (made of stone) to store the vats. Prevails, as already mentioned, the type of building with double pitched roofs perpendicular to the façade, that we called in this catalog, basic variant 1 (Fig. 7). This model presents size variations of rooms dimensions (4.00 × 6.00 m), a window in the lateral facade and sometimes have a small opening above the front door of the house. The position of the access door is generally not in axis with the building, in order to optimize the management of the internal space; in fact, the prospect becomes symmetrical as soon the longitudinal dimension of the room increases. A second type identified is the basic variant 2 (Fig. 8), obtained by the lateral juxtaposition of two basic variant 1, with the ridge of the double pitched roof parallel to the main facade; the local access is provided with stone benches (that we can also find outside together the stone rings for tethering animals) while in the next room there is a built in wardrobe partially realized using the thickness of the wall mass.

Another case study is the basic variant 3 and 4 (Figs. 9 and 10) that presents a sloped roof, parallel or perpendicular to the main facade. The design of this models is been realized through a lateral juxtaposition of two rooms connected by an opening without infix and windows on the long side of the building.

Fig. 8 Basic variant 2 in locality *Agra* (Magomadas)

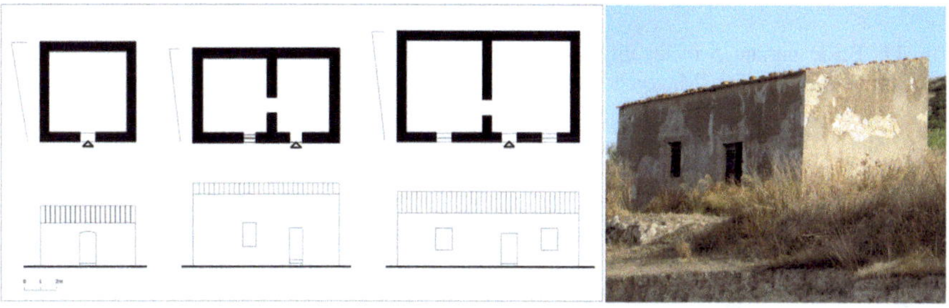

Fig. 9 Basic variant 3 in locality *Giagonia* (Magomadas)

Fig. 10 Basic variant 4 in locality *Sorrighes* (Bosa)

An increase in depth is been identified and defined basic variant 5 (Fig. 11); this model—obtained by juxtaposition of two module—shows six interconnected rooms used as local work, kitchen and bedrooms.

The repertoire is completed with the variant 5.6 (Fig. 12) and 7 (Fig. 13, casa Prunas) that represent some solutions with unique characters, rooms with a biggest

Fig. 11 Basic variant 5 in locality *Crastu Ruggiu*. In this case the border line of the property follows the railway line (Magomadas/Bosa)

Fig. 12 Basic variant 5 in locality *Abba Mala* (Bosa). and *Santa Lughia* (Magomadas)

Fig. 13 Casa Prunas: localization of "casa Prunas" in the historical cadastral map (courtesy of Archivio di Stato di Nuoro) and view of the house in the summer of 2007

dimension than others basic variant; this models are characterized by worked stone and a symmetry that demonstrates a particular attention to architectural forms and constructive details. The materials employed are the late Miocene sedimentary lithologies or ignimbrites of the cast-alkaline cycle oligo-Miocenic (trachyte) or the Pliocene basalts [9]. The materials used for the masonry structures are—close to the plateau—marly tuff and basalt, while the structural elements of the vertical openings are realized with trachyte and basalt, often accompanied by discharge arches made with roughly cut stones or blocks of tuff. The lintels of the internal openings are generally made of wood with two elements arranged side by side.

As we approach to Bosa and the coastline, and move away from basaltic plateau there is a progressive increase in the use of trachyte for elevated structures and the panels of the doors and windows, which are embellished with simple decorative elements. The walls of the house (average thickness of 50 cm) are plastered and inside the rooms we find shelves, cupboards and a fireplace located in the entrance room, that sometimes result without the wall structure and flue pipe; in this case the removal of and flue gas directly moves through the roof that usually present two inclined pitches realized with a main reinforcement of wooden beams that do not exceed 5.00 m of length; the secondary frame is made of beams (arranged in a transverse direction respect to the main reinforcement) is completed by an overlying structure that supports the external finish rods realized with a curved roof tiles. The houses are usually provided with a cistern located under the ground floor; this system collect the water from the double-pitched roof, but sometimes we can find flat roofs with collection function; this solution was used to obtain a water "cleaner" with the aim to avoid the runoff on the mantle of tiles. The rainwater, which preferably was accumulated in the winter months (because it was considered of better quality) was subject to a filtering phase that preceded the use. The windows are wood double shutters; external ones have an average size of (1.00 × 2.00) meters, the internal ones are smaller than the external and sometimes are absent, as happens in the variant model with lateral juxtaposition. The variant base 1, realized without a window opening, present at the top of the door a little frame (about 40 × 40) cm that replaces it (Fig. 7).

Inside the building are present the housings for the vats, realized in masonry as well as the benches and the ring used to bind the donkey. Regarding the spatial organization of the fund, we find the owners' house tile or adjacent to the wine cellar. When the fund was entrusted to the care of a sharecropper (*mezzadro*), he stayed in a house next to the owners, and if possible in the closest position to the entrance of the property, as well as the wine cellar, in order to facilitate working operations.

The home of the sharecropper (*mezzadro*) usually was a part of the building, in particular one of the cells; differently happened in the properties of wealthy landowners who permanently resided in the fund; in which case the home of the sharecropper is separated from the main building.

1.3 Architectural Survey and Representation: The Case of "Casa Prunas"

Architectural survey and landscape drawing [8] have been offered an interesting repertoire of historical building models realized in perfect harmony with materials and shapes of places. The aims of the research has required an analysis of historical maps, landscape drawing, direct and photogrammetric survey which have product a first documentation of this important "abandoned" heritage and a first catalog of plan models integrated with constructive details.

Furthermore, the state of ruins that characterize the historical buildings has made necessary a quick and no-contact procedure already employed in similar case study

Fig. 14 Casa Prunas: survey database (spring of 2018)

[10]. The “casa Prunas”, located in locality *San Giovanni* (Magomadas) has been selected to test this methodology; direct and photogrammetric survey (supported by drone flights) has defined a digital model from which obtain aerial views and orthophotos useful for different analysis; the aerial view allows an observation of the architecture and its landscape context, the orthophoto lead a rigorous analysis of geometrical matrices employed during design solutions and constructive phases (Fig. 14).

Acknowledgements The author would like to give a special thanks to Salvatore Ganga (Archetypon) for his precious and qualified help in surveys operation and in particular during photogrammetric survey with drone flights. Furthermore author wants to give a special thanks to Vincenzo Bagnolo, Giovanni Sistu and Ezio Pirinu (my father) for their participation during field operation and remember Gianni Fois (†) for his useful and passionate information about country life and traditions.

References

1. Le Lannou M (1941) Pâtres et paysans de la Sardaigne, Tours Arrault, Traduzione italiana (a cura di Manlio Brigaglia): Pastori e contadini di Sardegna. Cagliari, Edizioni della Torre 1979
2. Day J (1973) Villaggi abbandonati in Sardegna dal Trecento al Settecento. Parigi
3. Ortu GG, Sanna A (2009) Atlante delle culture costruttive della Sardegna. Le geografie dell’abitare. Department of Architecture of the University of Cagliari, Sardinia Regional Office of Planning
4. Corda A M, Mastino A (2005) Contributi all’epigrafia d’età augustea. In: Actes de la XIII rencontre franco-italienne sur l’epigraphie du monde roman, a cura di Gianfranco Paci, Macerata, pp 277–314
5. Milanese M (2006) Vita e morte dei villaggi rurali tra Medioevo ed Età Moderna in Quaderni del centro di documentazione dei villaggi abbandonati della Sardegna, Borgo San Lorenzo (FI)
6. Pirinu A, Sistu G (2015) Paesaggi imperfetti. Il DI-Segno del paesaggio della Planargia Costiera. (Sardegna Centro – Occidentale), Editore Edifir (FI), p 128
7. Atzeni C (2009) Architetture delle colline e degli altipiani centro-meridionali, I manuali del recupero dei centri storici della Sardegna. Regione Autonoma della Sardegna – Università degli Studi di Cagliari, Dipartimento di architettura – DEI Tipografia del Genio Civile
8. Pirinu A (2008) Il modello insediativo dell’ager bosanus: gli insediamenti sparsi in atti del Convegno internazionale Le case e i luoghi del lavoro: letture e confronti, tenutosi a Fisciano (SA) il 28–29 settembre 2007, pp 635–644, luglio 2007
9. Carmignani L, Oggiano G, Funedda A, Conti P, Pasci S, Barca S (2008) Carta geologica della Sardegna (geological map of Sardinia), scale 1:250000. Litografia Artistica Cartografica, Florence
10. Pirinu A (2017) Pinnettas: traditional Shepherds’ Huts of Sardinia. Geometry, Shape and Materials, in putting tradition in to practice: heritage, place and design. In: Proceedings of 5th INTBAU international annual event, a cura di Giuseppe Amoruso, Editore Springer, pp 467–474

Riverscapes and Watersheds: Cultural Heritage Layers Along the River Guadalbullón (Jaén, Spain)

Pilar Chías and Tomás Abad

Abstract Humboldt went beyond the old concept of landscape as a mere visual scene. The modern notion aims to find a landscape's inner structure by means of dynamic visions provided by panoramas and itineraries. These prove to be particularly useful when applied to the analysis of riverscapes. According to Braudel, a landscape is a cultural construction in the guise of something natural, and this is particularly evident in the proposed case study. From the perspective of natural processes, the River Guadalbullón flows along a narrow valley with a variety of land forms and topographical features. From the cultural point of view, the river was the borderline between the Muslim territories and the Christian kingdoms throughout the Middle Ages. They were linked by means of an ancient royal road that was recently transformed into a highway from Madrid to Málaga, thereby subjecting the landscape to considerable strain. But the narrow valley still conserves old archaeological sites that date back to the Bronze Age. The old route is still in use and dotted with lodgings, bridges and watermills as described by the travellers since the 17th century. The castles that controlled passage through the valley also survive, as do the *atalayas* which sent out visual signals alerting to dangers. All of them share the same space as ancient crops like oil groves, elements of vernacular architecture such as *alquerías*, and ancient opencast mines. This study aims to prevent the disappearance of all these structures, to avoid the extinction of species, and to preserve the memory of territory and landscape.

Keywords Landscape · Sustainable development · Territory · Regional planning · Local identity · Ancient cartography

P. Chías (✉) · T. Abad
School of Architecture, University of Alcalá, Alcalá de Henares, Madrid, Spain
e-mail: pilar.chias@uah.es

T. Abad
e-mail: tomas.abad@uah.es

G. Amoruso and R. Salerno (eds.), *Cultural Landscape in Practice*,
Lecture Notes in Civil Engineering 26,
https://doi.org/10.1007/978-3-030-11422-0_3

1 Introduction. Concepts and Goals

Ever since Humboldt, the idea of a landscape as being merely a visual scene has been superseded by a modern conception which aims to discover its internal order. To do so it turns to the dynamic vision afforded by panoramas and itineraries, which prove to be of particular use when studying landscapes with a linear development, most especially, riverscapes.

A complementary approach tackles the evolution of the idea of landscape from a heritage-centred concept tied to monuments—to what is singular and exceptional—but also in terms of its broader territorial dimension [1], thereby going beyond aesthetics to include ethical, scientific, social [2] and pedagogical dimensions.

On the basis of this encounter between the natural and the human, thinkers like Braudel [3] regard landscape as a cultural creation in the guise of something natural since the direct or indirect impact of human presence has transformed unspoilt natural areas into areas at risk, when it has not effaced them all together.

Today it is this view of the land as a constructed or produced element [4–6] which prevails. Landscape is a palimpsest requiring an analysis which attends to the natural geographical support as well as the elements that have been built over time, and which assesses their degree of insertion and extension, the ways in which the new is grafted onto the old, and the extent to which the landscape has been transformed during the process of absorption or being put to new uses. To this end, methods must be used which adopt the essential evolutionary view, which is now commonplace in urban studies but has only recently started to be applied to territorial analysis [7, 8].

Consequently, we approach territorial analysis as the key component of local identity, as a deconstructive process requiring deep knowledge of a landscape's internal structures and thereby throwing light on long-past modes and conditions of life, skills and capacities, and assesses resources and the multiple ends to which they are put. Thus, territorial analysis becomes a source of historical knowledge about societies of the past [9].

The goal of this research is to prevent the territory from disappearing and to recover its memory as the first step towards developing it respectfully and sustainably for tourism, thereby regenerating the local economy.

1.1 The River Guadalbullón Valley: Historical and Geographical Context

Focusing on the how a territory and a landscape are constructed is particularly apt in the case of the Guadalbullón valley. Its river a tributary of the River Guadalquivir, the valley has been inhabited continuously since the Bronze Age (2nd millennium BCE) [10] and traversed since time immemorial by an important communications route joining the Mediterranean coast and the Central Plain (Figs. 1 and 2).

Fig. 1 Location of the Río Guadalbullón Valley in Andalusia, Spain. *Source* ESA/NASA Spain, Copernicus Sentinel Data, 2/3/2016. The red rectangle corresponds to the area studied here

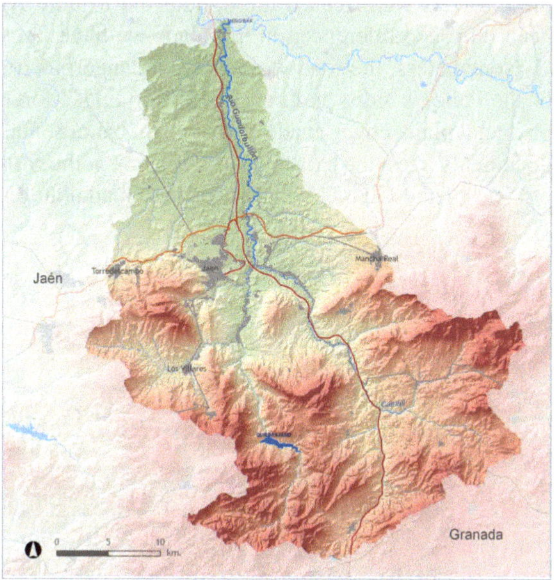

Fig. 2 The Río Guadalbullón Valley. *Source* The Andalusian Government's Centre for Landscape and Territorial Studies, 2012

It was when settled by the Iberos (7th-1st c. BCE) that the first territorial structure began to take shape along the lines of a *pagus*, with the building of a shrine at the river's source and the foundation of various *oppida* wherever there was fertile land for farming and ground of sufficient elevation to control the pathways and tracks [11].

After the Roman victory in the Second Punic War in 201 BCE, the main Iberian *oppida* were transformed into Roman settlements: *Iliturgi*—situated where the Guadalbullón flows into the Guadalquivir—was re-founded and converted into one of the area's principal markets; similarly, Puente Tablas, 18 km further upstream from Iliturgi, was inhabited almost without interruption until the Islamic period.

In order to safeguard communications between the upper Guadalquivir and the intra-Betic region, under the reign of Augustus Caesar a road was built between 8 and 7 BCE which followed the course of the Guadalbullón downstream towards the sea [12]. Evidence of this are the four milestones which have been found along the valley. These not only prove that it was part of the *Viniolis-Mentesa* section of the *Acci-Castulo* road, but also testify to the successive restorations and improvements carried out under the emperors Hadrian (136 CE), Maximinus Daia (305 CE), Constantine (307–317 CE) and Crispus (317–326 CE) [13].

The valley's strategic location led to the reinforcement of its system of vigilance. To this end, a line of 11 watchtowers was built, each four km from the next and visible the one from the other. This way, not only was transit along the valley controlled but visual signals could also be sent to transmit messages. The line stretched from Castillo de Arenas in the upper stretch of the river to Torre Bermeja, in the lower lying land around Jaén. A number of archaeological finds show that these towers were reused throughout the Middle Ages (Fig. 3). These finds have been duly geo-referenced in the maps we have drawn.

Nonetheless, the road was always a dangerous route because of the steep slopes of the valley's sides and occasional flooding. Moreover, the higher land's orography and plant cover made it home to bandits and renegades. As early as 765, documents attest the existence of uprisings in the Muslim district of *Wadi Abd Allah* or "Río of the Guard"—that is, of the Guadalbullón [14].

Fig. 3 Landscape where the highland and lowland zones meet in the River Guadalbullón valley. The tower "Torre de la Cabeza" in the village of Pegalajar, aligned with the Royal Road or "Camino Real". *Photograph* The authors

Between the second half of the 13th century and the late 15th century, the valley remained a frontier zone with shifting limits which separated Christian territory from the Nazari kingdom of Granada. It was the scene of constant skirmishes which forced the settlers to take refuge in the impressive fortresses of Jaén and Puerta Arenas. These skirmishes also led to a considerable slump in population and the appearance of large areas of wasteland, the valley's fertility notwithstanding. As a consequence, the centre of population that lay on the border enjoyed certain royal privileges in matters of taxation and franchises, and that transformed them into genuine emporia and encouraged resettlement.

Otherwise, the highland areas that spread out to either side of the valley were a mixture of woods with dense undergrowth, making them perfect hunting grounds, with, here and there, meadows for grazing.

When truces permitted, the valley was a thriving commercial hub between the Guadalquivir valley and Granada. Markets were held frequently and their locations survive in the place names of today. In fact, by the late 15th century two roads ran in parallel, one of either side of the river: the road on the left bank was known as the "Road of the Dale of the Gate of the Tower of the Star" (Camino de la Cañada del Puerto de la Torre de la Estrella), where the toll was charged; this was the route taken by the *almayales* or cross-border traders. The road on the right bank climbed up to Cambil—today's Carretón—Pass and was used by the Catholic Kings as a military route when engaged in the conquest of Granada, whose fall would culminate the Reconquest (Fig. 4).

Peace brought appreciable changes to the valley's physiognomy. The economy and people's ways of life were changed, while the re-population of either old or newly founded settlements entailed significant changes to the landscape, which translated chiefly into the disappearance under the plough of large areas of woodland and meadows. New socio-economic ties were struck between the settlements in the valley which led to the legal independence from Jaén of other neighbouring settlements which had thitherto formed part of its "alfoz", or dependent villages.

The roads were subject to intense traffic in the following centuries, the one on the left bank becoming a Royal Road and, later, the main N-323 road, many stretches of which follow the course of the original. During the 18th century, King Philip IV and a host of other travellers to Granada—among them François Bertaut in 1659, who was followed by many more until the late 19th century—left records of their journeys through the valley.

As for geography and natural processes, the Guadalbullón boasts a number of fascinating singularities in so far as the valley is part and parcel of the territory and has a very varied geomorphology and geology (Fig. 5).

A look at the longitudinal cross-section of the river's bed shows how its drop in altitude—almost 500 m in Little more than 70 km—is significantly greater than those of the Guadalquivir's other tributaries. Yet in terms of its longitudinal gradients, it is a most homogeneous river.

As for the course of the Guadalbullón, it can be divided into three landscape types: the upper section, about 37% of its total length, from the source of its

Fig. 4 Gaspar Salcedo de Aguirre 1587: *Geographia o description nueva del Obispado de Jaén*. Biblioteca Nacional de España, Sign. MR/42/639. Manuscript map. The north is towards the bottom, the edges are graduated in leagues; approximate scale 1:350.000. The rectangle corresponds to the area studied here, with the city of Jaén as the centre-point

Fig. 5 The upper section of the valley seen from the road to Carchelejo. *Photograph* The authors

different branches to La Guardia is a riverscape typical of high mountains and highlands; in its middle section, about 36% of its total length, from La Guardia to Jaén, the river flows through lower-lying land, between hills or hillocks; the lower course may be subdivided into an undulating riverscape, 21% of its total length, until it reaches the floodplain of the Guadalquivir, and its final section of 1 km as far as the village of Mengíbar [15].

As for its physiognomy, the valley is notable for its irrigation agriculture—records of which survive from the 15th century—chiefly in the alluvial valleys and the floodplains to either side of the river between La Guardia and its outlet at Mengíbar. The uppers section of the river flows through natural landscapes with abundant trees and bushes. In this connection, the most extensive transformations have affected agricultural uses due to changes in the relative amounts of land devoted to irrigation and cultivated species.

Finally, the valley is home to unique aquatic and terrestrial fauna and flora, which need to be protected and are to be fond as well in the neighbouring Natural Park of Sierra Mágina and the Peri-Urban Park of Monte la Sierra.

All the foregoing is evidence of the valley's wealth of cultural strata, of that palimpsest drawn and built over a period of four thousand years, which more than justifies its analysis.

2 Methodology

Territorial studies are approached from different perspectives, among them those of geography, ecology … and territorial organisation and transformation, the perspective which concerns us here in view of our stated goals of recovering the memory of the territory as the first step towards developing it respectfully and sustainably for tourism, thereby regenerating the local economy.

In our case we have applied a methodology that has already been tried out successfully in other geographical areas of the Iberian Peninsula [16]. This methodology makes use of information and communication systems and other digital means offering ways of approaching heritage and its imaginary on different scales [17]. We implemented a geographical information system for heritage and organised the methodology into various steps.

As a starting point we delimitated the area of study with precision and decided to include in the research noteworthy features of physical geography—for example, points of high ground with invisibility the whole length of the valley, such as La Serrezuela (1127 m), the Sierra de Grajales (1660 m) and Monte Prieto (980 m)—, together with elements built by man through the ages. This was done by means of a thorough inventory of geo-referenced resources which was cross-checked with other inventories and existing plans.

The next step, consisting in the collection of data, entailed exhaustive documentary research encompassing written, graphic and cartographic resources, as well, of course, as the obligatory fieldwork which enabled us to verify the existence or otherwise of the documented features and elements, to record their current characteristics, and to geo-reference them. Reminiscences, the collective memory and traditions were also considered. All this data is reflected in the relevant sketches at different scales and is stored in multiformat relational databases (Fig. 6).

The digital cartography on which the SIG was based was carried out on the National Topographical 1:50.000 Map Series-sheets 926, 947 and 969. The oldest

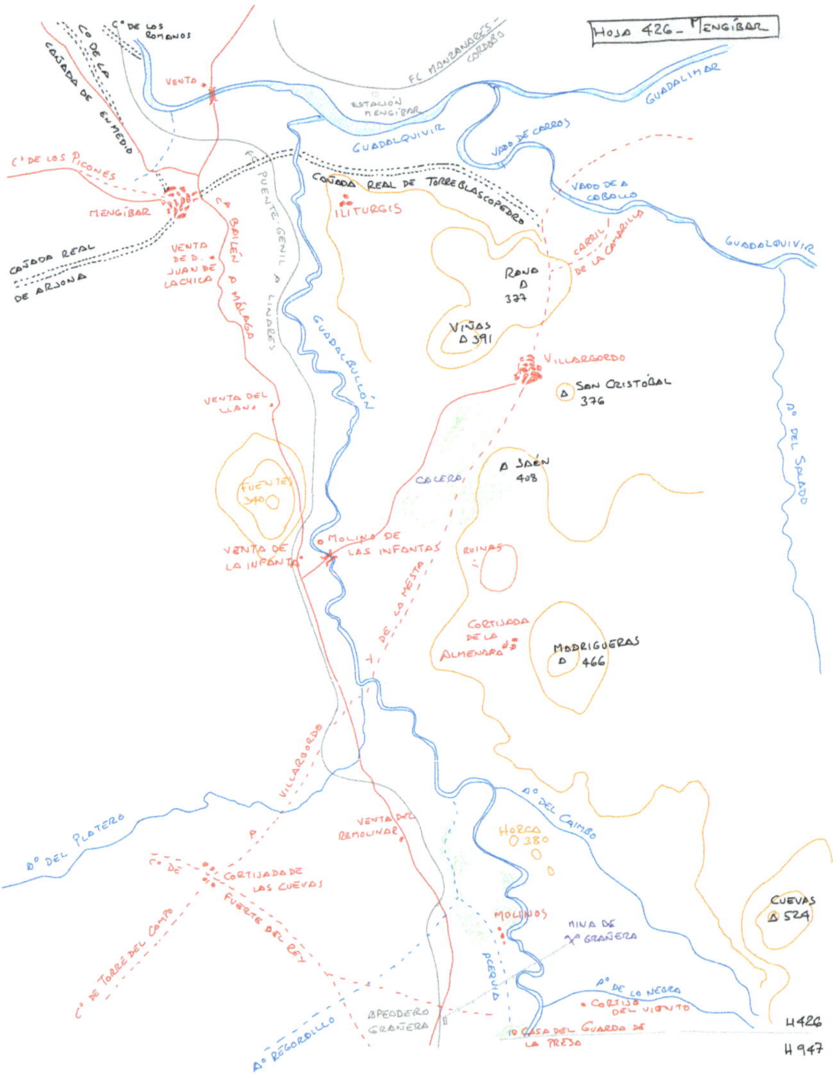

Fig. 6 P. Chías y T. Abad 2017: field sketch showing the heritage elements in the middle-lower course of the River Guadalbullón. The north is towards the top. Original scale 1:50.000

editions available were used with a view to recovering the toponymy. We then incorporated into the map the data collected from the different sources and the field work [18].

After referring as and when necessary to the SIG databases, we managed to obtain the information and to perform the analyses required to frame a regeneration proposal.

3 Analysis and Action Proposal

Our analysis was focused on locating and defining the weaknesses of the structured and dynamic system which makes up the valley, as well as on identifying its sources of potential.

Its most egregious weakness is the impact of edification since the river is under considerable strain around the built-up areas of Cambil, Jaén and Mengíbar, where spaces traditionally devoted to agricultural uses have been occupied by buildings which at times actually encroach upon the riverbeds. It is particularly urgent that the over-exploitation of aquifers should be avoided if the springs are to be preserved—a case in point being the unique hydraulic system of Pegalajar—, and that crop transformations be controlled in order to weaken the monopoly of the olive grove and its pernicious effects on market gardens and orchards.

One of its sources of potential include a group of features which make it unique and worth preserving, reinterpreting and revaluing. Moreover, the river enjoys great ease of access, and that smooths the way for proposals for recreational and leisure activities such as fishing in the lowland trout preserves around the headwaters or bird-watching in the area of Puerta de Arenas-Santa Lucía.

Other noteworthy public uses and festivals of interest to tourists include the Santa María Magdalena de Mengíbar pilgrimage, which takes place on 14 July and includes an open-air celebration on the riverbank. In this connection it might be remarked that in an effort to support tourism, the European Union, the Ministry of the Environment and the Government of Andalusia have published a guide to the region's gastronomy and designations of origin, not to mention other local and regional tourist guides, and that there is an ample supply of accommodation, including campsites and hotels in the main centres of population.

On the basis of our analyses and inventory of heritage resources, our conclusion is that the most valuable action for regenerating the valley should focus on proposing the creation of a heritage park capable of uniting the different periods of the river's shared history and of providing a coherent and attractive interpretation of it, which is borne out by the available resources [19].

As in the previous steps, we carried out exhaustive documentation of the historical periods which are best represented in the valley before defining a series of thematic sub-areas with a common link serving as the backbone to a specific narrative.

Fig. 7 Correspondence concerning the 1774 suit over the demarcation of the vale of Cazalla, Jaén. Archivo Histórico Nacional, Madrid, Sección Nobleza, Bookmark BAENA C.254, D.159–176

Thus, we were able to define clearly distinct sub-systems in a territorial area contained within the valley as a whole and the park. Among the sub-systems we propose are: 1/the sub-system formed by the group of fortifications and defensive towers that punctuate the valley; 2/the hydraulic system of Pegalajar, comprising the Fuente de la Reja, la Charca and la Huerta, which, organised into terraces supported by dry-stone walls or *jorfes* and prolonging the urban areas, constitutes an ecosystem and irrigation scheme which is unique in the Mediterranean; 3/the ranches, farmsteads and other types of popular architecture in the Dehesa de Cazalla, which together conserves an age-old system of land ownership (Fig. 7).

To define the project, we used the elements recommended by Lynch [20], namely *landmark*, *region*, *node*, *path* and *edge*, all of which have the additional advantage of lending themselves to SIG representation with the aid of dots, lines and areas, as well as to topographical verification.

All the foregoing has allowed us to devise paths to be followed using forms of transport which are related to the history as it is told—on foot, on horseback, by boat or by bicycle—since it is essential to get a feel of the route. The aim of all these paths is to enhance the image and uniqueness of each place while at the same time maintaining a design in which everything fits the overall proposal. Finally, the resources inventoried which are not important to the sub-narratives were joined together later by means of secondary paths.

4 Conclusions

Although the River Guadalbullón valley is included in the Andalusian Watersheds Master Plan, and despite its unique collection of heritage and the need for economic development in the upper stretches, there is still no regional development project in place.

In this regard, one of our studies main conclusions is that the time is ripe for the creation of a heritage park in view of the results of our analysis of weaknesses and sources of potential.

A heritage park would attract visitors and investment alike, creating opportunities for further actions and areas for new projects.

And, with strict respect for the territory's features, the park should integrate the preservation of heritage resources by means of education, leisure, tourism and the economic development required to make living in the area a feasible and attractive option.

Finally, the adequate solution to each landscape lies in its very essence.

References

1. Sereni E (1989) Storia del paesaggio agrario italiano. Laterza, Bari
2. Lefebvre H (1974) La production de l'espace. Éditions Anthropos, Paris
3. Braudel F (1985) La Mediterranée: L'espace et l'histoire. Flammarion, Paris
4. Muratori S (1967) Civiltá e Territorio. Centro studi di storia urbanistica, Roma
5. Caniggia G (1967) Strutture dello spazio antropico. Uniedit, Firenze
6. Cataldi G (1977) Per una scienza del territorio. Studi e note, Uniedit, Firenze
7. Chías P, Abad T (2011) La Bahía de Cádiz: Territorio fortificado y paisaje/The Bay of Cadiz: Fortified territory and landscape. In: Chías P, Abad T (eds) El Patrimonio fortificado. Cádiz y el Caribe: una relación transatlántica/The Fortified Heritage. Cadiz and the Caribbean: a transatlantic relationship. Publicaciones de la Universidad de Alcalá y Fundación Jose Félix Llopis, Alcalá de Henares, Madrid, pp. 18–172
8. Chías P, Abad T (2012) El arte de describir el territorio: Mapas y planos históricos en torno al puente de Alcántara (Cáceres, España)/the art of describing the territory: historic maps and plans of the brigde of Alcántara (Cáceres, Spain). Informes de la Construcción 64:121–134. https://doi.org/10.3989/ic.11.071
9. Ortega Valcárcel J (1998) El patrimonio territorial. El territorio como recurso cultural y económico. Ciudades 4:33–48
10. Escobedo E, López JA (2013) Asentamientos de la Edad del Bronce en el Alto Guadalbullón. Sumuntán 31:79–98

11. Lechuga MA, Bellón JP, Rueda B (2015) Nuevas propuestas de actuación para el estudio del oppidum de Iliturgi desde la arqueología del territorio. Revista Atlántica-Mediterránea 17:211–221
12. Sillières P (1976) Un grupo de cuatro miliarios en La Cerradura (Pegalajar, Jaén). Boletín del Instituto de Estudios Giennenses 90:55–70
13. Olmo A (2004) Mentesa romana, visigoda y musulmana. La Guardia fronteriza. Sumuntán 20:13–28
14. López Cordero JA (1996) El valle del río Guadalbullón en la Baja Edad Media. Una frontera entre Castilla y Granada. Jaenseñanza 9:19–30
15. Zoido F et al (2012) Los paisajes fluviales en la planificación y gestión del agua. Elementos para la consideración del paisaje en la Cuenca Hidrográfica del Guadalquivir. Junta de Andalucía, Universidad de Sevilla, Sevilla, pp 513–533
16. Chías, P, Abad T (2004) Los caminos y la construcción del territorio en Zamora. Catálogo de puentes. CSIC (Instituto de Estudios Zamoranos)/CEHOPU—CEDEX. Madrid. 2 vols
17. Salerno R (2015) City ideologies in techno-urban imaginaries. Urban 508(9):185–192
18. Cardone V (2015) Modelli grafici dell'architettura e del territorio (a cura di Salvatore Barba). Maggioli editore, Santarcangelo di Romagna
19. Galindo J, Sabaté J (2009) El valor estructurante del patrimonio en la transformación del territorio. APUNTES 22(1):20–33
20. Lynch K (1960) The image of the city. The MIT Press, Cambridge, Massachusetts and London, England

Indian Villas in the Valencian Landscape (Spain): *Casino del Americano*

Pablo Rodríguez-Navarro and Sergio Estruch González

Abstract After the phenomenon of emigration from Ultramar, a new architectural style, know as *Indiana* architecture, appears mostly located in the Canary Islands and the northern provinces of the Peninsula. This particularity makes of our case of study an exclusive construction of this emergent social class in the Valencian landscape. *La Quinta de Nuestra Señora de las Mercedes*, also known as *Casino del Americano,* is a recreational villa that since 1869 forms part of the Benicalap neighbourhood, and whose architectural and construction characteristics make it a property of great heritage value, and one of the last Indian villas in the city of Valencia. Due to the lack of bibliography, information will be mostly obtained through the study of the building. The graphic surveying is the main working method, which has an intrinsic value because of the danger of collapse of the building. The main aim pursued by this research is to present a historical, stylistic and constructive study, as well as a definition of the main characteristics of Indian architecture, through the most representative case of Indian villas in the city of Valencia.

Keywords Recreational villa · Graphic surveying · *Casino del Americano* · Indian villa · Valencian landscape

P. Rodríguez-Navarro (✉)
Architectural Graphic Expression Department, Institute for the Heritage Restoration of the UPV, Universitat Politècnica de València (UPV), València, Spain
e-mail: rodriguez@upv.es

S. Estruch González
Conservation of Architectural Heritage, Universitat Politècnica de València, València, Spain
e-mail: sergioestruch90@gmail.com

G. Amoruso and R. Salerno (eds.), *Cultural Landscape in Practice*,
Lecture Notes in Civil Engineering 26,
https://doi.org/10.1007/978-3-030-11422-0_4

1 Bourgeois Villas and Casinos in the Valencian Landscape. Indian Architecture: Indian Villas

The historical landscape of Valencia's orchard has its origin in the configuration generated during the Islamic era, characterized mainly by a complex network of ditches and roads. However, the territorial organization has undergone changes throughout history, being able to highlight three major periods that have determined different operating schemes associated with an architectural typology, its form of population and territorial distribution. Originally, the *andalusí* type orchards, (8th–13th centuries), where great extensions of orchard predominated, linked to each one of the *alquerías* (farmhouses) representative of a population unit and associated to a transit or irrigation channel. After the Christian conquest, the period of feudal orchards began (13th–19th centuries), in which a social transformation was made in terms of the water management and control, a fact that affects the territorial physical arrangement and their habitable spaces. And finally, the bourgeois gardens (19th–20th centuries) where fragmentation of the extensive territories took place to give birth to small land fragments, originating in this way the great dissemination of small residential architectures [1].

The 18th–19th century industrial revolution is associated with a change in the social model of the time, which is implicitly reflected in the architectural type. The landscape of the Valencian orchard before the 18th century, under the dominion of a feudal model, is characterized by the presence of two types of traditional architecture, such as the *alquería* and the *barraca*, whose origin was linked to the exploitation of the landscape. However, with the bourgeois revolution, new archetypes flourished in the Valencian territory, such as the country or orchard house, as a consequence of the fragmentation of the feudal latifundia, and the bourgeois village or pleasure village. These last models arose before the desire of the high bourgeoisie to have large gardens surrounding their residence. Depending on the size, cost and appearance, we can identify them as suburban mansions, casinos or villas and recreational villas.

Recreational villas and casinos are defined as holiday period residences of families with greater economic capacity and social position, in the city's surrounding areas, bordered by orchards and an idyllic landscape. The versatility of the architects, the different stylistic influences, the margin of creativity that the architects have due to the high possibilities of their clients, their personal whim and the freedom of the municipal norms as regards to these buildings, provoked a great variety of types.

These are exempt buildings, around which there is a private environment based on fenced gardens for closing off the estate. Their main features are: the square layout or Greek cross plan, the basement accessed by a staircase, its organization on one or two stories, an attic if they have sloping roofs, in addition to a tower or *miramar* for the contemplation of the landscape [2].

Our case study presents a unique characteristic of a movement that emerged in Spain towards the mid 19th century and the emigration to the Americas. Emigration to Ultramar is a phenomenon that is registered in the period between 1855 and 1930. The underlying reasons for this tendency resided in a period of conflicts, governmental instability and economic crisis that, along with the population surplus, promoted a set of measures to favor emigration. These new policies had an impact in very specific areas of the territory, in particular, the Canary Islands and the provinces of the north and northwest of the Peninsula such as Asturias, Cantabria and Galicia and to a lesser extent Barcelona, Madrid, etc. [3]. This social class, after obtaining fortune in Ultramar and returning to its native town, will be known as Indian or "American", and will seek to capture its new social status and acquired success through the construction of their new residences; it is what will be known as Indian architecture. This is defined as: "*The set of buildings whose construction (...) has been financed with money of American origin (...) are intended for stable housing of their owners and in cases of migrants with a more solid financial position, the villas respond to a scheme of second residence or recreational house, of temporary and summer occupation, to share with another main house, located in the urban nucleus*" [4, 5].

The majority of these constructions were solved with the participation of the masters of works, since the services of an architect were demanded by great fortunes as a sign of high social position and thus counting on the guarantee that the house presented an exclusive design, far from the catalogued constructions (Fig. 1).

Regarding the architectural style, numerous studies conclude that there is no particular language attributable to Indian architecture, that is, this group will

Fig. 1 *Casino del Americano.* Current status

configure their residences on the foundations of the native bourgeoisie in which it is inserted, incorporating connotations, customs and constraints due mostly to the rural location of these villas. Standing out among the most characteristic elements the presence of galleries and viewpoints, arcades and verandas, façades and roofs crowned by attics, ornate combs where the construction date was inscribed, initials of the owner's name or allegorical reliefs to his origin, garret and turrets.

However, in these recreational villas, if there is a characteristic element, that would be the garden. The Indian garden is generally for recreational use and contemplation, but sometimes complemented for productive purposes by planting fruit trees and crops. This is defined by the combination of three elements: the symbolic tree, garden and orchard. The symbolic tree of the Indian condition is the palm tree. They are arranged in pairs presiding over the front space between the residence and the main road, although they can also be planted isolated and integrated with other trees or defining an avenue. Other exotic species that can be found are araucarias, thujas, ligustrums, magnolias, cedars, chestnuts and pines.

These residences surrounded by gardens, tree masses and orchards create a unique environment around the main building capable of producing shade in summer, purifying the environment, creating a place of relaxation and being an element of comfort and prestige. Associated with this landscape is the enclosure of the property, constituting the external element of the property, proof of the success and quality of the house, hence it becomes one of the most ostentatious elements.

The garden, in imitation of European styles, follows two trends: the garden of French origin or the romantic English garden, although we find mixed solutions in which the French garden coexists in the front facade with the picturesque garden with abundant and exotic species in the posterior (Fig. 2).

Fig. 2 Elevation drawing of the *Casino del Americano*

2 Case Study: *Casino del Americano*

El Casino del Americano, also known as *La Quinta de Nuestra Señora de las Mercedes*, is a recreational villa that since 1869 forms part of the Benicalap neighborhood, in Valencia's North orchard, and whose architectural and construction features make it a building of great patrimonial value, and one of the last Indian villas in the city of Valencia.

Valencia's orchard has been morphologically estructured by means of four elements: the plot of ditches, the road network, the population centers and the parcel organization. That is where *El Casino del Americano* gets integrated, coexisting in its immediate surroundings with historical and traditional buildings, such as the *alquerías de la Torre* and *dels Moros*, remains of the *Molí de Pallús* and the layout of the Tormos ditch, but also with others examples of casinos or villas such as the well-known *Palacete del Rosal*, constituting a consolidated and characteristic image of this neighborhood, as impressions of the architectural, cultural and historical past.

So, this emblematic construction, defined within the scope of the typology of summer residence of the high bourgeoisie, with the connotations and characteristics of a villa, along with the conditioning of its owner's origin, make it an exclusive work of great interest, which is in an advanced state of deterioration despite being one of the last examples of this architecture in the city of Valencia (Figs. 3 and 4).

Fig. 3 Valencia's city map in 1883 [6]

Fig. 4 Print from the magazine "El Artífice" 1869

Historical analysis: Family chronological axis

Originally named *Quinta de Nuestra Señora de las Mercedes*, it is known as "Casino del Americano" due to the origin of its wealthy primitive owners, from the Americas. Joaquín Megía y Ortega, a native of Granada, was a soldier assigned to the Island of Cuba, where he married a wealthy Havanan, Mercedes González Larrinaga y Cruz. After retiring from the army, Joaquín Megía along with his wife Mercedes and her son Jacinto, moved to Valencia in 1865, where they bought three hectares of land in the Benicalap orchard and ordered the construction of a recreational village (1869), surrounded of beautiful and exotic gardens so that his wife felt like in her native land. The master builder Manuel Piñón was responsible for building this residence.

In April of 1876, the proprietor Dª Mercedes dies, beginning the discrepancies between her husband and her son, which lasted for three years, until in May 1879 the division and liquidation of assets of Mercedes' inheritance was executed, bequeathing *El Casino* to her husband. However, he had decided to retire to his birthplace in Granada, and therefore resoluted to sell his belongings in Valencia. After the sale of a first batch of land plots at the beginning of 1881, Jacinto Gil de

Fig. 5 Canvas print by José Peris in 1890

Fig. 6 Postcard 1965 (Ana Mª Reig)

Avalle (son of Dª Mercedes) acquired from his stepfather the part of the remaining lands, which he could barely enjoy, since a couple of years later, in July 1885 he died, and the property of this estate passed into the hands of his widow. *El Casino* then passed into the hands of Mercedes' granddaughter, Ángeles Gil de Avalle, who, after not leaving any descendants, decided in 1941 to sell *El Casino* to the Valencian industrialist Plácido Navarro Pérez [7]. During the 1980s, it was rented as a school center. And from the 1990s, it was transformed into a pub-disco. Finally, after its closure, it entered a process of abandonment and deterioration as a result of multiple lootings. It was in 2011 when it became a municipal property of the Valencia City Council, after an expropriation worth 3.5 million euros. The mansion was occupied on many occasions suffering from damage and fire, which is why most windows and doors have been bricked over (Figs. 5 and 6).

Formal description: The garden and the villa

As a result of the new trends and tastes towards the end of the 19th century, there is a variant of the orchard-garden associated with the villas of the bourgeoisie, whose changes reside in the following premises: "*(…) the main variation lies in the change of the house from the perimeter fence to the interior, occupying a position centered on the plot. It maintains the axial composition, the relationship between the house and the pond, the presence of ornamental species around it and the access road, although other elements change their orientation, such as the pergola and the garden elements that come to occupy the front (…) Towards the 80 s, aesthetic gardens with new exotic species will be strengthened, and the main axis and the interior roads become promenades with benches*" [8] (Fig. 7).

Fig. 7 Turrets of the perimeter fence

Considering the definition of bourgeois style orchard-garden, different elements can be identified in the plot, such as:

- The fence or perimeter fence. The villa has a symmetrical composition around a longitudinal axis in a south-north direction. It is formed by openings made of cast iron iron gates on an ordinary masonry sill with a lime mortar lining on which a mock ashlar is made. Each vain is articulated through small casting columns. In addition, located in the longitudinal axis, the access door is arranged, with an elaborate metal lock between pilasters of solid brick clad, and topped with an ornament, with two lions as guards, that at present have disappeared.
  The primitive enclosure was more delayed, so that it was crowned at the corners by two turrets. These hexagonal layout towers have an ordinary masonry wall with lime mortar and reinforcement in corners with solid bricks. The rest of the body, based on a solid brick factory, has a cladding and simulated ashlars following the style of the rest of the enclosure and a top finish with a crenellated body. In its interior small arris vaults are highlighted.
- Main axis and paths. They are what define the formal composition of the garden. The main path constitutes the symmetry axis of the composition. The geometric garden is organized on each side of the axis, based on the traditional division into squares, crossed by orthogonal axes of secondary paths in whose intersections there are little squares or roundabouts, as haven spaces. In addition, the paths used to be slightly raised to produce the sensation that the side vegetation was a tapestry on which one walked.
- Vegetation. In a mixture between the native and exotic styles of plants, we would find a garden invaded by Valencian species such as orange, lemon, pine, cypress and palm trees as well as low plantations such as roses, thyme and rosemary, and a place reserved for exotic species such as araucarias, magnolias and chestnut trees to satisfy the wishes of their owners.
- Characteristic elements: these were the pergola, animals, and water. In the main axis, stands out the presence of the pergola, formed by eight vains based on casting columns over brick masonry dies, with vaulted shapes and invaded by trellises creating an illusion of light and shadows.

The third element was the presence of animals. In our study case, we observed the existence of two birdhouses or aviaries of polygonal plant with a cover based on palm leaves and protected with wire mesh.

On the rear façade, we find a patio with a square floor plan with lateral accesses to the garden and, in the center, a water cistern for the collection of rainwater or fountain. In addition, there was water in continuous movement through the ditches that crossed the property and allowed irrigation.

Behind the patio, the residence is completed with a series of buildings or auxiliary pavilions such as stables, yards and a pigeon house or henhouse.

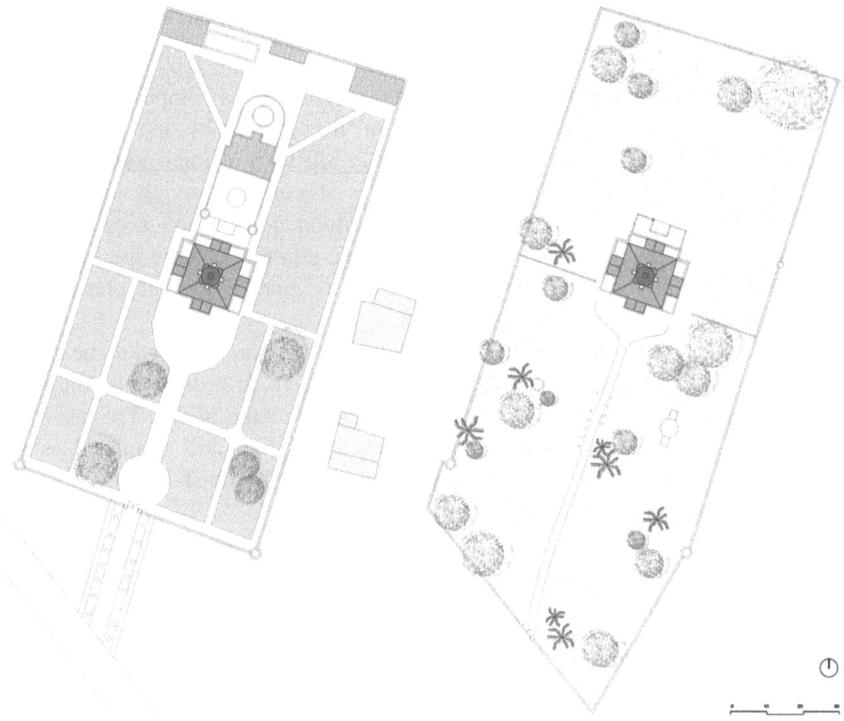

Fig. 8 Comparison of the organization of the plot and garden between the hypothetical state (left) and current state (right)

In the representation of the plan of the municipality of Valencia of 1929–1944, in the final part of the property, appear a series of agricultural buildings such as the gardener's house, *alquerías* and ponds (Fig. 8).

With respect to the residence, the first graphic reference is found in the magazine *El Artífice*, number 36, published on December 5, 1869. In this publication, in addition to the representation of the plant and façade of the project, there is a description that reproduces the article "Valencia in progress" published in the *Prophetic illustrated calendar, previously Pata de Cabra, for the ancient Kingdom of Valencia 1870:* "*(…) everything is prevented in its distribution despite not being able to be counted among the number of the great palaces; and without enumerating the halls, rooms, bedrooms, private living room, dining room, kitchens, library, waiting and weapons room, galleries and service stairs, which constitute the owner's home, contains, a billiard room, two private living rooms, and several departments for the maids on the second floor; and in the ground floor, quarters for the doorman, gardener and other servants, cellar, tack room, hall and bathrooms, coach house, kitchen, etc…; a large inner patio with a water cistern in the center, elegant and spacious stables, and a yard for the breeding of several domestic animals*" [9].

Volumetrically it consists of a ground floor, noble or main floor, a second floor, and a loft or attic under the sloping roof, being crowned in the center by a lookout tower or *miramar*. This tower is topped by a walkable terrace, from which the sea level horizon may be observed.

The residence has a square floor plan in which another is superimposed in the form of a Greek cross. The floor composition is based on the nine-square system.

On the ground floor, it is known that there was a Corinthian oratory on the west arm and the main staircase on the east arm, and completing the distribution we find rooms for the doorman, gardener and servants, cellar, tack room, living room and bathroom, coach house and kitchen.

From the noble floor we have more detail, since its distribution is shown in the original plan. In the arm looking out to the main façade is where the great hall would be located, at the end of the rear façade the dining room with exit to the roof that overlookes at the patio, and next to this the kitchen. At the east end, we find the main staircase. It is an imperial staircase with a first wide flight that rises and then divides into two symmetrical flights going back. On the landing, a service staircase starts, of a single section, which connects to the first floor, and from which a passageway is taken over the main staircase, which faces the start of a second staircase with three sections that connects with the second or servants floor. The anteroom on which the staircase disembarks, presents a dome with a lantern over four pendentives and a small "drum" (Fig. 9).

On the pendentives we find some pictorial decorations based on oval portraits of relevant characters like the painter Diego Velazquez; Christopher Columbus,

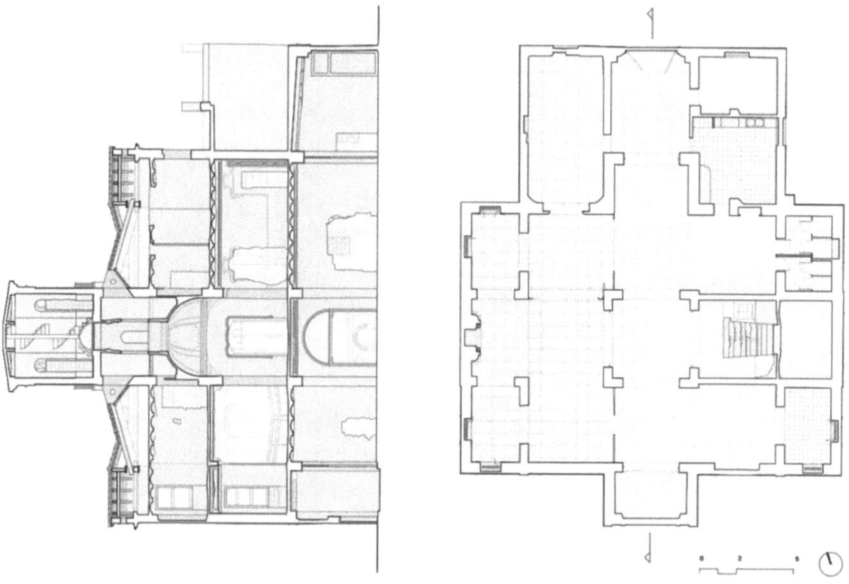

Fig. 9 Ground floor and section. *Casino del Americano*. Current state

discoverer of the Americas and land of origin of the owners; Gonzalo Fernández de Córdoba, a nobleman, politician and military Andalusian; and the writer Miguel de Cervantes.

The rest of the dependencies of this floor are occupied by the bedrooms located in the four corners, in addition to other rooms, anterooms, private sitting rooms and galleries.

The second floor is intended for the living areas and bedrooms of the servants, in addition to presenting a billiard room and two private sitting rooms. From the south chamber, a linear staircase starts, running above the spherical dome and disembarking in the central core where the lantern is located. This small room with a square floor plan has openings in the form of a window with semicircular arches on each of its sides, which can be seen on the roof as skylights. To this element is superimposed the *miramar,* of octagonal layout, which is accessed from a staircase that follows the same footprint as the previous one. In this last level, we found a double spiral staircase that allowed access to the *miramar's* upper passable terrace.

Finally, we would like to point out that in the *Casino's* printed image, the floor appears flanked on the back by two circular towers at the corners, inside which the word staircase is written. At present, there is no trace of their existence, and the only representation we have of them, is in the 1929–1944 map of the municipality of Valencia of the Geographic and Cadastral Institute (Figs. 10 and 11).

Fig. 10 Imperial staircase. Ground floor

Fig. 11 *Miramar's* double spiral staircase

Regarding the configuration of the building envelope, the element of which greater documentation is available is the main façade, finding the oldest image in the magazine "El Artífice" of 1869.

In this façade plane, all the elements are ordered from a vertical symmetry axis. In the composition, a basement consisting of the ground floor, a body formed by the main floor and the second floor and a top level formed by the attics and the gable roof, on which the *miramar* is found. The floors are highlighted by a superposition of Corinthian pilasters in the corners, which support entablatures that reveal the horizontal division.

The central composition shows a plinth with a mock ashlar work and large vains with basket-handle arches, except for the second floor, whose vains are materialized in the form of a shield. As for the ornamentation, "simulated" niches stand out with the classic sculptures of the noble floor, a medallion or rosette where appear the year of construction of the residence (1869) and the presence of sculptures crowning the pediments. The sculpture of the central façade features a weather vane, while those of the lateral façades have pulleys that raise and lower the awning of the terraces.

However, on the sides of the façade, where the presence of exposed brick stands out, there are some windows with shields topped with the iconography of a helmet, and terraces with rails in the shape of "pointed umbrellas" that are covered by the awnings.

The roof has two dormers for lighting and ventilation of the space bellow. From the *miramar,* a spire shaped mast supporting an awning stands out.

Finally, on the description found in the magazine "El Artífice", the comment made about the color is of particular interest: "*(…) the color of the* façade *is somewhat bold, lacking therefore the smoothness that would much enhance a remarkable building (…)*" [9].

3 Methodology: Obtaining Three-Dimensional Model in Heritage Conservation

In order to develop this research, scientific and rigorous guidelines and phases of work were established, with the aim of obtaining a high-precision three-dimensional model on which to develop graphic documentation as an inverse project before the likelihood of collapse of this property, currently in an advanced state of ruin.

In addition, due to the scarcity of documentary sources and publications on the property, we decided to resort mainly to the information provided by the main source, that is, the building itself through graphic surveying as the most important analysis tool.

Subsequently, with the graphic documentation, we have been able to develop the historical, stylistic-compositional, constructive-material and pathological analysis in order to obtain a thorough diagnosis of the property, and to successfully face the subsequent phases of rehabilitation and enhancement. This work methodology has been addressed in three phases:

1. Historiographic research. In this first phase, a review of all existing documentation has been carried out through the bibliography (magazines, books, articles…) related to *El Casino del Americano* and its architectural context, and through the study of documentary sources, information screening from the archives available in the city of Valencia, and the review of historical cartographies and aerial photographs.
2. Field work. After obtaining the necessary permits to access the interior of the municipal property, during the first visits, a photographic catalog was made, in addition to obtaining sketches and drawings, and carrying out a general data collection through direct measurement.
   After this first approach, we proceeded to analyze the characteristics of the property, selecting the method through which to carry out the graphical survey of the current state. We decided to apply 2D photogrammetry and image rectification since it is one of the most economical and effective methods to obtain

Fig. 12 *Casino del Americano*. Point cloud

drawings at scale of facades with flat geometries, while we used a 3D laser scanner to obtain the three-dimensional model due to its interior spatial complexity [10].

We must highlight the difficulties encountered during the lifting phase, due to the poor state of preservation of the building, discovery of fallen slabs, interiors with insufficient lighting, shores that generated shadows during the sweeping of the scanner and trouble accessing some rooms (Fig. 12).

3. Processing of data and information. In the phase of registration and data management, we proceeded to the alignment of all the stationings made, that is, the processing of the point clouds to obtain the building's 3D model, using Autodesk ReCap 360 software. From this model, we were able to draw the two-dimensional plans to scale that will provide us with accurate information regarding the current state of the building (Figs. 13 and 14).

4 Conclusions

The architecture carried out during the 19th century and the beginning of the 20th century in the surroundings of Valencia, stands out as a reflection of the changes taking place at the economic and social levels. A clear example of this is *El Casino del Americano*, though it can not be conceived without the landscape that surrounds it, both the one in which it is inserted and the one that it is associated to it. This is what makes up the Valencian rural landscape of the time, a characteristic landscape that unfortunately no longer exists and this fact is also reflected in the property itself.

The recovery of the original image where these architectures were conceived is not feasible since it has been absorbed by urban expansion. Not withstanding,

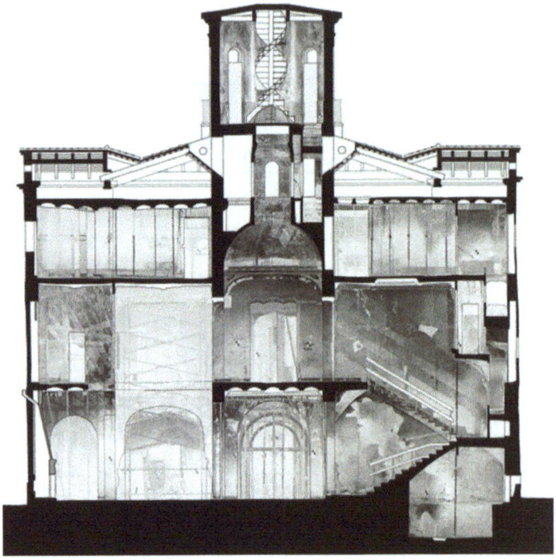

Fig. 13 Textured point cloud section

Fig. 14 Photogrammetric elevation

the immediate environment or landscape which, along with the architecture still present, could recover the essence of a particular time and its historical and cultural values.

Examples of similar recoveries are the gardens and villas of Monforte and Ayora, which with similar characteristics, and having been inserted into the urban fabric of the city, have recovered their essence.

References

1. Pérez A (2017) Arquitectura tradicional en la huerta sur de Valencia. Final master's project. Valencia, Máster en Conservación del Patrimonio Arquitectónico, Universidad Politécnica de Valencia
2. Benito Goerlich D (1992) La arquitectura del eclecticismo en Valencia. Vertientes de la Arquitectura Valenciana entre 1875 y 1925. 2nd edn. Ayuntamiento Valencia, colección "estudis" 3, Valencia
3. Álvarez C (1986) Emigración Asturiana a Ultramar y arquitectura. Parte I, la emigración. In: Boletín del Instituto de Estudios Asturianos. Año XL, nº. 117, enero-abril 1986, pp 243–270
4. Álvarez, C. (1986) Emigración Asturiana a Ultramar y arquitectura. Parte II, la casa indiana. In: Boletín del Instituto de Estudios Asturianos. Año XL, nº. 118, mayo-agosto 1986, pp. 525–562
5. Bores, F. (coord.) (2000) Casas de Indianos., ed. Xunta Galicia, Galicia
6. Llopis A, Perdigón L (2016) Cartografía histórica de la ciudad de Valencia (1608–1944), 4th edn. Universidad Politécnica de Valencia, Valencia
7. Reig AM (2011) El Casino del Americano y el Palacete de Pescara: villas indianas en la ciudad de Valencia. In *Boletín de la Sociedad Castellonense de Cultura*, Volumen LXXXVII, Enero-Diciembre 2011, pp 397–419
8. Besó A (2016) Horts de tarongers: La formació del verger valencià. Institut Alfons el Magnànim, Valencia
9. Anonymous (1870) Casino del Sr. D. Joaquín Megía. In: El Artífice, volumen 36, Diciembre 1869, pp 282–283
10. Rodríguez-Navarro P, Verdiani G, Gil T (2015) Comprehensive methodology for documenting the defense towers of the Valencian Coast (Spain). In: Defensive architecture of the Mediterranean. 15th–18th centuries, vol I, Editorial Universidad Politécnica de Valencia

Traditional Identity and the Progressive Loss of Local Character in La Sagra Region (Toledo, Spain)

Alejandro García Hermida

Abstract This study presents the basic aspects of a research dealing with the identity features of the architectural tradition of La Sagra region, in the Spanish province of Toledo, and the needed review of the criteria applicable to its conservation in order to avoid its disappearance. The problems presented in this regional analysis are very similar to those in other places and, therefore, similar studies may be valid for them. For this research, the architectural tradition had to be studied first, as the existing diversity and complexity in the local traditions of this area were unknown before undertaking this study, all previous studies considering it as a homogeneous entity. Finally, towns have been studied building by building, recording the transformations operated in each one of them. This is, the evolution of the building elements which define local urban landscape. To this aim, the composition, volumes and various constructive solutions used both in the new buildings and in the interventions made in the last decades on existing buildings have been analysed, then contrasting them with the traditional local identity municipal regulations of each town seek for preserving. Finally, the results obtained have been compared with that stipulated by the regulations in question, trying in this way to identify their strengths and capacities.

Keywords Traditional architecture · Local identity · Identity preservation · Traditional building

1 Introduction

This study presents the basic aspects of a research dealing with the identity features of the architectural tradition of La Sagra region, in the Spanish province of Toledo, and the needed review of the criteria applicable to its conservation in order to avoid

A. García Hermida (✉)
Universidad Alfonso X el Sabio, Villanueva de la Cañada, Madrid, Spain
e-mail: a.garcia.hermida@gmail.com

G. Amoruso and R. Salerno (eds.), *Cultural Landscape in Practice*,
Lecture Notes in Civil Engineering 26,
https://doi.org/10.1007/978-3-030-11422-0_5

its disappearance. The problems presented in this regional analysis are very similar to those in other places and, therefore, similar studies may be valid for them.

Although it is a well-defined region, analysed in an autonomous way by the main treaties on Spanish vernacular architecture, it has not been the object of a detailed and documented investigation with enough amplitude. The practical homogeneity of this area, very well defined geographically by the valleys of the rivers that frame it, does not prevent there being numerous local peculiarities determining the unique identity of each of its towns.

However, the lack of knowledge on these traditions has burden the preservation and empowerment of these singularities, progressively leading to the homogenization of both the region in general and each town in particular. In the best cases, generically Toledan architectural features have displaced those determining the unique character of each of the existing urban ensembles.

Trying to better understand this problem and to be able to contribute to its future development, I am carrying out from the Polytechnic University of Madrid since 2010 a detailed investigation on this particular case, object of my doctoral thesis.

Towns have been studied building by building, elaborating a series of analytic sheets in which the transformations operated in each of them are recorded. This is, the evolution of the building elements which define local urban landscape. To this aim, the composition, volumes and various constructive solutions used both in the new buildings and in the interventions made in the last decades on existing buildings, then contrasting them with the traditional local identity municipal regulations of each town seek for preserving. Finally, the results obtained have been compared with what is stipulated by the regulations in question, trying in this way to identify their strengths and capacities.

Only some of the conclusions and the initial considerations of the mentioned study are presented here. They allowed to better understand and properly study the selected samples, which were treated until now as a homogeneous whole, while they present a diversity and complexity which were unknown before undertaking this study. To know more about the final results of this research, the very doctoral thesis will be soon available for anyone interested.

2 Historical-Geographical Context

La Sagra is a region of the Southern Spanish Subplateau. It extends northeast from the city of Toledo. In addition to Toledo municipalities, it also includes others nowadays belonging to Madrid, although in many of them the study of its traditional architecture is already practically impossible, since its urban centres have been systematically destroyed.

Its most defined limits are the Tajo and Jarama rivers, which border it by the southeast, and the Guadarrama valley, which runs by the northwest. BY the northeast, its borders are not that sharp, because they do not derive from physical hiatus, but political or cultural ones. At present, it could be said that this limit is located where the influence of the city of Toledo fades, and that of Madrid prevails. In any case, as Leopoldo Torres Balbás explained [11], "the natural regions, enormously complex, do not have boundaries fitting a well-defined reality, in which land, life and mankind coincide. Its limits, in many cases, are imprecise, not sharp lines, but broad areas of diffuse edges, the more faded the more natural the region is by itself, passing through imperceptible gradations from one to the other".

Whatever its exact extension is, the studied area is a wide flat plain dotted with small hills. It is placed about 600 m above sea level. Its climate is continental, with short autumns and springs and significant thermal oscillation between winters and summers. Precipitations are scarce and snowfalls rare. Their lands are sedimentary, rich in gypsums and clays, the main materials with which their image has historically been molded.

In spite of the intense processes of transformation undergoing there in the present, due mainly to its proximity to the Spanish capital, they are lands still maintaining an eminently agricultural character. Despite the real estate pressure, especially high in the municipalities best connected to the main communication routes, it is still a region of intense and old cereal exploitation, with a remarkable presence of vineyards and olive trees.

Located in what was once the territory of the Carpetan people, its history has always been linked to that of the city of Toledo. Thanks to the thriving situation of this city under Roman rule, La Sagra would become by then the main granary of the city. The aristocracy of Toledo exploited the region through a network of large rural villas, Carranque being the most notable of those which remains survived to date, but there are many other sites bearing witness to this historic role. After the definitive collapse of the imperial system, large late-Roman estates fragmented into a dense succession of villages and houses of labour controlled by local landowners [12]. This new model of settlement does not seem to have meant an absolute break with the precedent, being the new settlements sometimes placed on old villas, as it happened in Cabañas de la Sagra, although leaving aside their most representative architectural elements, which would be used only as a quarry for building materials. The transfer to Toledo of the Visigothic court would strengthen the agricultural role of La Sagra and would imply an important contribution to the population to the whole region, being often reflected in its toponymy. Faced with the temporary decline of other cities, in Toledo there was a progressive densification of what should have been the primitive Roman site, overcoming its limits by the significant demographic increase, what would also imply an intensive exploitation of this territory to ensure its supplies.

The Islamic occupation should not have meant substantial changes in its economic structure, but in the urban one. Between the 8th and 9th centuries the population diminished, and many smaller settlements were abandoned, their population being gathered already in a way very similar pattern to the present one [12]. For this reason, the Muslim cultural influence on the urban landscape these towns would present in the future may have been remarkable. With the taking of Toledo by Alfonso VI the area passed into Christian hands in the year 1085, when the important Mozarabic community of Toledo took possession of numerous properties in the region. Subsequent repopulation waves coming from the north made their population grow since then. However, the local architectural traditions of the old kingdom of Toledo would not change much during the Middle Ages, staying relatively indifferent to the cultured architectural currents of northern Europe, imposed with great difficulty in the area just in cases such as the Cathedral of Toledo. While the southernmost areas of La Mancha would be controlled mainly by military orders in charge of border defence, in La Sagra royal power predominated, as well as the mighty Archbishopric of Toledo. This situation would make these lands depend both culturally and economically from its capital city. The few existing noble dominions, like Seseña, Cedillo del Condado or Villaseca de La Sagra, were generally late and never came to constitute differentiated cultural entities from the rest of the region. Nevertheless, to some of them we owe the few castles erected in the region after the border moved to the south, dating mostly from the fifteenth century.

Since the sixteenth century, however, the Toledo decline against the rise of new capital city, Madrid, brought with it a period of progressive decadence for the old capital, which the towns of its jurisdiction generally solved by severing themselves from it to a greater or lesser extent, and growing with this process its relative power. It would be by this time when a more classical language would begin to hybridize in its most monumental buildings with the vernacular features of the place. Meanwhile, its popular architecture, which would live from this moment a period of intense renovation, would be progressively ennobled thanks to the enrichment of small local landowners.

Nowadays, largely thanks to its proximity to Madrid and the progressive rebirth of Toledo, La Sagra is one of the most important regions both demographically and economically in the Castilian-La Mancha area. This is a stage of great opportunities and, at the same time, of great risks for its cultural heritage.

A monoculture of single-family homes sustained its regional economy until 2008 and provided its present and future inhabitants just with a single, though copious, harvest.

The confluence of all these factors has shaped the sober landscape of La Sagra (Fig. 1), in which natural vegetation can barely find its way following the course of some meagre stream. Every corner of this territory, except for a few semi-arid hills, has historically been converted by mankind into a productive mechanism. This process has determined the traditional ways of inhabiting this land and from it derive the architectural and urban types it still houses today.

3 Traditional Urban Landscape

Azorín, as a result of the admiration he felt for Cervantes, who lived for a long time in these lands, was the first to pay attention to the towns of this region, dedicating picturesque descriptions to it. The architect Leopoldo Torres Balbás already worked specifically on the subject of this study [11], dedicating a section of his work

Fig. 1 Landscape of La Sagra (Bargas, Toledo)

La vivienda popular en España to the houses of La Sagra, though practically limiting itself to the old house of Cervantes in Esquivias and referring mainly the words previously written by Azorín. Fernando García Mercadal in *La casa popular en España* included only attractive drawings of traditional houses in one of its towns, Illescas (Fig. 2) [5]. Finally, both Luis Martínez-Feduchi [2] and Carlos Flores [4], in their monumental works on Spanish popular architecture, would emphasize the uniqueness of the houses of La Sagra within the general type of La Mancha, multiplying the examples and already referring some of the architectural aspects of the region in their respective works. Nevertheless, they were not able, given the national scale of their study, to treat the subject with the depth it deserves, nor could they therefore refer to the many existing local variants. Other studies would follow, not deepening significantly into the analysed topic beyond these early works [7, 8, 10].

Good examples of medieval monumental architecture are preserved, while the oldest surviving domestic buildings date back to the sixteenth and seventeenth centuries. The latter would suffer only slight typological variations until the arrival of industrialization. In addition, when compared to the most representative buildings, which, due to their symbolic and monumental character, show their vernacular features altered by the cultural currents of the time, popular residential architecture is more closely linked to the place and the existing ways of exploiting it,

Fig. 2 House in Illescas, by Fernando García Mercadal

which would hardly vary until the mid of the twentieth century. In any case, within the relative continuity of these local traditions, a slow but constant evolution can be identified, showing how cultural factors also condition this architecture, although they do so to a lesser extent than in the case of the main buildings.

La Sagra towns dot the monotonous landscape of the region. They are always presided over by the parish church, which generally stands on some elevation of the land and is rarely located in their main square. Their contours were traditionally well delimited, forming compact urban ensembles, with only some little hermitages beyond their limits. In them large urban farm houses and smaller and more humble houses are grouped together. The larger houses usually occupy their main squares and streets Both types are equally surrounded by ample patios and corrals. Long walls with few openings and wide gates delimit the extensive and irregular blocks. The main streets, unlike the secondary alleys, usually have well-defined routes, often almost rectilinear and, in general, following the directions of the most important communication routes crossing the town. Winding smaller streets depart from them, cul-de-sacs being also frequent. This type of layout provides some neighbourhoods with a vague Andalusi resemblance still surviving nowadays in some particular cases (Fig. 3).

Fig. 3 Alleyway (Bargas, Toledo)

Churches follow, regardless of their chronology, the Mudejar construction traditions characteristic of the old Toledan kingdom (Fig. 4). All of them are therefore built using *aparejo toledano* walls (a very local mix of brick and rammed earth walls), or in some cases *cal y canto* walls (a traditional type of stonework similar to rammed earth for its building process). Nevertheless, the diverse solutions for roofing them varied, for being adapted to the prevailing techniques from each period: from the Mudejar coffered ceilings, used mainly in the late Middle Ages, to the classicist brick vaults that would be used in most of the later periods, existing also some examples of the star vaults which were so popular in the late Spanish Gothic. They are, in any case, Latin cross plan temples with three bays and generally a single apse. They have a unique bell tower located at their feet. But there are also exceptional cases such as Illescas, Esquivias or Borox, where the bell tower position, next to one of the sides of the transept is due to the extension to the west of the original temple plan, keeping after it the primitive tower. The motifs embellishing elements such as cornices and openings varied in time as the prevailing styles did.

Convents, hospitals and little hermitages complete the picture of the monumental architecture of this region, showing similar continuities and variations to the ones explained for the churches. Except in the most recent cases, communal granaries and consistorial houses lacked special monumentality. The façades of the latter usually face the main square, having always some architectural feature to preside over the festivities and public events taking place there. In order to enjoy the same

Fig. 4 Parish Church of Carranque, by Luis Cervera Vera

events, the façades of the houses delimiting these squares were often opened to them by long wooden cantilevered galleries (Fig. 5), but many of them were replaced later on by equally long wrought-iron balconies or completely disappeared. Nevertheless, in these squares there are very few houses with wooden porticoes, so frequent in near regions, though a good example is still preserved in La Sagra today, in the main square of Villaseca de la Sagra.

In the large rural properties located outside the towns there are remarkable farm houses. Occupying strategic locations, they usually dominate the surrounding farming territory from some hillside or hillock. As they are not subject to urban pressures of any kind, the layout of their volumes is freer than the one of urban houses, but their main architectural elements and constructive solutions are very similar to those existing in the towns.

All large farm houses, being isolated or within the urban core, as the humblest houses, were complex organisms tending to self-sufficiency. They were at the same time places to work and to inhabit, having all necessary elements for storing and processing the agricultural products of their inhabitants, as well as the spaces for the animals both serving them in these tasks and complementing their diet.

In La Sagra, though there are also local variants diverse to this this type, the most common model of this kind of house consists of a front patio, more domestic, and a backyard or corral, more utilitarian, both delimited by high walls (Fig. 6).

Fig. 5 Main Square of Villaseca de la Sagra, by Luis Cervera Vera

One of the main peculiarities of the houses of this region within the general type of La Mancha area is it is not often accessed directly from the street, but can through the first and more domestic enclosed courtyard. This happens both in the noblest and in the humblest houses. Both courtyards or patios make up most of the lot's surface, which is why, as these houses use to be grouped together, they conform large-sized blocks, and imply a relatively low population density, what complicates the preservation of these areas today. All traditional courtyards, but specially the back or corral, have been inexorably disappearing when facing increasing real estate pressure.

The corral always has its own independent access, by large doors or *portadas*, as they are often locally called, designed to allow access by carts. They are mostly open to smaller and less important streets than the front patio doors, which usually have a more representative character. For this reason, in the noblest houses they are built with carved jambs and lintels in stone. Despite this, it is not uncommon either doors to both courtyards coexist in the same street, especially when the main building is parallel to the street.

In fact, the general configuration of the local house type depends on the position of the main building with respect to the adjacent street. They can basically be of two kinds: in perpendicular to it, the most particular of this region, or in parallel to it, facing the alignment of the street by its larger side. In both situations the main

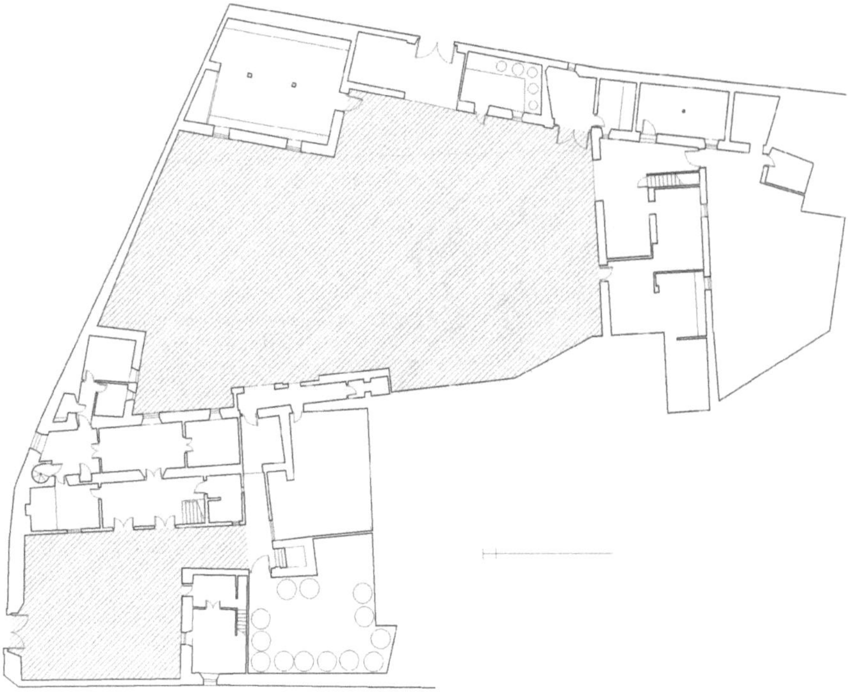

Fig. 6 Hypothetical original plan of the Cervantes House (Esquivias)

building layout is "L" shaped. The first occurs more frequently next to streets running approximately north-south. It is in this one where the openings to the outside of the block are the scarcest (Fig. 7). The second one is more typical from axes oriented approximately east-west.

The origin of this configuration is in the prevailing will to open the main façade of the house to the south, independently of the layout of the surrounding streets. The domestic patio was generally located to its south, leaving the corral in these cases to north of the house. In this scheme the roof slopes, generally facing the main elevation of the house, often pour the rain water towards the courtyards and not towards the street.

The southern façades of the main houses often had recessed porches and wooden galleries opened to the outside to catch the winter sunlight and warm the house in the colder months, while keeping its walls in shadow during the torrid summer months, when the light of the highest sun is intelligently avoided by traditional knowledge to create more comfortable spaces.

When these houses worked as relatively autonomous productive entities, the stables and chicken coops are located around the corral, and, in the largest farm houses, also the wine press and the olive mill. The size of the corral is usually

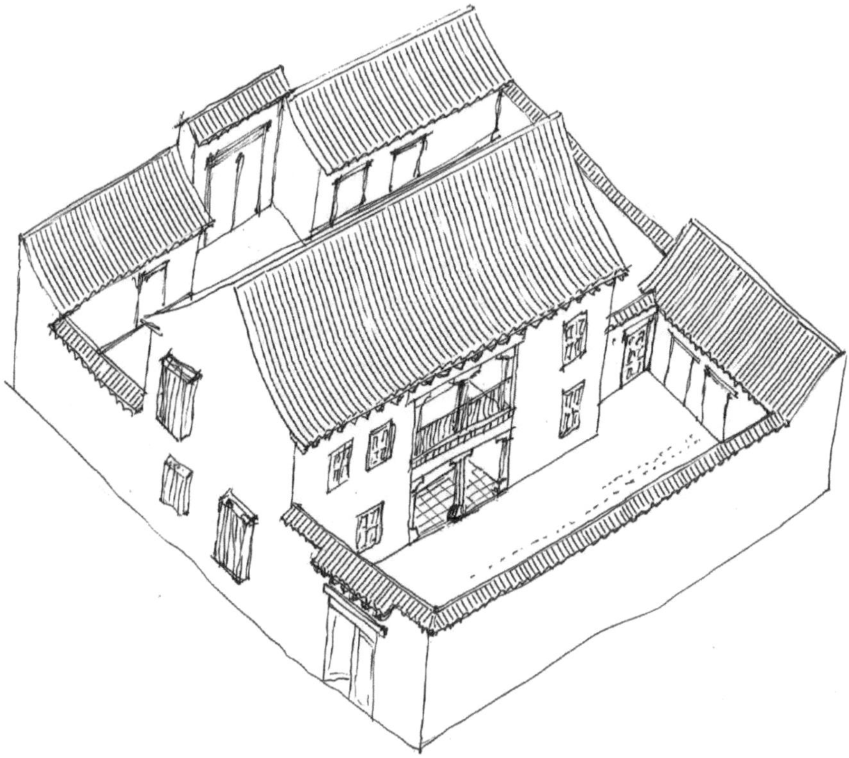

Fig. 7 Basic scheme of the traditional house of La Sagra

significantly larger than that of the front patios and, unlike them, usually lacks cobble stone pavements and has little vegetation.

The front patios, being mainly dedicated to leisure and representation, have trees and vine training forming arbors and providing plenty of shade in the summer. They are paved in cobble stone and there is usually where a well is located. In the main building there are numerous openings to this patio, and, in the large farm houses, also often a wide recessed porch and even, on the upper floor, a *solana* or wooden gallery, if there is more than one floor. As it has been explained, the relation of this patio with the house responds to needs of climatic control, allowing a greater sunlight in winter while ventilating and refreshing its interior in summer.

These houses always have one or two floors, both usually arranged in a two bays structure, and a garret, locally known as *cámara*, under the roof. The work areas sometimes have a higher building or turret often housing a dovecote. Thus, unlike in many other Spanish regions, the dovecotes of La Sagra were always integrated into the urban fabric.

The recessed porch generally gives way to an entrance hall where the main stairs are located. A hall, as well as the kitchen, are accessed from it. If there is a secondary stair, they are often spiral stairways. The kitchen was one of the main rooms of the house and was at the same time the dining room. If there is an upper floor, there is usually another small hearth or heater over the kitchen with a smaller fireplace to heat the upper floor. Until the nineteenth century, in the absence of corridors, each room was giving way to the next, being generally the bedroom or bedrooms located furthest away from the entrance hall. Later on, the interior layout of these houses would be adapted to the new requirements and more specialized functions.

The larger houses also frequently have caves, sometimes forming a long and complex network, since they generally produced and stored their own wine. In this aspect the last preserved caves in Esquivias stand out, honouring the ancient fame of its wines, as Esquivias was at times and thanks to this industry one of the richest and most populated towns in the region. Carved in soft sedimentary soils, they develop in long tunnels, often exceeding the limits not only of the plot where the house stands, but also its very block. Throughout them, paired cavities were built to place large earthenware jars. The structure of these caves is made of carved vaults, commonly barrel or pointed, as in the case of all those preserved in Esquivias, where the intersections with the lateral cavities were carved with the geometric rigor of being real groin vaults. To reduce moisture, which could deteriorate the soils where they are carved in, there are periodic vents opened to the patios and the cave doors also allow the air to go through, thus favouring cross ventilation. Today, unfortunately, these vents are ignorantly blinded, moistening the space and ruining the excavated tunnels.

Fig. 8 Traditional front courtyard

The described types of urban houses all together define a slightly inhospitable or introverted urban landscape at first sight. Only in its main streets the towns show some pomp. The other streets being enclosed by long walls, now whitewashed, with large doors and *portadas*, often distant from each other, and the houses showing few openings to the streets. However, these doors were always open, and, after these, one could contemplate an exuberant vegetation, a place rich in shade and water which still nowadays constitutes the greatest pride of its inhabitants, who dedicatedly take care of these patios (Fig. 8).

4 Traditional Building Techniques Shaping the Landscape

While the overall configuration of the buildings in La Sagra remained basically the same until the mid-twentieth century, as well as their structure and their main constructive features, other aspects were always changing, slowly but inexorably. This can be mainly understood in those concerning the external image of houses, where the representative function has had a greater importance. This is the case, for example, with the *portadas*, the carpentry works of the windows and doors, the wrought iron grilles and the exterior renders. In any case, also some interior renders, especially those of the most public rooms of the house, were renewed again and again, adapting them to the different fashions or to the changing tastes of their successive owners.

That said, foundations are usually in *cal y canto*, generally not very deep, given the scarcity of appropriate stone in most of the towns of the region. The walls stand on them either directly or with an intermediate course in masonry or brick to protect them from the possible humidity climbing from the ground. The noblest farm houses and the most representative buildings used to be built with mixed brick and rammed earth walls (the already mentioned *aparejo toledano*), or mixed brick and *cal y canto* walls, though there are also local exceptions. Nevertheless, most of the houses were built mainly with rammed earth walls. This rammed earth can be the regular one or, as it is the most common in La Sagra, *brencas* rammed earth, where there is a kind of lime or plaster mortar reinforcements, depending on the abounding materials in each place. These reinforcements are arranged forming very characteristic half-moon shaped rammed earth pockets in the walls (Fig. 9). There are several variants of this *brencas* rammed earth in the area. The most characteristic type is composed of almost semi-circular half-moon pockets vertically aligned, and whose vertical joints are misaligned by only about ten centimeters. But *brencas* rammed earth with more horizontal pockets is also common, when grouping several units of rammed earth are built one after the other within the same reinforcement. Therefore, in the majority of the municipalities of La Sagra, walls where the rammed earth was combined with brick forming the *aparejo toledano* type were rare and even more scarce those walls built entirely in brick, simply because it was a more expensive material than raw earth. The only exceptions to this are the southernmost villages of this region, those closer to the city of Toledo, where bricks were probably more available.

Wood was a scarce resource in these lands, so it used to come mainly from the distant pine forests of the mountains of Cuenca, taking advantage of the Tagus River for its transportation. It was used not only in doors, windows, roofs and horizontal structures, but also in the construction of porches and wooden galleries.

All houses have gable roofs, with their ridges running parallel to the main façades. The ridge usually rests on an intermediate wall, taking advantage of the double bay layout existing in all floors. In those farmhouses where grain was stored in the space under the roof, the *cámara*, the roof of the main building sometimes has hip roof dormers for ventilation. In the humbler houses and in all secondary buildings, having smaller dimensions, simpler couple rafters with collar and tie beams are frequent.

According to the answers given by the Sagrean towns to the questions raised in the 16th century by the crown for the national survey ordered by king Philip the II, known as *Relaciones Topográficas de Felipe II* [6, 13], thatched straw roofs predominated there by then, while Spanish tiles were only used in the noblest houses. This type of straw roofing has completely disappeared in this region, so that today in the totality of the traditional houses preserved, whether rich or poor,

Fig. 9 *Brencas* rammed earth wall in Carranque

all traditional roofing is done with Spanish tiles placed on wooden boards directly nailed on the rafters.

In the main houses eaves are usually wooden corbels of straight profile, sometimes topped with a simple quarter round. Frequently the corbels are arranged in pairs and they are always independent pieces both from the rafters and from the beams of the horizontal structure (Fig. 10). Cornice solutions in brick in various compositions are also common, though this detail was often linked to those few walls built in the same material or in brick combined with rammed earth, that is, in the richest buildings. The smaller buildings, on the other hand, show only a simple projection of the edge tiles, usually with lime mortar endings, or, occasionally, several rows of tiles in progressive projection one over the other (Fig. 11). On all eaves and cornices there is a change of the slope of the roof when it meets the wall, known as *silla* (chair) and generated by the horizontal distance between the head of these and the point where the rafters meet the edge beams. This traditional change in the slope is solved in this region progressively reaching the upper slope with a lime mortar and rubble filling.

The horizontal structures are always formed by beams or even logs placed close to each other to allow covering the gaps using little plaster vaults. To allow the plaster vaults to sit on the lateral sides of the beams, the adherence of these sides is improved by nailing ropes to them (*entomizado*). Besides, these sides are usually carved creating a linear grooving on its lower part to provide a little horizontal surface to support these little vaults.

As for the exterior openings, they are always protected by wrought ironworks (*rejas*), and, in case of having balconies, until the eighteenth century they were also entirely enclosed by *rejas*, with wide projections over the street. Nowadays, however, the most common balconies in La Sagra have wrought iron works only as

Fig. 10 Paired-corbels eave detail

Fig. 11 Simple eaves by projecting tiles, with endings in lime mortar

railings, usually composed by a single straight railing, and vertical bars of circular or quadrangular section, the latter type being always placed in a diamond shape with respect to the front of the balcony. The corner bars are usually a singular type, having a special section, often of a lanceolate kind. The most characteristic crowning of these corner pieces is a pointed ending, sometimes with a bronze ball on a cup base inserted on that point. But there are also some examples of flower shaped tops. The balconies are generally supported by curled wrought iron brackets with a growing section as they are progressively flattened towards their two ends.

There are also wooden galleries, as it has been said, both in upper floor openings and, in some of the houses arranged around the squares, projecting over the streets. Already lost other beautiful examples, a good sample is still preserved in Villaseca de la Sagra. In these cases, the banisters enclosing these galleries can be both in wood or in wrought iron, in both cases using balusters of square section and arranged in diamond shape with respect to the façade.

As for the finishings, the pavements were made on the ground floor by using square ceramic bricks arranged in a diamond shape with respect to the walls and framed by the same kind of square bricks but in this case parallelly placed to the walls. On the upper floor and in the *cámaras* located under the roof, gypsum mortar floorings of excellent making and pleasant touch were commonly used. Wooden floorings are also common in these upper areas. On the stairs, treads and risers are usually made of wood.

On the interior walls renderings are made in at least two layers of gypsum plaster, while the exterior renders were always in at least two layers of lime or earth and lime mortars. As they often were adapted over the centuries to changing trends and customs, sometimes the number of layers multiplied in time. Due to the higher cost of lime production, gypsum plaster or even slightly enriched earth mortars had to be used also for these exterior finishings in the earlier periods, but it is difficult to find traces of them, since from the 16th century on, when the oldest existing houses were built, the use of fine renders, rich in lime, seems to have prevailed.

As it has been mentioned, the composition, tonalities and ornamentation of these renderings varied depending on the place and date when they were made. It is significant in this respect how during the 18th century the temporary fashion of making renders imitating brickwork replaced the previous predominance of renders imitating ashlarworks. The dazzling whitewashed finishings nowadays characterizing these towns, and especially their humblest houses, had a fairly late emergence, probably generalized just from the 19th century on, in parallel to the cheapening of the lime production. Before their spread, the light tone coloured renders, tinted by the varying colours of the diverse local earths seem to have predominated.

5 Local Singularities

There are towns in La Sagra where their architecture meets all those typological and constructive traditions typical from this region, confirming practically all of what has been previously explained. Located in the heart of La Sagra, cases such as Esquivias or Illescas are paradigmatic in this sense. Maybe as a consequence, the region had been analysed in all previously published studies as a homogeneous whole. However, during the development of this research, unexpected richness and diversity were found in this tradition, with a number of local singularities. Its architecture had not been studied to date using any scientific method nor in-depth systematic procedures. Thanks to them, each municipality of La Sagra has proven to have its peculiarities.

In spite of the abundance of local constructive variants, there are also cases of very significant typological differences. This is the case of Borox, where the noblest urban farm houses have their own local scheme, closer to that of the Toledo houses [9]. Though constructively there are only occasional divergences, in this Borox type of farm house the layout of the domestic front patio is completely different, being located in the center of the dwelling, surrounded by it, and with a perimeter wooden gallery working as a distributing element to access the different rooms (Fig. 12). However, access from the street to the house is once more directly through this patio, in this case commonly linked to the main entrance through a covered passageway. The domestic patio becomes, therefore, the antechamber and heart of the house at the same time. The galleries of the central courtyard are porticoed, using wood both for its structure and for the flooring of its upper floor. The function of the spaces on the ground floor could be considered a transition between those linked to the rear corral and the noblest rooms, which occupy the upper floor. Servants' rooms and the main kitchen, as well as large warehouses of oil and other products and the access to one or several caves where the wine is kept are located around the court on that ground floor. Finally, there are also some very local constructive features, as the enormous timbrel vaults existing sometimes over the spaces on the ground floor, forming groupings of parallelly arranged barrel vaults.

In the case of Yeles, till very recently, almost all of the humblest dwellings were cave houses, with most of their rooms being underground. Nowadays, however, there is hardly any ruinous vestige of this ancient configuration.

In any case, these singular local features are more generally found in the materials or constructive solutions commonly adopted in each municipality, rarely affecting the architectural typologies.

We could analyse each village of La Sagra in particular finding similar local variants, so that, within an a priori homogeneous region, these local constructive features are important elements to define the identity of each specific town, the key of their uniqueness. Therefore, it is a fundamental step to previously identify them in order to find out the existing variables affecting the results of this study.

Thus, in municipalities such as Cabañas de la Sagra stone masonry walls are more abundant than any other type, as well as mixed brick and stone masonry walls, which are also common in other places such as Yunclillos.

In the preliminary studies carried out for this investigation, the transformations suffered by the urban landscape of a municipality where *brencas* rammed earth

Fig. 12 Central courtyard of a traditional house in Borox

walls predominate proved to be diverse to those suffered by those towns which walls are mainly made of brick combined with other materials (earth, stone, etc.), the latter being better preserved. The transformations suffered by those houses where main façade is apparent to the street when compared to those affecting houses where the main façade is hidden within the walls of the front patio have also proven to be totally divergent, the latter keeping many more of their traditional features.

These cases clearly show the eminently cultural motivation of most of the alterations suffered by the traditional urban landscape of the area. This finding challenges the cause usually argued as the main driving force for this kind of transformations: the changing functional needs, confirming what researchers such as Amos Rapoport or Julio Caro Baroja already pointed out decades ago. But it also means variations of this type can introduce important distortions in the results of this study if not taken into account. It is, therefore, essential for such a research to select a sufficiently homogeneous sample both at a typological and constructive level, so its alterations can be correctly analyzed. That is to say, by means of the selection of the study cases for the more detailed research the impact of those two variables must be annulled or minimized. Only in this way, the detailed results of a study of this nature can be contrasted with the existing municipal regulations for the protection of local identity without reaching conclusions which would be altered by these other variables.

This need does not preclude similar analyses may be undertaken in the future in order to compare the result of the transformations suffered by the urban landscape of municipalities with diverse constructive traditions. In that case, other variables affecting the studied problem should also be annulled or minimized, such as the diverse building types or variants in the applicable regulations themselves. This is, other variables should be sufficiently homogeneous in such a sample to avoid similar problems to those which might have altered the results of this study.

To comply with these premises, the chosen case studies for this detailed analysis have finally been Bargas, Olías del Rey, Mocejón and Magán, those municipalities forming the southernmost strip of the subregion known as Sagra Baja (Lower La Sagra).

Regarding other possible variables to be considered, its proximity both to Toledo and to the main communication routes connecting that city to Madrid allows supposing the real estate pressure being faced by the architectural tradition of this area should not be, leaving other factors aside, very disparate among these municipalities. The average rent in Bargas and Olías del Rey, according to data gathered by the National Tax Agency in 2014, was among the highest in the province of Toledo, while those of Magán and Mocejón were below the provincial average. However, evidences seem to indicate these figures have been distorted by the average data corresponding to the new suburban developments occupying the lands located between the urban centers of Bargas and Olías del Rey and the city of Toledo, not by those in their urban cores. Thus, the economic situation in the historic centers of both municipalities should be very close to that of the centers of Magán and Mocejón.

It is, however, the relative typological and constructive homogeneity of these four municipalities the main factor that has determined their selection among the rest of La Sagra localities.

The most common walls of the traditional houses of the four chosen municipalities differ from those used in most of the towns located further north. The *aparejo toledano*, combining rammed earth with brick reinforcements, is the most frequent solution in their noblest houses, unlike the scarce number of them in this fashion existing in other towns of this region. Besides, there is not a single example of *brencas* rammed earth walls, the walls of their humblest houses being mostly done in conventional rammed earth. Walls made of a very local combination of brick and stone masonry courses and rammed earth are also a common feature in these towns, especially in their smaller houses. This type of technique, moreover, is present in practically all the smaller houses forming the lowest part of the walls (Fig. 13). In those towns located immediately to the north from the four selected ones, as in the case of Villaseca de La Sagra, the conventional rammed earth walls become less common, as well as the *aparejo toledano*. Meanwhile the *brencas* rammed earth walls prevail, becoming more and more common while going north. In most of the northernmost villages of the region, those located in the area known as Sagra Alta (High Sagra), the *brencas* rammed earth is the basically the only technique used in all their houses, whatever their size.

Fig. 13 A wall in Bargas

The typical renders in the chosen towns follow in any case what has been exposed for the Sagra region as a whole, showing the same periodic variations, the same slight changes of tone depending on the available kind of earth in each place and the same late predominance of whitewashing, its whiteness sometimes enhanced by rendering the walls bases in bright light blue, using lime dyed with indigo. In the richest houses in this area, where the *aparejo toledano* is often found, only its rammed earth pockets were rendered, these renders usually simulating ashlar masonry. Nevertheless, there are also examples of renders designed to resemble brick works, generally later than those imitating ashlars, especially used on walls entirely built in rammed earth.

The traditional cornices and eaves of the richest houses in this area do not differ from those already described, though in Bargas and Mocejón there are less houses with wooden eaves than in other places such as Magán, Chozas de Canales, Borox, Carranque, Ugena or Illescas, including those paired wooden corbels so typical in towns such as Olías del Rey, Yuncos, Cedillo del Condado, Esquivias or Villaluenga de La Sagra. However, cornices in brick, more or less complex, more or less projected on the street, become very common in these chosen towns. Sometimes some of their brick courses are made of rotated bricks, placed forming a kind of sawtooth profile. In the humblest houses, as in the rest of this region, a simple projection of the last tiles of the roof is the most frequent design, often with their ends finished in whitewashed mortar. Some cornices are also formed by the progressive projection of several courses of tiles horizontally placed.

Wherever means for it were found, all openings have jambs and lintels made in brick. In the humblest cases, doors and windows are just mere interruptions in the rammed earth structure with wooden lintels. These wooden lintels were often simple uncut logs and the traditional *entomizado* (nailed ropes) was used to upgrade their adherence for being rendered in lime mortar.

Regarding the windows and doors, they follow what has already been explained for La Sagra as a whole, as it is the case with all traditional ironworks or with Spanish tile roofs.

However, among the existing singularities, it must be noted, especially in the case of Bargas, there is a common type of chimney which significant presence in the elevation of the house is very characteristic of the humblest houses. The proportion of its apparent volume to the street when compared to the relative low walls on which rises makes these chimneys an important feature in the urban landscape, more clearly than in other areas of La Sagra. In the best-preserved examples, their tops are always made by a pair of inclined bricks facing each other, forming so a kind of miniature gamble roof covering the chimney.

We could analyse each village of La Sagra in particular finding similar local features, so that within this a priori homogeneous region these local variants in its traditional houses are important elements to define the identity of each specific town, as well as important determinants to be considered for the study of the transformations being suffered by those identities.

6 The Will to Preserve Local Identity: Achievements and Failures

While monumental architecture is relatively well preserved in La Sagra, most of its vernacular architecture seems to be dammed to be transformed or disappear. The second of these options received a greater boost till the 2008 crisis as a result of the real estate pressure, when its increasingly decimated remains were seriously threatened. Local urban policies have contributed to increase property value in the existing town centers when allowing higher buildings to be built. The wide corrals and patios characterizing local architecture have been the first victims of this process and, especially in those urban ensembles where higher buildings can now be built, also the old one or two floors houses are gradually disappearing.

Particularly sensitive to change is also the interior of the houses. Many rooms have to be transformed to adapt them to current requirements, as it often happens, for example, with kitchens. These kitchens, traditionally one of the most representative spaces in each house, however, can be today preserved only by accepting the loss of their old function or significantly altering them.

In the last decade, even what has always been the most precious asset of this region: its rural landscape. Modelled by mankind during centuries to optimize its use, it can be seen now how its wheat fields, olive groves, vineyards, etc. have been succumbing to a monoculture of serial single-family homes (Fig. 14). This is, a kind of crop already proven to be unsustainable for producing one, even if copious for one generation, single harvest.

Fig. 14 New suburban developments in La Sagra

However, in the mid of this adverse situation, there is a remarkably strong will to preserve the traditional identity of these towns in both their inhabitants and their administrations. Most of them arduously try to keep at least some of their most characteristic features, what makes each of these towns a unique place to belong to. And these efforts are also generally shown by their very urban regulations.

Beyond the economic-tourist value of preserving their identity, this requirement has per se an outstanding value. Though it results from a complex psychological or cultural need, the very intensity of the will to keep these identities helps with understanding this is as important or even more for the inhabitants of this region than other concerns of a more pragmatic nature.

Despite this, urban planning regulations, which are a local competence in Spain, show a great superficiality and a very limited expertise on this matter. They often try to deal with this complex problem just by prescribing certain façade finishings and some volumetric limitations for new buildings and by protecting specific elements, selected on dubious criteria, from specific buildings. This problem by itself means this protection attempt is already generally limited to façades, what in buildings often humbly built barely allows the dissections this would have to produce. The superficial character of these regulations sometimes leads to the absurdity of protecting just a balcony or a stone coat of arms from a specific house, what, after its replacement by a new building, has to be inserted, completely decontextualized, in the new façade.

The case of the evolution of the front courtyards is a very significant one. They, as explained, are one of the most characteristic elements of traditional local architecture. Nevertheless, they progressively disappear, falling into disuse in new constructions. In the urban centers where the current urban regulations allow the setback of the new buildings with respect to the street alignment a very widespread will to get equivalent spaces to these old courtyards can be found, to keep their role as antechamber of the very houses. However, it is generally prescribed in all towns, as a rule, these new front patios must be now open to the public space. The consequent loss of privacy in these patios practically nullifies their use. They are thus transformed into transition spaces between the house and the street, and their only function becomes a merely representative or symbolic one. Besides, due to its limited usefulness, they are usually reduced to minor fenced setbacks, only a few meters away from the street alignment (Fig. 15). Interestingly, they keep part of what was its former look, and still are often very ornamented by flowerpots and vegetation.

In any case, practically all new buildings built in old urban centers and most of those designed in the most recent neighborhoods try to incorporate to a greater or lesser extent characteristic features of local tradition (Fig. 16). The same thing can be said of those existing buildings which have been renovated in the last decades. Most of them intend to resemble as traditional as possible. In the analysis of the recorded data from this research on each building of the selected area, in each one of the historic centers studied, conclusions are very enlightening (Fig. 17). In particular, about 75% of the buildings built in the last decades in the studied area try to follow local tradition to a greater or lesser extent, a percentage that escalates to

Fig. 15 Contemporary front yard, still open to the street, though very differently

about 80% in those built in the last twenty years. In addition, while a 20% of new buildings dealt with local character in a merely superficial way until then, in the last two decades that percentage has grown up to becoming a 60% of what has been built. This is, while the percentage of new works trying to keep the place identity is increasing even more, what it is growing in a much more significant way is the proportion of them just getting to unskillfully or very partially do it.

At the same time, there are other highly significant figures. In the four towns studied in more detail, while the existing humblest houses are being widely transformed, most of the old noblest houses keep most of their traditional features. Explanatorily enough, these widespread transformations in the humblest houses is not leading to a modernisation of their image. On the contrary, these changes seek for getting a closer image to that of the noblest traditional houses. For instance, those houses built in rammed earth are often hidden under a mockery of *aparejo toledano*, as if having been partially built in brick, a most expensive material.

Similar conclusions are reached when comparing the changes taking place in the exterior image of those traditional houses where their façades stand inside their courts to those where the façades face the streets. The latter have generally been significantly altered, also often trying to seem richer buildings, while the other ones have mostly kept their traditional image. This is, not only these transformations prove to be a cultural matter, related to the usual social perception of diverse

Fig. 16 Recently built houses in Bargas, showing unsuccessful attempts to imitate local *aparejo toledano* walls

traditional building materials or house types, but they also prove to be a problem which must be fought from a cultural perspective even more than from a practical one. Being so, education becomes a key tool still to be used in this field.

As explained, regulations aimed at preserving traditional local identity are increasingly concise in terms of façade materials and the image of particular building details. But what the gathered data show, though the proportion of new buildings trying to respond to local identity is growing, is they are not being very successful in reaching their aim. And this is probably linked to a general educational problem, especially serious in the case of those in charge of designing and building for this demand.

Many of the master builders and architects who work and have recently worked in the area, exclusively trained in Modernist design and in the more industrialized construction techniques, lack the necessary knowledge to satisfy this important demand, which proves to be so significant for local population. In the worst cases, they lack even interest in satisfying it, what also seems to be a consequence of their kind of training. This problem leads, in the first case, to the appearance of a large number of designs which approach tradition in a superficial and generic way, often replete with inconsistencies or linguistic errors. In the second, the case of architects or masters trained in a widespread disinterest in the studied question, resulting

ESTADO DE CONSERVACIÓN DE LA IDENTIDAD LOCAL TRADICIONAL
DEL PAISAJE URBANO DE LA SAGRA BAJA

BARGAS 0001

LOCALIZACIÓN:
- Calle Progreso 2

DESCRIPCIÓN:
- Tipología de la edificación: Vivienda unifamiliar
- Bajo comercial: No
- Alturas: B + 1
- Acceso: Directo
- Grado de conservación: Bueno
- Clasificación: Construcción histórica

MANTENIMIENTO DE LA IDENTIDAD LOCAL TRADICIONAL
Grado de mantenimiento de la tradición local:
- Alteraciones puntuales

Alteraciones sufridas:
- Constructivas

Alteraciones Constructivas:
- Estructura portante: Tradicional
- Solución de vanos: Imitación correcta de la tradición local
- Acabados: Imitación correcta de la tradición local
- Carpintería de puertas principales: Imitación correcta de la tradición local
- Carpintería de puertas secundarias: Ajena a la tradición
- Carpintería de ventanas: Imitación correcta de la tradición local
- Rejerías: Tradicional
- Cornisas: Imitación correcta de la tradición local
- Chimeneas: Inexistente / Desconocido
- Cubiertas Ajena a la tradición
- Balcones: Imitación incorrecta de la tradición local
- Galerías Inexistente / Desconocido

FOTOGRAFÍAS:

Fig. 17 Example of the data form used for this research

designs use to become tortured hybrids of modernist and traditional forms, showing the wrestle of the designers and builders with their clients, probably leading to equally unsatisfactory results for both (Fig. 18).

In spite of it, there are also artisans and architects who are very sensitive to this problem and who develop an important and quite silent work (Fig. 19). And, in some areas, there is still a remarkable knowledge on some traditional crafts among construction workers, as it is the case of traditional masonry in Bargas (Fig. 20). They seldom enjoy recognition and will often face their work being despised by those professionals who consider themselves more avant-garde people. All of this makes the result of their work perhaps even more meritorious.

However, the image of La Sagra, as it is happening in many other regions, is becoming increasingly less specific, more homogenized, precisely what so much effort tries to avoid. In spite of the growing concern of its inhabitants and the increasing number of regulations developed aimed at this problem, the cultural

Fig. 18 Recently built house in Bargas, awkwardly merging weird versions of "local" building details

Fig. 19 Recently built house in Esquivias successfully drawing on local tradition

aspects of this process have not been highlighted nor faced till now. The nature of this homogenization deserves scientific analysis to allow a better glimpse of the way in which it occurs and perhaps to tackle it more effectively in the future.

Beyond the emergence of foreign or global architectural trends, and the increase in the built volume in town centers and the real estate deals leading to clumsy suburban sets of serial homes, this homogenization can also be increasingly found in most of the very architecture which intends to meet local tradition. As explained, contemporary architecture which tries to respect local identity often does not achieve its objective either.

Fig. 20 Contemporary traditional masonry works in Bargas

Many of the traditional construction techniques which gave rise to those local features have been quite neglected for decades, and the last generation able to put them into practice is now on the verge of disappearing while no institution is focusing on their role in this process. This will make it difficult in the future both to study them and to implement possible initiatives aimed at its recovery. Therefore, most of the architectures today trying to preserve the local identity often have to limit their will to a sometimes imprecise and poor imitation of the exterior finishings of those buildings superficially interpreted as typical of this region. Not being previous studies on a local base, it is not even known if those models really belong or not to the particular town where they will be built. Thus, the streets of La Sagra are being filled with architectures trying to show a more or less traditional image, but they are increasingly unsuccessful in reaching their goal, often for being no longer a direct consequence of the constructive process but depending solely on the authors' knowledge on the original models and their expertise when trying to follow them. Despite the effort and good work of just a handful of great practitioners, those knowledge and expertise are becoming scarce in much of what it is being built.

In this context, even new traditional architecture is also contributing to homogenise the region, which occurs at two different levels: a local and a regional one.

At a local level, the historic hierarchy from monumental architecture to private architecture is being diluted. Constructive details previously almost limited to the richest public institutions are now used in all kinds of buildings (Fig. 21). The traditional gradation from the most utilitarian to the most representative architectures is thus annulled. This former gradation was largely due not only to their different scale, but also to the progressive enrichment of the architectural language, which usually reached its highest levels in religious and communal buildings. Those details previously mainly circumscribed to the noblest buildings are now profusely used regardless of the character of the buildings. This occurs both in new construction and in adaptive reuse works, as many of the old rammed earth houses are now covered by simulated brickwork, being this technique considered to provide a richer look. This also happens in elements such as chimneys or cornices, which progressively become more complex in all kinds of constructions. What were the richest details of the monumental architecture of La Sagra are now found or mocked in all smaller buildings. Thus, the richest brick cornice models, present only in the past in churches, convents or other singular buildings, now crown most

Fig. 21 Use of excessively monumental elements for a private house in Mocejón

Fig. 22 Contemporary houses showing a mixture of generic Toledan characters in the main square of Esquivias

of the most modest constructions. In any case, this progressive enrichment of local building culture has been an historical continuum, progressively happening since the early times of these towns. Besides, the former variety which was lost in this sense is of a relatively minor importance when compared to the second mentioned process: homogenization occurring at a regional level.

In a wider regional area, local tradition is being displaced by a generic Toledan architecture, unaware of what makes each local tradition unique and singular. The same models are indiscriminately used in almost any town, without paying attention to their distinctive features. Thus, for example, new houses in the very main square of Esquivias seem completely alien to the place where they have been introduced, while in some cases they would have relatively better fit if integrated in urban centres such as Bargas or Olías del Rey (Fig. 22).

7 Conclusions

It is still possible to make a well-defined typological characterization of La Sagra traditional architecture by its direct observation, though this is uncertain for the future, since, apart from its greater monuments, the short-term survival of even the best-preserved examples of its vernacular architecture is not at all granted.

Given the preservation of the peculiarities of local traditions goes through their detailed knowledge, it is urgent to develop a wide cataloguing and documentation work that will help with preserving them, or at least their memory, for future, and perhaps more sensitive generations. Traditional formulas from each town, including key aspects of them, such as the materials used in each place and the specific local construction techniques, must be inventoried and spread in detail in the very short term, as, at this point, the mere passing of a few years could make it already impossible. For this reason, it is also singularly important, before the last direct connoisseurs of this local knowledge disappear, to study and promote the construction processes that gave form to this architecture. Finally, according to the results of the study I am carrying out, an inexcusable step is the development of more useful planning tools to properly control this process, but even more important is training in the respect and understanding of architectural tradition and local identity all the agents who could be involved in their preservation and possible continuation.

Fortunately, as these traditional identities are more and more threatened, the growth and progressive generalization of the existing interest in preserving them can already be identified onsite. But upgrading the training of architects and building crafts people of the construction, meeting the current importance of this demand, with its economic, cultural and environmental benefits, as well as providing them with tools to be able to satisfy it, seems to be the key step to get a better future for these traditions.

The initiatives we are developing in Spain, not only from the administration, which has undertaken as important projects as the National Plan for Traditional Architecture, a fundamental conceptual framework, but also from the civil society. Good examples are those activities we are developing from Terrachidia or from INTBAU, the latter thanks to the support of the Richard H. Driehaus Charitable Lead Trust, aimed to help with alleviating these shortcomings: the Rafael Manzano Prize for New Traditional Architecture and its courses and seminars, the Richard H. Driehaus Architecture Competition, the Richard H. Driehaus Building Arts Awards and its scholarships for masters and apprentices, and the National Network of Traditional Traditional Building Masters. We hope we can notice its effect on the studied phenomenon in the future, and it is not already too late.

References

1. Cervera Vera L (1998) Plazas mayores en la comarca toledana de La Sagra. Real Academia de Bellas Artes y Ciencias Históricas de Toledo, Toledo
2. Feduchi L (1974–1984) Itinerarios de arquitectura popular española, vol 5. Blume, Barcelona, pp 152–153
3. Fisac M ([1985] 2005) Arquitectura popular manchega. Colegio de Arquitectos de Ciudad Real y Lampreave Ed, Ciudad Real
4. Flores C (1973–1977) Arquitectura popular española, vol 3. Aguilar, Madrid, pp 415
5. García Mercadal F (1930) La casa popular en España. Espasa-Calpe, Madrid, pp 70–81

6. López Gómez A, López Gómez J (1992) La vivienda rural toledana en el siglo XVI según las Relaciones Topográficas de Felipe II. Boletín de la Real Academia de Historia CLXXXIX (1):1–21
7. Miranda Sánchez A (2008) Casas de Toledo. Colegio Oficial de Arquitectos de Castilla-La Mancha Demarcación de Toledo, Toledo
8. Navajas P (1983) La arquitectura vernácula en el territorio de Madrid. Área de Urbanismo y Ordenación Territorial de la Diputación de Madrid, Madrid
9. Passini J (2004) Casas y casas principales urbanas. El espacio doméstico de Toledo a fines de la Edad Media. Universidad de Castilla-La Mancha, Toledo
10. Sánchez-Horneros Gómez A (1981) La arquitectura popular toledana. Colección Temas Toledanos. Publicaciones del IPIET, Diputación Provincial de Toledo, Toledo
11. Torres Balbás L ([1933] 1988) La vivienda popular en España. Folklore y Costumbres de España 3:137–502 (Ediciones Merino, Madrid, vol 150, pp 433–444
12. Vigil-Escalera Guirado A (2007) Granjas y aldeas altomedievales del norte de Toledo. Archivo Español de Arqueología 80:239–284 (Consejo Superior de Investigaciones Científicas, Madrid, pp 246–251
13. Viñas C, Paz R (1951) Relaciones histórico-geográfico-estadísticas de los pueblos de España hechas por iniciativa de Felipe II: Reino de Toledo, vol 2. Instituto Balmes de Sociología e Instituto Juan Sebastián el Cano de Geografía, Consejo Superior de Investigaciones Científicas, Madrid

The Liberty Network in Varese Province: Strategies for Its Knowledge and Enhancement

Claudia Caramel and Anna Anzani

Abstract Between the end of 19th and the beginning of 20th century, the research of a national style produced various architectural experiments in Europe including the development of Liberty. In the Province of Varese as well as in the adjacent Canton Ticino, the diffusion of Art Nouveau was fostered by the creation of new rail links that became a source of tourism development. The great potential of Liberty valuable built system, connoted by a stratification of both human and natural elements that blended into a whole, is currently extremely vulnerable, its assets being often classified as minor examples and not subjected to law protection restrictions. Sustainable strategies for its knowledge and enhancement worth be promoted, adopting a holistic approach that includes both the preservation of the built heritage and the improvement of life quality, in a continuous dialectic between memory and design, tradition and innovation. Awareness should be raised about the importance of beauty as a fundamental right of human beings, considering the aesthetic experience as a normal attribute of our daily life and understanding that space livability is based not only on rational and functional aspects, but also on values and a sense of belonging with places. The purpose of this paper is to suggest possible perspectives to improve the people daily experience by enhancing the identity characters of a specific urban environment, inevitably including the preservation of the ecological conditions. Solutions based on a "space outfitting" approach could be proposed, which can be a response to current social changes, but also a transdisciplinary perspective to address the relationship between present and memory, relying upon reuse instead of new construction, low environmental impact, no waste, energy sustainability, social network and shared economy, participatory design, attention to the details, comfort and wellness.

Keywords Liberty · Cultural heritage · Beauty · Ecology · Identity

C. Caramel · A. Anzani (✉)
Department of Design, Politecnico di Milano, Milan, Italy
e-mail: anna.anzani@polimi.it

C. Caramel
e-mail: claudia.caramel@polimi.it

G. Amoruso and R. Salerno (eds.), *Cultural Landscape in Practice*,
Lecture Notes in Civil Engineering 26,
https://doi.org/10.1007/978-3-030-11422-0_6

1 Introduction

Between the end of the nineteenth and the beginning of twentieth century, the emerging industrial bourgeois made the area of Varese, with its agreeable landscape characterized by the presence of several lakes and the Prealpine hills, one of its favorite holiday destinations. An essential support to this process was the construction of the railway that allowed the Milanese citizens to reach the countryside and spend their weekend time there. Many holiday homes, hotels, and leisure facilities were built according to the language of Liberty style that found here a perfect balance in the natural landscape [1]. Regardless its real influence to the Art Nouveau expression at a European level, this architectural heritage is strictly associated with the image of the city and still offers a significant contribution to its aesthetic quality despite the transformations suffered over time. In fact, beside few remarkable well-known buildings designed by influential architects, such as Giuseppe Sommaruga [2], Silvio Gambini, Ulisse Stacchini, today hundreds of noteworthy elements have emerged as a huge underestimated heritage spread in all 139 municipalities of the province of Varese.

However, in spite of its value, sometimes citizens and administrations seem not to be aware of the relevance of this little-known heritage to the quality of the urban space and the local identity. Many individual and social factors can prevent people to notice beautiful elements surrounding them. According to the American psychologist James Hillman [3, 4], a condition of "sensory anesthesia" can be pointed out as a result of fast movements, distraction, also due to the use of mobile devices together with our scarce aesthetic education and an increasing individualism, which leads us to focus on ourselves. In addition, the lack of protection laws and phenomena such as generational change, maintenance cost, etc. expose these buildings and artifacts to high risk of loss.

These little elements can also be considered from an ethical and ecological perspective, in terms of the impact of our everyday environment on the quality of life. In this respect, our city are not less significant in comparison to the scale of our relations. New technologies and the possibility of traveling low-cost make us perceive to be part of a global dimension in which the "proximity" does not seem to affect our well-being. Nevertheless, as Serge Latouche wrote [5], because of the lack of oil and the climate change, we cannot exclude that in the future our movements will be increasingly less distant, less frequent, less rapid and more and more expensive. However, independently from ecological and economic reasons, paying attention to our everyday environment is essential to feel at home in urban spaces and to improve our emotional attachment to places.

Appreciating the beauty of the little-known heritage asks for a particular awareness [6]. We need to look around with different eyes. How can we change our point of view? How can we enhance the beauty of the little things, to improve the quality of dwelling in the city? We believe that the best answer is to make places "significant". We live in the cities with our body but also with our mind, spirit, ideas, imagination, memories and all these aspects should be taken into consideration.

Above all, we believe that to properly enhance the less known heritage we have to design actions aimed to improve people's awareness. Indeed, making places liveable is not only a technical issue; it should involve the inwardness of human beings, leaving space to people emotions, imagination, and stories. Usually, we experience the heritage mainly through cognitive faculties, by learning its history, or through physical engagement, for example by visiting sites, museums or buildings. However, we should design strategies which allow us to experience it also through emotions and sensations. Indeed, the urban spaces are part of our everydayness, and their value grows in time according to a complex system of relations that involve personal experiences. Many elements are involved in the everyday experience of the city and we should be able to interpret them in their complexity. Considering all these contributions within a "holistic" approach we could improve our aesthetic appreciation of the little-known heritage and recognize the reality of our mental life, both in its rational and fantastic dimension. Thanks to a transdisciplinary approach, we may weave humanistic and scientific aspects of knowledge, combining solicitations that come from the anthropological and psychological culture and from visual arts, making beauty identity and memory become not only public goods, but also essential dimensions for the collective and individual well-being [7].

2 The Features of Varese Liberty Heritage

The landscape identity of the Varese area is characterized by a complex of Liberty elements that form a diffused system of great aesthetic value. With the aim to support its knowledge and provide public administrations with tools to protect it, a project was started in 2011 addressed to the cataloguing of buildings and artifacts in Liberty Style. During this research, assets that are not always accessible and form the Varese Liberty heritage could be analysed [8].

Beside some masterpieces, well-known at a European level (e.g. *The Grand Hotel Campo dei Fiori* and the *Palace Grand Hotel* designed by Giuseppe Sommaruga, *The Poretti brewery* by architects Bihl and Woltz), most of Liberty Style buildings are constituted by private villas, generally inaccessible to the public. Indeed, in many cases, these assets are characterized by architectural elements, like turrets, painted bands, concrete window frames decorated with floral motifs, wrought iron railings, etc. that could play the role of embellishing the urban space. Whereas in many private villas distances, fences, trees, and other physical obstacles obstruct their view, in the case of ordinary residential buildings decorative elements of great interest are visible, as well as in the case of other remarkable catalogued buildings like hotels, leisure facilities, factories, funicular, tram and railway stations built according to the needs of the emerging industrial bourgeoisie (Fig. 1).

Fig. 1 Window frames decorated with floral motifs, Gallarate (VA)

Fig. 2 Wall fence of Villa Mauri. Gallarate (VA)

In many cases, the main facades are the only parts of the buildings that the citizens can appreciate, although sometimes the buildings' height cannot be easily perceived in its complexity due to the small width of the roads. Several remains of precedent buildings can often be found, like for instance window frames, ceramic tiles, painted decorations, graffiti, wrought iron railings, plaques, ornamental bands, decorative cement, as well as scattered isolated artefacts such as sculptures, gates, lamps. These details are typical features of the Liberty Style in Varese area, and can offer unpredictable occasions to discover beauty in our proximity.

Noteworthy interior decorations have been discovered inside the buildings made accessible thanks to the collaboration of the public administrations. Painted ceilings, wrought iron railings, decorated radiators, concrete tiles, etc. are all typical elements of the Liberty style in Varese area that in many cases went lost.

Among the catalogued elements, also many gardens and yards have been included, which often hosted beautiful elements such as vases, sculptures, gazebos, etc. (Fig. 2).

3 The Little-Known Varese Liberty Heritage, Strategies for Its Knowledge and Enhancement

Considering the current ways of using private and public spaces and the chronic lack of economic resources of public entities, a reflection is proposed, indicating some suggestions aimed at achieving a greater knowledge and enhancement of the described valuable heritage. These proposals are mainly based on a range of reversible and low-cost tools that can be adapted to specific features of the buildings and artefacts here considered. Specifically, all the interventions have been designed according to the following principles:

- all solutions should be temporary, flexible and reversible [9] and should not damage the heritage and the urban environment;
- all solution should be based on low cost materials;
- all proposals should try to involve citizens to take an active part in the enhancement process;
- new elements introduced into the landscape should be all the same color to help people to recognize them as part of the same project.

On this base, ten main actions are suggested in the attempt to define an approach in which many factors involved in the aesthetic appreciation of the little-known heritage are taken into consideration. Indeed, the aim is to increase people's awareness of the real impact that the beauty of the little things can offer to their everyday experience of urban spaces. Not only top-down procedures are suggested but also proposals that encourage citizens to take an active role in the enhancement process to improve their sense of belonging to the places and develop a natural attachment to them. Above all, actions are aimed to awaken people from the condition of indifference in which sometimes they experience the city, helping them to notice and appreciate elements in their proximity. Below, all these suggestions are briefly illustrated.

3.1 *Look for Beauty*

To increase the citizens' awareness of the importance of remarkable buildings, artefacts, details hidden in the city, they could be invited to look for them through a playful approach. Actually, top-down processes aimed to improve their knowledge are not enough to help people becoming aware of the potentiality of their role to the quality of the urban spaces [10]. By cooperating with local public administrations, an interactive platform could be set up to geo-localize elements already catalogued and to collect new contributions created by citizens using social media. At the same time, stickers with a QR code connected to the platform could be freely offered in

Fig. 3 Looking for beauty

public libraries, schools, city halls, etc. or could be downloaded from the website. This could allow people to know the project and encourage them to place the stickers near elements they consider noteworthy (Fig. 3).

3.2 *Underline*

To help people noticing already identified buildings and artefacts across the city, solutions based on removable and low-cost materials could be adopted. For example, an ideal link with noteworthy buildings could be created just by highlighting their presence through colored adhesive tapes. In order not to damage the elements surfaces, the tapes could be placed directly on the structures facade if possible or, alternatively, on the urban pavement. In this way, a path could be created linking important elements and driving the citizens to their discovery. Similarly, where possible, remarkable details on building façades could be evidenced always using adhesive tapes. This kind of temporary installations, that could be carried out occasionally, are an attempt to awaken the citizens' attention. Also, to encourage people to take an active part in the enhancement process, they may be invited to mark elements that they consider attractive by using the same tools.

3.3 Change Perspective

To help people perceiving urban spaces from a different perspective, some temporary installations can be suggested. By placing mirror adhesive films on suitable surfaces, beautiful elements hidden in the city can be discovered indirectly. An attempt could be carried out to allow some views that normally cannot be appreciated from the sidewalks, due to the buildings height or to the presence of physical obstacles.

3.4 Break

In the attempt to allow people to stop and observe beautiful city elements, some areas could be temporarily identified as places in which citizens have the "right to enjoy beauty". They could be marked on the pavement by using adhesive tapes, chalk sketches or other provisional solutions. This proposal is aimed to raise people's awareness about the importance to pause, as an occasion to experience the beauty of little things. According to this approach, to increase citizens' involvement, they could be invited to identify other areas around the city in which they would like to pause (Fig. 4).

Fig. 4 Changing perspective

3.5 Bring Out

As observed during the fieldwork, some beautiful objects are kept in private houses preventing people from their knowledge and appreciation. In an attempt to overcome the hindrance posed by the private property and to encourage the owners to display their goods, a temporary structure (like a container) could be used as a "museum" in order to host movable objects in public spaces. According to their availability, private owners could exhibit here their goods allowing citizens to see them. This small "museums" aim to soften the boundaries between inner and outer urban dimensions and become an opportunity for people to meet and strengthen their sense of belonging to places and communities. Moreover, according to the specific features of the Liberty heritage of Varese, widely scattered in the landscape, these museums can be moved from time to time across the 139 municipalities of the area to host the local heritage. By becoming "familiar", the "museums" could contribute significantly to raise the people attention to the enhancement project (Fig. 5).

Fig. 5 Bringing out

3.6 *Let People in*

In the same attempt to connect private and public dimensions, owners could be invited to host occasionally an event in their gardens, allowing the outsiders to enter in secret areas. As emerged during the fieldwork, few owners of the catalogued buildings have taken part in similar initiatives in the past, for example hosting some vintage cars in their courtyards. Showing little-known elements and interior spaces of inaccessible buildings in private locations such as gardens, this initiative could offer a double support to establishing public-private connections. Considering the presence of noteworthy decorations in several buildings, the exhibition could also be housed in interior spaces according to their owners availability, allowing citizens to enjoy a secret heritage.

3.7 *Give Voice to the Invisible City*

Encouraging people to share their memories and stories about the little-known heritage and the city, some sketchbooks could be made available in public spaces. According to the book-crossing concept, people could find sketchbooks everywhere. Instead of reading a book, in this case, the users are encouraged to write or to draw their thoughts and interpretation of the urban spaces, also anonymously. Then, people are invited to leave the sketchbook in another place. The aim of this "sketchbook-crossing" is to increase people's emotional attachment to the places and to enhance the intangible city heritage. Periodically, these contributions could be collected and shared, for example in public readings, or using an interactive platform to signal remarkable elements identified by people. In addition, the most significant sentences could also be made public by writing them on posters affixed in public spaces.

3.8 *Use Differently*

In an attempt to enhance the little-known heritage of Varese through low-cost solutions, a proposal to use urban facilities could be developed referring to bus stops. Indeed, these structures form a system of elements scattered over the whole area of Varese that people are accustomed to see and use in their everydayness. The time spent waiting for public transport could become a precious occasion to catch people's attention towards the little-known heritage. In particular, some images of remarkable buildings or details placed in the proximity of the stop might be put inside these small structures, asking citizens if they could see them ("Can you see it?"). The attempt is to make people aware of close elements and to invite them to pay attention to their surrounding environment. Also, in this case, a playful approach might be useful to achieve the expected result. The proposal could be developed in collaboration with the local public transport system.

3.9 Make People Feel at Home

To increase people's 'attachment' to the little-known heritage, noteworthy buildings and elements could become the scenery of some pleasant moments. To this aim, a system of areas could be identified, allowing citizens to use them peacefully. According to the principals underlying this project, only low-cost and removable elements could be used to equip these areas and make it possible to play, sit, read, meet, etc. The idea is that by making people feel at home in the city a deep interest towards the surrounding environment can raise. In particular, specific proposals addressed to children could be developed. Allowing the future citizens to grow up surrounded by beauty may encourage them to claim it as a fundamental need for their well-being. Moreover, by making people aware of the importance of the aesthetic quality of urban spaces, we can raise a careful attitude that can become a real safeguard action.

3.10 Take Care

Besides beautiful buildings and artefacts, some negative situations have also emerged from the study. Indeed, several elements are in a state of decay, and their presence worsens the aesthetic experience of urban spaces. Making citizens aware of the importance of beauty means also encouraging them to report these cases. A photography contest could be proposed that invites people to pay attention to damaged structures, etc. The same platform created to point out beautiful elements can also be used for this purpose. In this case, no physical elements are added to urban spaces, but city users become protagonists and the agent factors that can improve the aesthetic quality of urban spaces by acting indirectly.

4 Conclusions

The actions proposed should not be intended as a series of steps to be followed in sequence. Indeed, the expected outcome is not only to improve knowledge but to raise awareness of the importance of its aesthetic and identity value for the quality of life. Only by making people aware, they may require the beauty of urban spaces as a citizens' right. Besides cognitive aspects, also people's interpretation and experience are essential. Some possible strategies and tools have been illustrated to offer practical ways which can be transformed into guidelines. The solutions has been outlined in the attempt to act immediately also in a period of economic and cultural crisis like the one that our society is going through. In fact, the heritage enhancement should not be a concern only in periods of well-being; on the contrary, it becomes even more necessary during difficult times, both to safeguard goods and to improve the quality of life. Indeed the perception of beauty is a real contribution to human well-being. The proposal shows some simple and low-cost solutions that, despite being tailored on the specific features and problems of the local heritage, can be easily replicated in other contexts, to enable people to appreciate the unique beauty of their everyday scenery. Consequently, each case becomes significant only within a more general vision, aimed to improve the quality of living in the city as a right for all. To achieve this goal, we need to open a perspective in which not only tools and strategies but possible actors—citizens, owners, local associations, institutions, etc.—should be involved according to their skills and responsibilities.

Acknowledgements The project, in collaboration with several institutions including Politecnico di Milano, Ente Parco Naturale Regionale Campo dei Fiori, Ordine degli Architetti, Pianificatori, Paesaggisti e Conservatori della Provincia di Varese and several municipalities, started in 2011. The team was composed by A. Baila, A. Anzani, E. Guglielmi, C. Caramel, M. Guarisco, and D. Niglia. E. Lonardo is gratefully acknowledged for his collaboration to create the pictures based on D. Niglia's photographs.

References

1. Anzani A, Baila A, Caramel C, Guarisco M (2013) Beauty and sustainability—survey on liberty heritage in Varese and Ticino area. In: Proceedings of XI Forum Internazionale di Studi Le Vie dei Mercanti, Heritage, Architecture, Landesign, 13–15 giugno, Aversa, Capri. La scuola di Pitagora editrice, Napoli
2. Baila A, Anzani A (eds) (2013) Palacento 1913–2013. Maggioli Editore, Santarcangelo di Romagna
3. Hillman J (1999) Politica della bellezza. Moretti & Vitali Editori, Bergamo
4. Hillman J, Truppi C (2004) L'anima dei luoghi. Conversazione con Carlo Truppi. Rizzoli, Milano
5. Latouche S (2009) Farewell to growth [Petit traité de la décroissance sereine] (D. Macey, Trans.). Polity Press, Cambridge (original work published 2007)
6. Anzani A, Caramel C (2017) Design and restoration—a holistic approach. In: Crespi L (ed) Design innovations for contemporary interiors and civic art. IGI Global, Hershey, pp 241–257
7. Anzani A, Baila A (2015) Enhancing the Ch: psychological and aesthetic issues. In: Proceedings REUSO 2015 - III Congreso Internacional sobre Documentación, Conservación, y Reutilización del Patrimonio Arquitectónico, Valencia, 22–24 Ottobre 2015
8. Baila A, Anzani A, Caramel C (eds) (2015) Nutrendo Anima e Corpo. Itinerari di delizia tra le ville del Sacro Monte sopra Varese/Feeding Body and Soul. Delightful journeys among villas in Sacro Monte above Varese. Maggioli Editore, Santarcangelo di Romagna
9. Crespi L, Anzani A, Caramel C, Crippa D, Di Prete B, Lonardo E (2017) Designing remains. In: Amoruso G (ed) Putting tradition into practice: heritage, place and design. Springer, Berlin, pp 1473–1482
10. Caramel C (2017) The beauty of little-known heritage as an aesthetic experience of urban spaces. In: Biamonti A, Guerrini L (eds) Polimi design phd_017. 10 Ph.D. thesis on Design as we do in POLIMI. Editore Franco Angeli, Milano, pp 56–66

RETRACTED CHAPTER: Survey and Digital Representation of an Architecture in a Landscape Between Karst and the Sea

The authors have retracted this chapter because it contains substantial overlap in sections 2.3 and 2.4 with the thesis[1] written by a different author. Additionally, the chapter contains Figures 8,9,11, and 12–17 for which the authors lack appropriate permissions.

All authors agree to this retraction.

Reference:

[1] Villa Alberi di Romano Boico a Duino: Tecniche avanzate di rilievo e rappresentazione della casa nella roccia", graduate student Roberta Purinani, supervisor prof. Alberto Sdegno, co-supervisor arch. Paola Cochelli, University of Trieste, Department of Engineering and Architecture, Three-year degree course in Architectural Sciences, 2015–16.

G. Amoruso and R. Salerno (eds.), *Cultural Landscape in Practice*,
Lecture Notes in Civil Engineering 26,
https://doi.org/10.1007/978-3-030-11422-0_7

RETRACTED CHAPTER

Creating a Map of the Underground Heritage in the Mediterranean Area: A Visual Representation for a Comprehensive Research

Beniamino Polimeni, Roberto Bixio, Carla Galeazzi, Carlo Germani, Mario Parise, Stefano Saj and Mariangela Sammarco

Abstract Interest in man-made (or artificial) cavities in the countries of the Mediterranean Basin led the Commission of the Italian Speleological Society to study and catalogue some of the most common troglodyte types in the region. Since 2000, the Commission has drawn up a study of the geographical distribution of rock-cut structures through a project initially developed by Mario Mainetti and Erica Besana in 1994. This geographical catalogue has been realised through integrating the information available in the international bibliography with the scientific research carried out by the Commission and the research groups affiliated therewith. The result of this work is a list of 1948 rupestrian sites distributed throughout 31 countries and represented on a general map. In this chapter, a general overview of the project is presented, along with a description of some case studies from different countries, including Tunisia, Libya, Turkey and Italy.

B. Polimeni (✉)
School of Architecture, De Montfort University, Leicester, UK
e-mail: beniamino.polimeni@dmu.ac.uk

R. Bixio
Centre for Underground Studies, Genova, Italy
National Artificial Cavities Commission—Italian Speleological Society, Bologna, Italy

C. Galeazzi · M. Sammarco
National Artificial Cavities Commission—Italian Speleological Society, Rome, Italy

C. Germani
National Artificial Cavities Commission—Italian Speleological Society, Egeria Centre for Underground Research, Rome, Italy

M. Parise
Artificial Cavities Commission—International Union of Speleology, University Aldo Moro of Bari, Bari, Italy

S. Saj
Centre for Underground Studies, National Artificial Cavities Commission—Italian Speleological Society, Rome, Italy

G. Amoruso and R. Salerno (eds.), *Cultural Landscape in Practice*,
Lecture Notes in Civil Engineering 26,
https://doi.org/10.1007/978-3-030-11422-0_8

Keywords Artificial cavities · Rock-Cut architecture · Mediterranean area · Maps · Cave dwellings · Underground defensive systems

1 Introduction

Artificial caves are human creations of historical, archaeological, anthropological and geological significance which are spread throughout the world and different in respect of their typology, construction period, and techniques. During the last 30 years, several interdisciplinary studies and research works have been conducted to create a homogeneous body of knowledge and compare experiences deriving from different areas. The most significant among them is the catalogue of the rupestrian sites of the Mediterranean area, an ongoing project—initially developed by Mario Mainetti and Erica Besana in 1994—in which 1948 rupestrian sites (Table 1), distributed throughout 31 countries, have been listed, analysed and visually represented. This investigation originated from the idea of creating a set of maps and datasheets to describe the distribution of underground architecture and the typological facets of the basic units. Subterranean heritage consists, in fact, of a diverse spectrum of architectural types, including dwellings, churches, mosques, minor places of worship, oil mills, dovecotes, and defensive structures. The majority of these examples are part of a vernacular language designed to respond to local needs, which perfectly exemplifies the interrelationships among cultural aspects, environmental control, and source sharing. The analysis and the adaptation of these ideas to contemporary represent an opportunity to develop new directions in sustainable techniques, and to promote diverse cultural and historical identities. In this chapter, different architectural types existing in some specific regions of Italy, Libya, Tunisia and Turkey are considered, with the aim of providing a broad overview of the different existing forms. References to historical events and geological features that determined settlements' distribution

Table 1 The number of settlements in the countries studied

Country		Country		Country		Country	
Albania	5	Macedonia	7	Malta	3	Serbia	15
Algeria	15	Georgia (eastern part)	4	Morocco	19	Montenegro	3
Armenia (western part)	1	Jordan	9	Moldova (southern part)	3	Syria	13
Bosnia and Herzegovina	2	Greece	157	Palestine	31	Spain	326
Bulgaria	9	Israel	9	Portugal	6	Switzerland	13
Cyprus	5	Italy	678	Gibraltar (UK)	1	Tunisia	161
Egypt (northern part)	4	Lebanon	18	Romania	10	Turkey	191
France (southern part and Corsica)	156	Libya	63	San Marino	1	Ukraine (Crimea)	10
Total							1948

and architectural features will also be included. Finally, a description of research methods and future directions will be identified, with the aim of increasing awareness of the heritage through data dissemination.

2 Research Method and Initial Results

The main idea of realising a map—and the related inventory—of the rock-cut sites in the Mediterranean area has been driven by the desire to create a compendium of all the information collected by different research groups and scholars in the field at different times.

In particular, the map has been originally created including the data available in scientific literature and those deriving from architectural surveys conducted by the National Commission of Artificial Cavities of the Italian Speleological Society (SSI). Contributions of other institutions and scholars have been included at a later stage to increase the level of awareness of rupestrian sites in countries not adequately investigated previously.

As mentioned previously, the project commenced in 2000 as a continuation of the experience of Mainetti and Besana, with the aim of providing a visual representation of a phenomenon which, with different forms, covers all of the Mediterranean Basin. Countries that do not border the Mediterranean Sea, such as Switzerland, Ukraine and Georgia, have been, however, included in this research to establish a cultural continuity among them and the bordering nations.

The visual material produced in this project has been realised while following a hierarchical structure. The first cartographical representation of the rock-cut sites of the Mediterranean Basin is a grey-scale index map in Mercator Projection (Fig. 1).

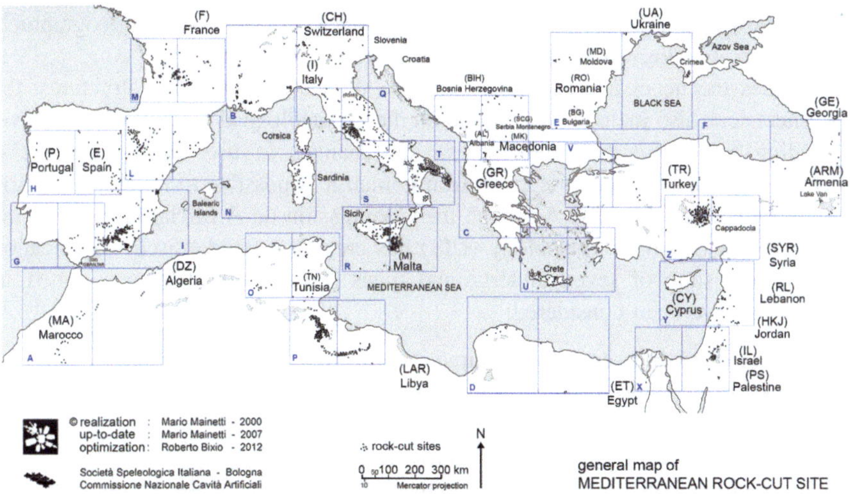

Fig. 1 General index map of the rock-cut sites in the Mediterranean area

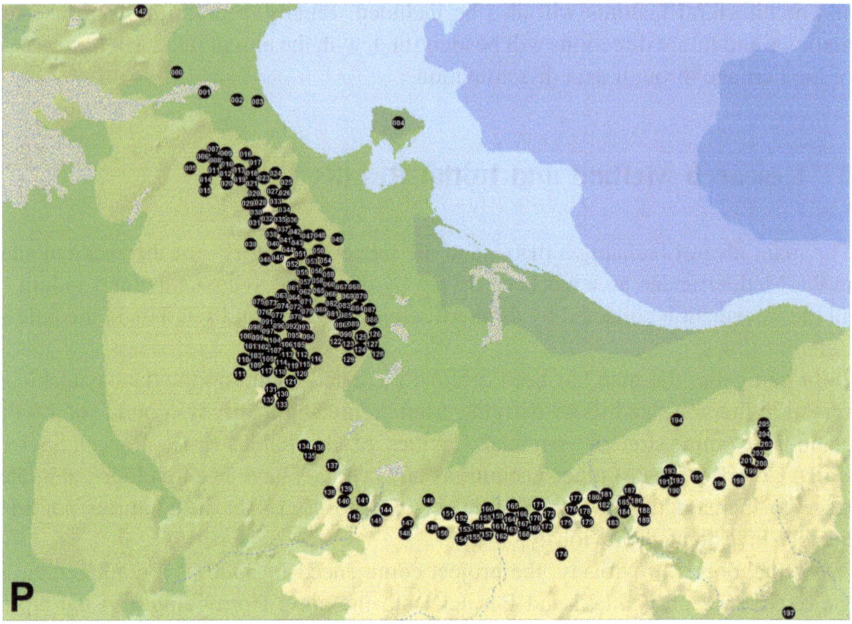

Fig. 2 Example of the map indicated with the letter "P", including Libya and Tunisia

This chart represents the distribution of the sites and provides an index of the relevant map sheets covering their regions of interest. The areas covered by the project amount to 23, with each one defined by a specific letter and represented through physical maps. In these drawings, colours are used to show changes in elevation and a set of black dots with numbers indicate the underground sites (Fig. 2). Each site has been associated with a set datasheet consisting of various information, such as geographical coordinates, typology, and the relevant bibliography if existing.

Besides the necessary information to pinpoint and classify the structures, the datasheets can also include information on the archaeological features of each cave, including transformations, extensions, and changes in use. Each datasheet is identified by an alphanumeric code, from which it is possible to know its location and typology immediately. Although the Commission on Artificial Cavities of the International Union of Speleology (UIS) has created a standardised classification system consisting of seven underground types [29], in this catalogue only four categories have been considered:

- Underground and semi-underground dwellings;
- Places of worship (sanctuaries of different types, including reused natural caves);
- Temporary shelters and underground defensive systems;
- Underground structures with utilitarian functions (barns, baths, animal shelters).

Underground structures such as burials, hydraulic works, mining works, transit routes as well as underground masonry works have been excluded from the present research.

Availability of the current data had a strong influence on the distribution of sites on the map. Differences in the number of sites for each country are not always dependent on the real quantity of real artificial caves or settlements, but they are somewhat related to the availability of such data. Accessibility of some of the geographical areas investigated is another essential element influencing the data distribution. Examples of this disproportion are represented by countries such as Armenia and Georgia, which are particularly abundant in artificial cavities, but rarely displayed on the map due to a lack of systematic studies.

The definition of "site" has been another crucial element which has influenced the distribution of underground settlements throughout the area. In fact, each site has been identified as a group (or a set of groups) of cavities on a specific piece of land. For example, Sassi of Matera has been reported as one site, even if the settlement consists of thousands of small dwellings dug in the tuff, subdivided into various groups and units. At the same time, other examples of sites can correspond to a single group of cave structures isolated on the territory.

In the next chapters, some models of the subterranean heritage of Italy, Turkey, Tunisia and Libya will be presented. These regions are representative of the varieties of cultural landscapes of the Mediterranean and are an expression of the extensive and robust relationship existing between peoples and their natural environments.

3 Case Studies

The case studies of the next paragraphs exemplify the diversity regarding the scale, geological features and climate of the regions analysed during this research. Different cultural, social and religious aspects will be described as determinants of traditional habitats and settlements.

3.1 Italy: Medieval Rupestrian Sites in Apulia

The Middle Age rupestrian phenomenon in Apulia represents the primary expression of a civilisation which has characterised for many centuries the history of the region. Apulia, an NW–SE elongated peninsula, which is surrounded by the seas and representing a natural bridge between the Italian Apennines and the other side of the Adriatic Sea, presents a peculiar morphology which significantly influenced the birth and development of the rupestrian civilisation.

At the same time, organisation of the human settlements developed according to the strategic, religious and political needs of the different ethnic groups.

Scientific interest in the rupestrian sites was initially focused on the caves of the most significant artistic and architectonic values, generally worship places such as rock-cut churches [7, 14]. It commenced during the last decades of the 1800s, and was dominated by the "pan-monastic" theory, according to which all Byzantine art in Southern Italy had to be considered the work of Greek monks, who arrived in Italy after the Arab invasions and the related iconoclastic oppressions. In this very general view, all cavities and hypogean settings were exclusively classified as Basilian settlements and hermit crypts.

With time, during the 20th century, a less closed vision started to develop. This new approach considered a "rupestrian culture" not limited only to the hermit and monk dimension, but essentially dedicated to "living in caves", in an attempt to understand this culture within the natural context offered by the local karst landscape. This latter, geologically characterised by carbonate rocks affected by karst processes, resulted in paucity of water at the surface (a serious problem for settlers), but, at the same time, offered many sites on which to carve and dig artificial cavities, thanks to the presence of soft rocks such as calcarenites [10]. These rock masses could be worked easily, simultaneously providing excellent mechanical properties to sustain the excavations, thus guaranteeing stable conditions for a variety of purposes [29]. Moreover, deep karst valleys and gorges, locally called gravine [28], offered along their steep flanks good sites on which to create settlements, which were hidden in respect of the view from the surface and, therefore, successfully protected from enemies' attacks.

Studies and researches on the rupestrian culture came to define a comprehensive chronology for the "life in caves", covering a long timespan, from prehistory to the Early Middle Ages. However, it was in the first two centuries of the second millennium that a real rupestrian culture could be identified, and represented a valid alternative to the classically built settlement. The rupestrian architecture was, in fact, linked to the masonry buildings, for both plans and volumes of construction, and the architectural elements, to recreate the illusion of a built-up setting [6]. In particular, rock-cut churches show many characters and schemes very similar to those of the built churches of the same age [25]. From the purest forms, consisting of a quadrangular room, an apse and niches on the walls, and adding to these further elements, the architecture of rock-cut churches reaches very complex planimetric shapes. A typical fan-shaped plan characterises the most complex and monumental architectural church types [7].

As aforementioned, the entire rock-cut village is strongly dependent upon the natural morphology of the site: the steep to vertical rock wall allowed excavating cavities, often dislocated in different stories and connecting one to the others through external stairs carved in the rock, or using internal links. Furthermore, road systems within the settlements allowed reaching the closest communication routes of greater importance. Water resources were provided through complex networks of surface channels to collect and transport the rainfall, and to store water in underground cisterns and tanks [24, 27, 30]. Some villages were also provided with defensive systems consisting of controlled accesses and sighting sites. In addition to civil dwellings and working places, many villages also hosted worship sites, located in particular areas of the settlements. Churches were situated on the outskirts of the

Fig. 3 Examples of crypts from Apulia: on the left, the crypt of S. Angelo at Otranto (Lecce province); on the right, a view from within a crypt in the Lama d'Antico site, at Fasano (Brindisi province)

town, since they were realised in a second moment, following excavation of the civilian dwellings, or merely to ensure the sacrality of the site. On the other hand, with regard to the settlements in the gravine, the crypt is typically located at the top, on the high plain or in the central terrace but without any other cavity above, so as to maintain and respect a specific hierarchy of values. The house of the priest or the custodian was often nearby, while tombs were typical in the area in front of the crypt.

The diversity of the forms and solutions to be found in this subterranean environment of Apulia represents a practical and efficient expression of a state of necessity, characterised by significant values. Its cultural, social and institutional aspects are surely comparable to those of the built-up villages (Fig. 3).

3.2 Italy: Rock-Cut Sites of Latium

In Latium, the rupestrian phenomenon is mainly concentrated in the volcanic formation areas of Volsino, Cimino-Vicano, Sabatino, and Vulcano Laziale (Alban Hills). This specific geological composition, consisting of tuffs, has fostered the development of different rock-cut structures.

In the area of Viterbo, after VII–VI AD, the Etruscan and Faliscan dug into volcanic tuff, creating vast necropolises. Among these tomb monuments, the necropolis of San Giuliano, located on a tuff cliff and already inhabited during the Bronze Age, is the best example of the variety of cave types (Fig. 4). Other underground sites in the Etruscan Latium include Blera, Castel d'Asso, Norchia, San Giovenale, Sutri, Tarquinia, Tuscania, Vetralla, and Vulci.

The fall of the Roman Empire and the subsequent barbarian invasion generated a progressive depopulation of the rural areas of Latium, although small rupestrian settlements connected to agro-pasture arose in the High Middle Ages. Reuse of old subterranean Etruscan or Roman structures was frequent during this period: the settlement of Castelnuovo di Porto in Belmonte [17, 36], the one called fosso Formicola in the area of Marcigliana and the structures found in the Alban Hills are all examples of such practice [8]. During IX–X centuries AD, many hermitages and

Fig. 4 The Necropolis of San Giuliano, Barbarano Romano (picture by C. Germani)

monasteries, often decorated with rupestrian paintings, were dug into the tuff and limestone, in the areas of Viterbo and the Apennines respectively (Valley of the Aniene River; [12]). In 2009, Tuscia University [1] carried out a research project comparing the historical heritage of Cappadocia and Latium. These two geographical areas present many similarities regarding architectural types such as churches, stores, cisterns and dovecotes, although subterranean hydraulic and drainage works—widespread in Cappadocia—are much less present in Latium.

Other significant studies, mainly focused on the sepulchral architecture of Etruscan Latium, have been conducted by Raspi Serra [33] and a group of speleologists who carefully analysed all of the sanctuaries in the region [13]. Further studies have been carried out on almost unknown sites such as Belmonte [17] and San Lorenzo Vecchio [16].

Finally, additional experiences in the analysis of medieval underground structures have been carried out during the previous decade [8] in some specific parts of the region. Dovecotes and pigeonholes, widespread throughout the region, are part of this diverse rupestrian heritage. These structures have been probably used since Etruscan times to collect both the guano and potassium nitrate to be used as natural fertiliser in agriculture [32].

Dwellings realised through adding new rooms in existing cavities have prolonged the rupestrian phenomenon to the 20th century, generating a series of structures in which new and old structures are part of the same system.

3.3 Tunisia and Libya: Berber Underground Settlements

Southern Tunisia and Western Tripolitania are the areas of North Africa characterised by significant examples of traditional architecture encouraged by the combination of Ibadism, Nafusi Berber culture and the peculiar topography of this territory. These types arose from the adaptation to climatic conditions and in response to the collective needs of the communities who have inhabited this area for centuries which encouraged a highly distinctive identity. This uniqueness is evident in landscape transformations, architectural solutions and the urban "organisms" which are entirely representative of the social aspects and perfectly adapted to the environment [31].

Fortified granaries and cave dwellings are the most remarkable examples of this landscape transformation which are spread throughout a semicircular mountain chain which extends from the Matmata region near Gabes, Tunisia, to the city of Gharyan in Tripolitania. The western part of this mountain chain, oriented along a north–south axis and bisecting the south of Tunisia, is called Djabal Dahar, with the eastern part crossing Tripolitania instead being called Djabal Nafusa.

The origin of the underground settlements, in both regions, is dated between the VII and XII centuries AD [19], between the first and second Arab invasions. However, the migratory process following the Hilalian invasion in the 12th century has finally defined the structure and configuration of the settlements located on both the northern and the southern parts of the mountain border.

Many scholars have tried to classify the caves of this vast region into different types. In this chapter we will use the classification proposed by Jean Despois [11], one of the earliest scholars who studied and analysed the underground structures, identifying four main types according to their historical evolution.

The first type, probably the oldest, is characterised by one or two horizontal rooms and can be found throughout the Djabal Nafusa (Fig. 5) and in the adjoining regions of Tunisia around Tatawin. Houses are excavated along the mountain slopes with horizontal bedding of alternately hard and soft layers.

Rooms are excavated in sequence, creating a system in which an area, generally used as a living room, is followed by an area used for sleeping.

Fig. 5 The construction of a single room in Kabaw, Libya (Picture by B. Polimeni 2011)

Some of these habitations were transformed with external spaces in front of the entrance in order to form a specific urban configuration that in the Tunisian towns of Douiret, Chenini and Ghomrassen is called Kalaa.

The main design feature is a central rectangular or squared courtyard surrounded by rooms that are long and narrow. Golany [19] reports that in 1972, in the city of Matmata, there were 600 underground houses, 60 of which were abandoned, connected by a system of pathways on the surface.

The third type of habitation is the "pit-in-depth", a cruder construction dug deeply into harder rock or earth. These are frequently found under modern houses and are used as cellars or for storing grain. Their original purpose was undoubtedly that of habitation. In the region of Kabaw, Libya, such pits are often used as olive presses.

Finally, the fourth type combines a cave dwelling with some other form of habitation, as can be seen in the Arab village of Tigrinna in the Djabal Nafusa. The plan is similar to the grand court type, but the patio contains only two or three caves, covered by a facade of stone walls which, in some cases, are connected to external rooms.

3.4 Turkey: Artificial Cavities in Asia Minor

In the Mediterranean Basin, Turkey is the country with the highest and most varied number of locations wherein rock-cut structures are present. One hundred and ninety-one, distributed throughout the country, have been included on the map, although the real consistency of existing artificial caves is undoubtedly much higher [4]. It should also be taken into consideration that, in some cases, significantly large areas are listed as a whole: in the region of Cappadocia, for instance, only the rock-cut churches comprise more than 600 [21].

The Anatolian region, enclosed with the provinces of Nevşehir, Kayseri, Kırşehir, Aksaray and Niğde, has the highest concentration of underground sites and has represented, for its geographical position, a cultural bridge linking the East and West since ancient times.

The presence of different cultures and religions is testified by the presence of some of the most significant archaeological, architectural and artistic heritage of the world. Among these, evidence related to the use of the underground for carving structures is not minor.

In this particular geological environment, often consisting of volcanic tuff, it is possible to identify evidence and strategies in respect of the lives of different populations. From Hittites to Urartu, Phrygians and Lycians; from Greeks to Romans; from Armenians to Byzantines; from the Mongols to Turkish ethnic groups, including Seljuks and Ottomans, many civilisations have succeeded and overlapped in Asia Minor.

Several Hittite terms, present in the oldest sources, allude to places or buildings excavated in the rocks [26]. Among these, the most significant example is probably the sacred spring cave called ^{D}KASKAL.KUR, which is located on the ancient site of Troy and mentioned in a treaty of 1280 BC.

The innermost part of the hypogeum was entirely artificially excavated using the well-known technique of opposing fronts, although this sector is probably more recent [15, 23] and personal communication, January 1, 2005. Realistically, it was needed to create a "tunnel-cistern" connected to the city overhead with supplying wells inside of the walls [5].

Other references to waterworks can be found in Herodotus [20] (Herodotus, I-75), which in the 5th century AD describes a bypass made by Thales of Miletus on behalf of Croesus in a war against Cyrus, king of the Persians, to divert the Halys River, today Kızılırmak. It is likely that this work corresponds to a tunnel excavated in the rock, located in the middle course near Sarıhıdır, and used to create a ford for caravans and commercial purposes [18].

The oldest historical sources mentioning underground structures used as residences date back to the 4th century BC and refer to an area not precisely identified in today's Eastern Turkey (Senofonte, L.IV, V, 25–28) [35].

The Middle Ages is probably the historical period in which most of the rock-cut architecture has been realised in various Anatolian regions.

Several examples show the massive diffusion of artificial cavities during this period: from the hundreds of cavities explored in the underground of Ani, in the Kars district [2], to those investigated at Ahlat, shared by Urartians, Armenians, Seljuks and Mongols [9].

Studying and cataloguing these examples has allowed the authors to understand the historical evolution of different rock-cut types, their state of conservation and distribution throughout the land. From this point of view, Cappadocia preserves the most remarkable repertoire of architectural forms. This area comprises thousands of churches, monasteries, tombs, dwellings, dovecotes and apiaries present, along with war shelters and water systems, excavated in the tufaceous cliffs and subsoil between the 4th and 13th centuries AD [3, 22, 34].

The first systematic study on this vast hypogeal heritage dates back to the beginning of the 20th century, conducted by an archaeologist, epigrapher and Byzantine scholar, the French Jesuit Guillaume De Jerphanion (1925–1942). At that time the structures had long since been abandoned (although some cavities are still inhabited today, or used as warehouses or workshops) and strongly compromised due to natural degradation, which has resulted in the disappearance of large volumes of rock and the loss of spaces excavated therein.

Since then, many scholars (art historians, archaeologists, architects, geologists, geographers and, in particular, speleologists, through their specific research and topographic survey methods) have been working to locate, explore and document this type of heritage, producing a remarkable body of literature. However, much remains to be done: the growing knowledge of rock-cut sites has indeed increased the awareness that the structures investigated thus far represent only the tip of a "stone iceberg" which has yet to emerge.

Fig. 6 Cave church of El Nazar in the Zemi Valley. On the left a picture of S. Saj, 1991; on the right the same church after the restoration in 2013 (Picture by R. Bixio)

At the same time, it is evident that the scientific community should implement the research into this topic, as it is exposed to the risk of rapid and widespread decay, until its inevitable disappearance.

The reasons for this threat are twofold: firstly, the persisting degradation is produced by natural agents on rock formations: erosion, rock falls, freeze-thaw cycles, flooding, etc. On the other hand, decay is a consequence of the increased impact of human factors (outside: the building of dams, roads, car parks, hotels, balloon areas, etc.; inside: non-archaeological sediment discharge, lowering of ancient tunnels, construction of masonry walls, gates and lighting systems for the ever-increasing number of tourists, the drainage of sewers to quickly solve the problems of local urban services, etc.). Such alterations are evidently due to the low sensitivity and knowledge of the same communities that today still exploit the economic potential of some of these anthropic cavities inherited from antiquity, thus having a substantial cultural and historical value, while others are subject to total carelessness.

Recent interventions of consolidation of the most important monuments (Fig. 6), thanks also to UNESCO patronage, are nothing but a drop in the ocean in respect of the rock-cut evidence of these places and, although necessary and commendable, still constitute an alteration of the original morphology of the sites, further complicating their historical understanding.

4 Conclusions

In this chapter the richness of the underground landscape of the Mediterranean regions has been described using four case studies, along with methods and strategies used thus far, to document and share knowledge acquired. This reference frame, created by mapping some specific aspects of the landscape, has delineated a substantial coincidence between the geographical space and the cultural environment. The rupestrian culture of this large geographical area testifies, in fact, to the birth, growth and development of several ancient civilisations, in which modalities of settling in the natural environment have left a significant mark in many countries of the region.

Although compiling a comprehensive and detailed inventory of the underground heritage appears to be complicated due to the multitude of sites, logistics and political difficulties, this first contribution can represent a solid base for further investigations which, in the future, might involve other scholars and institutions. To do so, an open-source geographic information system (GIS) accessible online, and including information deriving from different scientific fields, is the next step of this research.

The goal is to make this information freely accessible to a large community of people involved in the process of understanding and preserving cultural heritage, providing them with a powerful tool with which to foster the exchange of methodologies, case studies and best practices.

References

1. Andaloro M (ed) (2009) Terra di roccia e pittura. La Cappadocia e il Latium rupestre. Gangemi Editore, Roma
2. Bixio R, Caloi V, Castellani V, Traverso M (2009) Ani 2004. Surveys on the underground settlements. British Archaeological Reports-BAR, International Series 1944. Archaeopress, Oxford, England
3. Bixio R, Castellani V, Succhiarelli C (eds) (2002) Cappadocia. Le città sotterranee. Istituto Poligrafico e Zecca dello Stato, Roma, Italia
4. Bixio R, De Pascale A, Mainetti M (2012) Census of rocky sites in the Mediterranean sea. In: Crescenzi C, Caprara R (eds) The rupestrian settlements in the circum-Mediterranean sea. Università degli Studi, Firenze, Italia, pp 89–93
5. Bixio R, Parise M, Yamac A (2017) Idraulica rupestre in Turchia. Geologia dell'Ambiente, suppl. n. 3/2017, Convegno "Tecnica di Idraulica Antica", SIGEA, Roma, 18 Novembre 2016, pp 145–151
6. Caprara R, Dell'Aquila F (2004) Per una tipologia delle abitazioni rupestri medievali. Archeologia Medievale 31:457–472
7. Dell'Aquila F, Messina A (1998) Le chiese rupestri di Puglia e Basilicata. Adda, Bari
8. De Minicis E (2008) Metodi e strategie d'indagine per lo studio degli insediamenti rupestri nel Latium. In: Insediamenti rupestri di età medievale: abitazioni e strutture produttive, a cura di E. De Minicis, Atti del Convegno di studio, Grottaferrata 27–29 ottobre 2005, Centro It. di studi sull'Alto Medioevo, Spoleto, 2008, pp 293–314

9. De Pascale A, Bixio R (2011) Under and inside Ahlat: the KA.Y.A. (Kaya Yerleşimleri Ahlat) Project. In: Baş A, Eravşar O, Duran R, Dursun S (eds) Proceedings of the XIV. symposium in Medieval and Turkish excavation and art history studies, 2010 (pp 173–190). Selçuk Üniversitesi, Konya, Türkiye
10. Del Prete S, Parise M (2013) An overview of the geological and morphological constraints in the excavation of artificial cavities. In: Proceedings of 16th international congress of speleology, vol 1, pp 236–241. Czech Speleological Society, Brno
11. Despois J (1935) Le Djebel Nefousa (Tripolitaine): étude géographique. Larose, Paris
12. Dobosz T, Galeazzi C (2005) Dimore celesti per santi e briganti. Alla scoperta di tre eremi fra Latium e Abruzzo. Speleologia 50(1–2004):46–51
13. Felici A, Cappa G (1992) Santuari rupestri in provincia di Viterbo. Informazioni, Anno I, n. 7, Luglio – Dicembre 1992, 120–127
14. Fonseca CD (1980) La civiltà rupestre in Puglia. In Aa.Vv., La Puglia fra Bisanzio e l'Occidente (pp 37–116). Milano
15. Gabriel U, Treister M (1999) Lower City, spring cave and vicinity. Studia Troica, 9, 23–25. Universität Tübingen, Tübingen, Deutschland
16. Galeazzi C (2011) L'antico insediamento di San Lorenzo (Viterbo-Latium). In: Proceedings of Convegno Nazionale di Speleologia in cavità artificiali, pp 45–60. Opera Ipogea 1/2-2011, Società Speleologica Italiana Ed
17. Germani C, Galeazzi C, Dobosz T, Galeazzi S (2014) Cavità artificiali nell'insediamento medievale di Belmonte (Castelnuovo di Porto, Roma). Opera Ipogea, 2/2014:45–60
18. Gilli E, Yamaç A, Tok E (2014) Halys deviation tunnel and cliff dwellings of Sarıhıdır (Cappadocia - Turkey). Opera Ipogea, 2/2014:29–36
19. Golany G (1989) Earth-sheltered dwellings in Tunisia: ancient lessons for modern design. University of Delaware Press, Newark
20. Herodotus (2008) The histories. Waterfield R (trans.), Dewald C (ed), Oxford University Press, Oxford
21. Jolivet-Lévy C (1997) La Cappadoce. Mémoire de Byzance. Paris Mediterranée/CNRS Éditons, Paris, France
22. Jolivet-Lévy C (2001) La Cappadoce médiévale. Zodiaque, Saint-Léger-Vauban, France
23. Korfmann M (2003) Troia in light of new research (Reden an der Universitat). Universität Trier, Trier, Deutschland
24. Laureano P (2001) Water atlas. Traditional knowledge to combat desertification. Bollati Boringhieri, Torino
25. Lionetti G, Borneo V, Santarcangelo S, Pelosi M, Viva M, Parise M (2015) The San Pellegrino rock-hewn complex at Matera: a magnificent example of the rupestrian culture in southern Italy. In: Proceedings of the international congress in artificial cavities "Hypogea 2015", pp 41–52. Hypogea, Rome
26. Mora C, Balza ME, Bixio R, De Pascale A (2017) A link between "ancient words" and the "underground world": Cappadocian landscape, rock-cut structures and textual evidence from Hittite documentation. In: Parise M, Galeazzi C, Bixio R, Yamaç A (eds) Proceedings international congress in artificial cavities 'Hypogea 2017'. Cappadocia, Türkiye, pp 65–76
27. Parise M, Sammarco M (2015) The historical use of water resources in karst. Environ Earth Sci 74:143–152
28. Parise M, Federico A, Delle Rose M, Sammarco M (2003) Karst terminology in Apulia (southern Italy). Acta Carsologica 32(2):65–82
29. Parise M, Galeazzi C, Bixio R, Dixon M (2013a). Classification of artificial cavities: a first contribution by the UIS Commission. In: Proceedings 16th international congress of speleology, vol 2, pp 230–235. Czech Speleological Society, Brno
30. Parise M, Marangella A, Maranò P, Sammarco M, Sannicola G (2013b) Collecting, transporting and storing water in karst settings of southern Italy: some lessons learned from ancient hydraulic systems. Water Sci Technol: Water Supply 13(3):674–682
31. Polimeni B (2017) Notes sur le paysage culturel et l'architecture vernaculaire du djebel Nafûsa. Horizons MaghrÉbins Le Droit à La MÉmoire 33(76):135–142

32. Quilici Gigli S (1981) Colombari e colombaie nell'Etruria rupestre. Rivista Istituto Nazionale d'Archeologia e Storia dell'Arte, S.III, IV:105–175
33. Raspi Serra J (1976) Insediamenti rupestri religiosi nella Tuscia. In: Mélanges de l'Ecole française de Rome. Moyen-Age, Temps modernes, tome 88, n°1, 27–156
34. Rodley L (2010) Cave monastery of Byzantine Cappadocia (Paperback ed.). Cambridge University Press, New York, N.Y.
35. Senofonte (1984) Anabasi (Ravenna, E., Italian trans.). Mondadori, Milano, Italia
36. Stiesdal H (1962) Three deserted Medieval villages in the Roman Campagna. In: Analecta Romana Instituti Danici, vol II, Ed. L'Erma di Bretschneider, Roma, pp 63–100

The Challenges of the "Divided" Heritage of Cyprus

Kokan Grchev and Ozgur Dincyurek

Abstract By opening the issues of "divided" heritage of Cyprus, this study intends to discuss different continuities as well as internationally established consensuses on cultural heritage and its identity. Heritage is discussed as representing and evolving contemporary culture, problematizing human existence and guiding the future of the societies and cultures towards valuable identities. Perceived heritage of the island as a specific resource is questioned theoretically and by following the activities of the "Technical Committee on Cultural Heritage in Cyprus", in their efforts to bring together some practices aimed for heritage positioning in the focus of contemporary culture. Specific discourse is open towards the evident need of monuments on one side, and the social integration of heritage in cultural and development activities on the other. Unique emergencies are recognized as challenging the existence of both.

Keywords Cultural heritage · Challenges · Monuments · Identity · Emergencies

---

The Technical Committee on Cultural Heritage (in the text, we will use this title as TCCH).

As explained in detail in the booklet published January 2015 as part of the Partnership for the Future UNDP Program funded by the European Union: "The Agreement of 21 March 2008 reached between Greek Cypriots and Turkish Cypriots under the auspices of the United Nations, paved the way, among others, to the establishment of the Technical Committee on Cultural Heritage which is dedicated to the recognition, promotion and protection of the rich and diverse cultural heritage of the island. The Committee is supported in its work by an advisory board composed of archaeologists, architects, art historians and town planners from both communities" [1].

K. Grchev (✉) · O. Dincyurek (✉)
Department of Architecture, Eastern Mediterranean University, Famagusta, Cyprus
e-mail: kokan.grchev@emu.edu.tr

O. Dincyurek
e-mail: ozgur.dincyurek@emu.edu.tr

G. Amoruso and R. Salerno (eds.), *Cultural Landscape in Practice*,
Lecture Notes in Civil Engineering 26,
https://doi.org/10.1007/978-3-030-11422-0_9

1 Introduction

Increased interest for heritage nowadays effects with varieties of perceptions covering different domains as a result of processes of positive cultural contaminations with the globalized phenomena. Related profoundly with the aspects of identity (national identity), heritage and culture are taking the main role to determine the mainstreams of any culturally mediated transformations in the societies today [2] (p. 74). Critical discourses very often point to heritage as a promising issue to varieties of cultures and their attributes today, but at the same time, it is perceived as an unstable and submissive category to influences from outside were not reliable and unable to give back the promised results by default. Mainly because of the threat that over-emphasized heritage can oppose to objective cultural realities, very often instead of having a feeling of deeply rooted belonging, heritage is criticized for creating virtual realities, instant results, attractive products almost equal to any market-based values [3].

Nowadays, distinctions of Cyprus heritage are opened to be discussed in different contexts. Being aware of the potentials of heritage if focused on common objectives of social and economic well-being, future moves can affect the perceptions of cultural differences and how the actual realities are intertwined with the Cyprus conflict [4]. Relating it to Cyprus as an inevitable point of further development, another positive concern is raised about "heritage as the space in which futures are assembled" [5]. This is occupying our intentions not only because the extended field of heritage recognizes how heritage is more about involvement of practices that designs the future instead of being attached to the past; Moreover, it gives another perspective about possible relations and conditions in which tangible and intangible traces of the past are going to create new reality with an outcome to constitute specific resource "in and for the future" [5]. It can be expected, that the threat of experiencing Cyprus heritage as over-attached to the past or over-sensitive to localizations of different kinds, can be avoided by acceptance of new standpoint: translating heritage with a meaning of conversation about the values of the past, into activism where heritage is re-conceptualized so to take responsibility for future.

Contemporary conditions show how it can be followed that both cultural and natural heritage are addressed with attention as part of global efforts to put a focus on them in the creation of any strategy related to future of community, society or culture. The ongoing tendencies show even establishment of new collaborative research programs focused on assembling alternative futures for heritage; it is becoming obvious that the field and domains of heritage practices are expanded in this search for modalities of caring for the future. [6, 2] So far, the broader context and meaning of heritage is probably starting to dissolve or to transform its social and complex ingredients that create concepts of culture in the second decade of 21st century. There was a serious shift in understanding the identity and formation of nation-states ethics, especially in the last decades behind. Usage of heritage in that process of maturing became of immense meaning and importance. Exactly that feature of heritage will bring this term to the front line of identity defense, or in many other cases, it will contribute to the profiling of national identity mainstreams. Past and history through this prism are getting new social and cultural positions: "Modern nation-states use images of a chosen past to construct a national identity" [7].

At the other side, multiple challenges of various sources are changing the perceptions of heritage in contemporary cultures, mostly because of the shift in already achieved consensuses being unsustainable. In this transformative process, together with the evident progress in recognition, evaluation, and usage of tangible cultural heritage, goes a process of definitive loss of many aspects of the intangible, affecting the perceptions and experiences of the tangible heritage and participating in changing the definitions of cultural heritage identities and existence. Moreover, in the attempts to reorient and re-conceptualize heritage today, some researchers are willing to shift from the idea of the universal and inherent value of the heritage towards heritage as a collaborative, dialogical and interactive process. Through that understanding of relationships with heritage, it is becoming possible to open further domains and relationships of heritage, sustainability, resilience [6]. Enhancing the cultural dimensions of sustainable development further on, will be targeted in the definition of "cultural territorial systems" where the specific place-based approach is emphasizing the coexistence of efficiency and equity dimensions in development policy. Through the focus on minor historic centers and their natural and rural landscapes, heritage identity and cultural perspective are projected as a future of places and local populations [8]. Targeting the local sustainable development with a cultural perspective, heritage identity (together with natural resources) becomes significant ingredient and potential for economic development.

So, sensitive approach towards heritage issues in Cyprus has to be applied as multilayer installation. In this sense, the case of Cyprus deserves exceptional attention since its division produced varieties of side-effects, reflecting its reality in forms of existence that are unique by many different points. Even before historical events in Europe (and in broader context) in early 1990s that were marked with the enormous inner energy and the urge for instant change, division of Cyprus in 1974 precedes these events, taking its own mainstream (keeping it isolated and out of the main shifts till the recent serious political moves taken in a way to possibly re-unify the Island). In the light of such an approach, Cultural Heritage Technical Committee in Cyprus is established to respond to the newly developed mainstreams, supported with the belief that by recognizing intrinsic values especially in the architecture monuments recognized as important symbols for both communities it will be emphasized as a need and primary responsibility. In that sense protection of the endangered heritage of the island (considering its integral territory) points especially to its common recognition as "endangered". Solving these issues has meant not only to both communities but for humanity too [9].

2 The Cultural Heritage of Cyprus

With a proper historical distance from the events that left evident scars in the tissue of cultural heritage on both sides of Cyprus, it is possible to give another light to the heritage realities by fading partially the previous concerns and dilemmas. Especially after 2003 and opening the checkpoints that brought back many people to their familiar spaces of memory, places of cultural or religious significance and importance, some aspects of heritage started to be developed more intensively in different directions [4]. Both political interests on one side and experts opinions on the other, will bring together the significant efforts of both communities confronted with the challenge of everyday life management. Overlapping the differences in understanding, interpreting and presenting cultural heritage concerns is showing wide differences and close similarities at the same time. All of it seems enough to create a questionable atmosphere where both sides are trying to cope with the challenges so to establish and accept the new global realities of the world they live in.

After it became possible to travel to both sides and visit spiritually significant places and buildings, the neglected, damaged heritage (being a kind of spiritual bond for both communities) became a more sensitive trigger for a series of public reactions. Well known by the generations before the conflict, and transmitted to the new generations in a different global spirit of cultural understanding, heritage started to give new impulse especially in the treatment and care of religious buildings. This change in the general attitudes and policies is clarifying the reality of heritage sites and especially religious buildings associated and belonging to the "other", having consequences that will reveal some public and official anxieties [4]. Being emphasized mainly with different political agendas, there is an evident and constant pressure on both communities that additionally creates a specific platform on the struggle over heritage (and its identity) since the division of the island. Cultural matrix and changed social context of the territories after 1974 will directly affect the public (common) perception of heritage. In that sense, heritage becomes recognized as "divided" not by being a logical historical consequence or as territorial re-distribution but as real condition targeted by the impact of the ongoing political, cultural and social developments. At the other side of the ongoing processes of preservation and maintenance of the culturally significant historical buildings, there is a threat of this process being covered with multiculturalism and politics of tolerance presented to public (internationally and domestic) but not being efficient regarding some tendencies to bring back and to restore the function as well as the form of the buildings [4].

Which are the modalities of these restorations and how it can be understood in more complex social and cultural conditions today is difficult to be predicted. Following the fact of rapid change of the cities, landscapes and vernacular environments in Cyprus under influence of global trends, there is an evident shrink of the content of these items being recognized as specific "cultural containers" related to the cultural memory and identity formation. By using the term "container quality of the city" and underlying the necessity of witnessing its deterioration, this translation can be followed as a kind of alarm pointing to the evident rapid loss of authentic values [7]. We are recognizing the same concern about evident policies that are limiting the concept of heritage in a way of "monumentalization" of the buildings representing the past but far from being recognized as integrated with the everyday life and living culture. So it is becoming possible to recognize how the values of the cultural heritage need to be institutionalized, changed and culturally translated under such conditions and difficulties. Concern about possible "monumentalization by restoration" of some historical buildings, relates us to many historical assumptions of different kinds (used widely in the political rhetoric mainly). By confronting the consequences of the division with the heritage identity formation, both communities will be brought to the frontier where common cultural investment in the restoration of some particular sites of cultural significance becomes a real showcase of reconciliation efforts at many different levels [4]. Being confronted with the reality of the existence of the cultural heritage destructions that happened as a logical consequence of the turbulent history gives directions for possible solutions and future projects.

However, after decades of inevitable common existence with a lot of efforts invested in between, it seems that finally, communities have a role in the development of initiatives that involve their heritage. This is how it becomes of great importance to have a cultural heritage in the agenda that supports their involvement aside from the possibilities of direct political influence. [5] But still, in some cases it is even impossible to match and to balance differences in perceptions of heritage realities having a negative understanding of some localized problems seen from distance.

Because of those almost virtual cultural distances, even precise academic discourses of cultural heritage discussions contain a threat of not being understood and perceived under these circumstances. This introduces discomfort and sometimes misunderstanding of the perceived and analyzed conditions in specific cultural environments: instead of being integrated through commonly applied theories, methodology and legislation, cultural heritage is getting local, isolated and difficult to compare features. There are many examples showing the difficulties of the ongoing political situations confronted with the existing (accepted) international legislation that creates serious difficulties in the attempts to preserve cultural heritage in contested lands [4, 5, 10, 11]. For the professionals on both sides appears a new challenge: to work not only on recognition and evaluation of historic sites and buildings from different periods, but on the sensitive and intangible layers of heritage-related with the social and spiritual dimension of them and the specific influence they had in the collective memory formation. In most of the cases, such an approach can solve the possible lack of empathy for the spiritual values of some relevant historical buildings, although it can be agreed about the high degree of professionalism in heritage management and conservation on both sides since 1974 [11]. The level of commitment in this domain can be confirmed by following the project developments of already mentioned institutions and committees [1, 9, 12–14].

3 Theoretical Discourses—Divided Heritage

Questioning cultural heritage in Cyprus in varieties of theoretical aspects is not new. Different studies, materials, comments, and publications of different kinds are going to be presented to the public following the reactions of both sides to their "own" cultural heritage. In some of them, examinations of the conflicts and politics of heritage within communities and across ethnic divided Cyprus will be done by considering some implications of project activities upon different layers of cultural heritage (religious, antiquarian and modern) [4]. Further on, we are confronted with the threats to cultural heritage in the Cyprus conflict encompassing many aspects of the destruction of cultural properties and considering possible damages done to cultural heritage during and after the conflict. These studies are giving some aspects of political violence even analyzing it from the point of Islamic cultural heritage vs. Christian cultural heritage both on North and South part of Cyprus, including even vernacular architecture in its focus [15].

Analyzing problems of archeological sites as an important factor of cultural heritage identity, other studies are showing and explaining the so-called "social geography of difference" having it explained as a contested cultural reality of Cyprus. At the same time, using the same rhetoric it will be emphasized how in contemporary and ongoing political processes the practice of archeology is becoming redefined as political expression and cultural representation [16]. Being influenced by the current sociopolitical context, this part of the cultural heritage agenda is considered as constantly being vulnerable to the mainstream of international or domestic policies. Specific dichotomy is witnessed in interpretations of the facts about cultural heritage being endangered as a result of the mainstreams of political agendas: some will accuse of the "loss of civilization" by destruction of cultural heritage in occupied Cyprus [17]; still others will comment about common heritage and reconciliation, recognizing architecture, built heritage and specific buildings as representing common heritage. Reconciliation, restoring the past, avoiding ignorance and healing a division of four decades is recognized as the highest standard of benefit for both communities [18].

This enormous production of different attitudes supported by recent political mainstreams and interpretations seems almost never-ending: recognizing threats in recent urban developments affecting archeological sites and endangering this sensitive part of cultural heritage, asking for long-term decisions that will affect the future of archeological remains [19]. Others will continue the criticism about the potential looting of cultural heritage in "occupied Cyprus" by recognizing illegal excavations in different districts [20] or commenting about the potentials of some endangered sites. These observations will bring in front suggestions and proposals for positive future impacts, especially because some signs in that direction are already on the way under the umbrella of Europa Nostra, a pan-European federation for cultural heritage [21, 22] and later the European Union's Partnership for the Future Program implemented by the United Nations Development Program [9, 11, 12]. As a positive sign it can be stated that social and economic wellbeing is becoming the main common objectives based on co-operation between both communities all related to cultural heritage with common significance.

As a consequence of the political and territorial division of Cyprus, the sum of "heritage" is divided: intangible facilitated by migrations of people, tangible and immovable by being neglected in artificially created environments and not being attached to it logically. Since the intangible cultural heritage presents a fundamental element of the identities of the ones who create them, it refers to a further examination of the relevant implications in terms of human rights [23, 24]. Relations between cultural heritage and human rights present another field of specific and profound examinations exploring especially the impact of globalization on cultural heritage. This marks Cyprus on the list of such explorations not only because of the richness of the heritage: a witnessed struggle of the communities to survive their cultural identity and their way of life as part of their heritage is evident resistance against negative processes of cultural globalization. So far, it is evident that the richness and uniqueness of this heritage raise awareness and defines communities and societies, but at the same time, it is a potential cause of conflict [23]. That's why more profound touch into this matter (cultural heritage and human rights), will bring in front relations of the political aspects of heritage preservation and management relating it to human rights. Since the heritage is necessary for the preservation of the cultural identity, any progress towards a further display of heritage monuments can be on the way to balance minorities and majority. As a consequence, this is questioning the right of free expression (on one side) and further preservation of cultural heritage (on the other), together with the possibility of articulating and protecting that right [23]. However, with the scope of this research, the issue of heritage and human rights will not be examined in detail, although as part of this methodology, some of the relevant and shared questions about: possibilities of heritage to divide or unite; alternatives of control, defining and benefiting from heritage etc. will be opened and discussed in other contexts.

One of the important theoretical discourses considering heritage that is based on the fear and danger about maintenance and transmission of heritage in the future is especially targeting religious heritage and separate monuments, being aware the unpredictable mobilization of heritage for possible unknown purposes [25] (p. 7). As already mentioned in the text, monuments detached from their localized and spiritual existence are quite unstable in their real cultural impact to the communities; moreover, it is even more vulnerable by being dependent on varieties of political moves (local and international) distancing the real heritage problems to the unsafe ground.

Especially with Cyprus, after the last circle of political negotiations being postponed to undefined term, it is becoming obvious that even world is divided regarding the answers to be given to various global challenges within the separate heritages. In that sense, a division is not possibly recognized at the level of territories, people and identities only; it is referring also to the heritage being divided or detached by force from its origins. Being divided from its native environment, heritage becomes representative of negative rather than positive trends and mainstreams, with a risk to be managed in a utilitarian way [25] (p. 7).

Regarding this shift in understanding monuments, in the general discussion about the uses of heritage, for example, some interesting discourses about the period after the dissolution of communist countries in Europe will be opened referring to the general rejection and anxiety for the monuments as they were marking territories [25] (p. 7). With the Cyprus case, can this be indicating some of the unwanted consequences of the so far promoted use and understanding of the heritage emphasizing buildings and monuments rather than cultural heritage in general (without any attempt of direct analogy)? More specifically, in the case of Cyprus, it is becoming interesting to follow: how the logical shift from architectural to cultural (heritage) was successfully done worldwide, still in Cyprus case, what we can observe is that conservation is still about (separate) monuments and still trapped in architectural heritage conservation, having cultural heritage conservation without strong and influencing position for the society. Are these projects creating emergencies for the heritage of unique nature and character by reversing the process: instead of conservation dealing with the complexity of cultural heritage, it might be perceived as conservation of separate monuments, without expected serious impact on re-defining the cultural matrix of acceptance of this heritage and culturally balancing both communities? Conservation projects and accompanying developments are becoming new, undefined and difficult category/typology/of risk. Evaluation and management of these risks are probably new issues that need special care and attention.

4 Challenges of Identity (Memory and Heritage Identity)

As it can be perceived from the complexity of on-going processes in the domain of cultural transformations influenced by the current political mainstreams in Cyprus, we are introducing specific challenges that profile our perspectives on identity, culture, memory, heritage identity, cultural identity. Through identity as part of cultural heritage meanings and importance and the complex set of phenomena associated with, we are eligible to open discussion about culture and contemporary contexts within. Identity is referring to collective memories, nourished by the communities on both sides as contemplative, emotional input into contemporary cultural settings. A great part of newly defined emergencies in the domain of cultural heritage lies exactly in this part of identity understanding. This is even more complicated nowadays in the light of the status-quo political situation of Cyprus. Having in mind the historical necessities emerging from the actual situations in the newest history of Cyprus, it can be questioned about the possibility how to create more tangible feeling about the new landscapes and environments (including all aspects of heritage) and which are the possible mechanisms of making them new social interaction points for the new communities? Is it possible to keep the meaning of them to the level of personal identification and creation of new identities? Many different disciplines are going to profile the meaning of the term of identity and its sense, mainly because this abstract phenomenon is capable of change within the scope of any separate research and according to the changing parameters through time [26].

Introducing heritage in realization and conduction of various and recent socio-cultural developments is challenging us with a complex set of theoretical and practical mainstreams. Accordingly, in the frame of this research, some "emergencies" will appear as logical consequences of the gradual transformation and understanding of identity applied in various manners depending on the required contextual explanations. Referring to the cultural identity (identities) makes it feasible to easily recognize values of cultural traditions, contributing to better understanding of those aspects of the heritage which can bring improvement of lifestyles in different cultural environments [2] (p. 75–77). Thus, the case of Cyprus can be seen as an exceptionally interesting phenomenon, but observation of the profile of the cultural identity applied upon the integral territory of Cyprus can be seen as an extremely important and sensitive category. Referring to possible changes in the attitudes towards heritage, this also refers to the urgent necessity to understand the real dimensions and to determine the adequate cultural values within all aspects of its material and spiritual existence in the course of their turbulent historic development.

We can also argue that as a sequence in the understanding globalized world, conservation of identity appears as a special key for accepting the sum and varieties of localized values especially its historical, physical, socio-cultural and many other characteristics [26]. Some authors will question identity together with a place, landscape and heritage by arguing the possibilities of approaching to the phenomenon of identity; it is referring to varieties of discourses such as recognizing differences, place-making, reacting to traditions and representations [27] (p. 8). However, since these discourses belong to the field of social theory (social identity) it is becoming evident how new horizons are opened towards an understanding of landscape, place, and heritage, where heritage is getting in special relation to identity. Heritage identity, as it was mentioned before, is proving to have real potential for further economic development firmly rooted in places and local populations [8].

In this specific follow-up, when the politics of identity leads to petrified and non-negotiable issues, then culture, identity, and past are becoming non-negotiable leading to different divisions and formed boundaries, instead of unifying with co-existence. In that sense, people, things, places, buildings are being "monumentalized" as active creators of identity.

In other analysis, a phenomenon of "memory" will be perceived as extremely important and dangerous at the same time. But memory is related to specified cultural environments, where architecture (monuments) represents just one side of the coin; in the cases when the environment and even territories are changed (divided, contested), monuments lose their charming seduction already witnessed in part of their place-time existence. Exactly at this point, we are concerned about already mentioned shift from cultural to architectural, to "monuments". Identification at the level of cultural recognition is always difficult in confronted situations and complex historical transmissions giving an outline to our problem of defining emergencies in the light of the active ongoing processes. In contemporary conditions of globalized culture transmission of values, new generations are witnessing the past as a complex reflection of "time" and "place"; since the monuments are presenting the past, they are transmitting significant energy already mediated by several generations [28] (p. 75), [25] (p. 7).

5 Conclusions

The uniqueness in the approach towards divided heritage in Cyprus is related with important guidelines and dilemmas in a condition where common, utilized understanding of the values and identities of cultural heritage is grounded in a setting of the constant presence of the territorial and cultural division. The controversy appears with the listing of the monuments and hierarchy for interventions done by the Technical Committee: are they becoming interested in achieving harmony and balance between communities, rather than prioritizing the real heritage conservation demands? Is it possible that real hierarchy of needs for interventions are becoming intangible and not related to the objective condition of the heritage by fostering common and mutual understanding?

The decision of bringing monuments in the front line might be seen as a limited possibility in the given conditions, but in another way, it makes a distance from the more deep and profound definition of cultural heritage. In this sense, if insisted in the wrong direction, is it possible under (virtually) created circumstances to have the same expected "integrating" heritage in a different role: to divide, to mark, to determine? This is a possible risk and threat where conservation projects can produce "monuments" rather than "common spots for cultural identification".

Following the given methodological mainstreams of the TCCH, especially after it was established in April 2008, we can perceive that the Committee goal was to provide a "mutually acceptable mechanism for the implementation of practical measures for the proper maintenance, preservation, physical protection and restoration (including research, study, and survey) of the immovable cultural heritage of Cyprus." [1]. Further on, it is expected that fostering, cooperation and confidence building between Greek Cypriots and Turkish Cypriots will be provided under the umbrella of UNDP PFF as leading coordinator. Since the establishment of the TCCH in 2008, it is clarified that its mission is directed to the immovable cultural heritage of Cyprus [12].

The follow up of these events and establishment of the methodologies leads to the next source of our questioning: In the Phase 1 (Support to cultural heritage monuments of great importance for Cyprus) is emphasized that project aims at "contributing to the confidence-building process through the preservation of Cypriot cultural heritage and the implementation of emergency measures on a list of high importance monuments island wide according to the strategy of the Technical Committee on Cultural Heritage". Later in this document, it is emphasized the usage of the Study on Cultural Heritage in Cyprus and its completion is a prerequisite for the further actions [12]. Purpose of this study is to give light to the condition and estimated costs of restoring the immovable cultural heritage of Cyprus in order to contribute to its protection and preservation [13]. Again, instead of the sum of values and components of cultural heritage by the scope of the term, only partial actions are underlined as important considering immovable cultural heritage, and following inconsistent list of monuments, built heritage, cultural heritage, restoration, conservation, preservation, physical protection etc.[1]

So, opening new possibilities by leaving "open doors" for interpretations and giving legitimacy to the division as one of the alternatives, somehow shows that even in divided communities, things can be quite normal and acceptable making division durable. These notions are giving an appropriate framework to establish the genuine of the "emergencies" originality seen from inside. But being considered as an integral entity, cultural and natural heritage, including living culture of Cyprus, have to be evaluated for further developments throughout varieties of activities. In this process, it is of great importance to follow and create systematization and processing of relevant studies and projects (including ongoing and recent). Field research of different profile applied on the integral territory of the island can open different subjects of observation and inventory through an already established set of values. By bringing this data to the public, it is expected to become more effective in changing the attitude towards the heritage, in view of their activation through various activities. Profound re-direction of the mainstreams can probably participate in setting new directions for further developments, relying on a good and effective application of the entire know-how.

In this complex process of dealing with the layers of a divided heritage of Cyprus, intentional "balance" of the interests given by experts from both communities is surely good ground reflected through an affirmed term of "common heritage". But, is it possible in such environment to arrange and apply such evaluation that will be based on well known, established methodologies and theories, so to prioritize really those heritage units that urgently need experts touch, attention, and action? At the same time, all of the conservation projects to be done with a lot of emphasis on professionality expertizes and high expectations about the possible feedback of the communities as a direct response of the whole of heritage in new conditions (after restoration and conservations are done). Special evaluation methodology needs to be

[1]Developments in the domain of heritage as discussed in these documents are divided into phases [14]:

Phase 1: contributing to the confidence-building process through the preservation of Cypriot cultural heritage and the implementation of emergency measures on a list of high importance monuments island wide;

Phase 2: supporting confidence-building measures through conservation and emergency measures for additional high importance cultural heritage sites;

Phase 3: support the reconciliation process by increasing trust between the two communities through the implementation of confidence-building measures;

Phase 4: supporting the efforts of the bi-communal Technical Committee on Cultural Heritage to contribute to the ongoing peace and reconciliation process by bringing communities closer to their shared heritage through conservation and emergency measures projects.

established to measure this impact of projects on the real life. This urge is the point that refers to the new emergencies understanding. However, it is already witnessed how heritage objectively infiltrates people's lives, becoming a language that expresses their appreciation of objects, places, and practices. Through critical and a creative engagement with the various fields of study, further approaches should involve in critical heritage studies for the future [6].

References

1. UNDP Cyprus: Technical Committee on Cultural Heritage (2015) Available online: http://www.cy.undp.org/content/cyprus/en/home/library/partnershipforthefuture/the-technical-committee-on-cultural-heritage–2015-.html
2. Janet B (2000) On defining the cultural heritage. Published online: 01 Jan 2008. 49(1):61–85
3. Lowenthal D (2017) The heritage crusade and the spoils of history. Aug 2012. Cambridge University Press, 04 Dec 2017
4. Costas M (2012) Constantinou, Olga Demetriou & Mete Hatay. Conflicts and uses of cultural heritage in Cyprus. J Balkan Near East Stud 14(2):177–198
5. Lucas Lixinski (2011) Selecting heritage: the interplay of art, politics and identity. Eur J Int Law 22(1):81–100
6. Rodney H (2015) Beyond "natural" and "cultural" heritage: toward an ontological politics of heritage in the age of anthropocene. Heritage Soc 8(1):24–42
7. Munasinghe Harsha (2005) The politics of past: constructing a national identity through heritage conservation. Int J Heritage Stud 11(3):251–260
8. Rotondo F, Selicato F, Marin V, Galdeano JL (2016) Cultural territorial systems: landscape and cultural heritage as a key to sustainable and local development in Eastern Europe. Springer, 19 April 2016
9. UNDP (2015) The technical committee on cultural heritage in Cyprus. United Nations Development Program, Partnership for the Future, UNDP
10. Şevketoğlu M, Tuncel R, Şahoğlu V (2015) Protecting the cultural heritage of Cyprus: international laws and concerns. J East Mediterranean Archaeol Heritage Stud 3(2):141–148
11. Susan B (2007) Managing heritage in the wake of war and conflict in Cyprus. Deakin University, Melbourne, Australia. In: ICOMOS conference extreme heritage, James Cook University, Queensland, Australia 19–21 July 2007, pp 1–14
12. UNDP Cyprus: Partnership for the future. Available online: http://www.cy.undp.org/content/cyprus/en/home/operations/projects/partnershipforthefuture.html
13. UNDP Cyprus: Study of cultural heritage. Available online: http://www.cy.undp.org/content/cyprus/en/home/operations/projects/partnershipforthefuture/study-of-cultural-heritage-in-the-northern-part-of-cyprus.html. Accessed on 05 Nov 2017
14. UNDP Cyprus. Available online: http://www.cy.undp.org/content/cyprus/en/home/
15. Sam H (2014) Threats to cultural heritage in the Cyprus conflict. Heritage crime. Palgrave Macmillan, Macmillan Publishers Limited
16. Knapp AB, Sophia A (2002) Archeology, politics, and the cultural heritage of Cyprus. Archaeology under Fire. Nationalism, politics, and heritage in the Eastern Mediterranean and Middle East. Taylor & Francis e-Library, pp 13–43
17. Zaphiriou L, Nicolaides C, Miltiadou M, Mammidou M, Coufoudaki V (2012) The loss of a Civilization, destruction of cultural heritage in occupied Cyprus. Nicosia
18. Malcolm B. A crumbling symbol of a divided isle. Available online: http://news.bbc.co.uk/2/hi/europe/7596392.stm

19. Cyprus—ICOMOS, available online: https://www.icomos.org/risk/world_report/2006-2007/pdf/H@R_2006-2007_14_National_Report_Cyprus.pdf
20. The destruction of occupied Cyprus' cultural heritage. Available online: http://www.mcw.gov.cy/mcw/DA/DA.nsf/All/5C63072411078AB9C22572750055D67D?
21. Endangered site: Famagusta Walled City, Cyprus, Available online: https://www.smithsonianmag.com/travel/endangered-site-famagusta-walled-city-cyprus-54478493/
22. The buffer zone of the historic center of Nicosia, Cyprus, Europa Nostra/EIB Institute: The seven most endangered heritage sites, Nov 2013. Available online: http://7mostendangered.eu/2013/06/12/the-buffer-zone-of-the-historic-centre-of-nicosia-cyprus/
23. Silverman H, Ruggles DF (eds) (2007) Cultural heritage and human rights. Springer-Verlag, New York
24. Federico Lenzerini (2011) Intangible cultural heritage: the living culture of peoples. Eur J Int Law 22(1):101–120
25. Poulot D (2016) Uses of heritage. Cultural base, social platform on cultural heritage and European identities, pp 1–17
26. Rifaioğlu MN, Şahin Güçan N (2017) The concept of identity and its identification process in urban conservation projects. Jan 2007. Research Gate. 05 Dec 2017. https://researchgate.net/publication/284038848
27. Tilley C (2006) Identity, place, landscape and heritage. J Mater Cult 11(1/2):7–32
28. Bell Duncan (2003) Mythscapes: memory, mythology and national identity. Br J Sociol 54 (1):63–81

Conservation Issues on UNESCO World Heritage Sites in Russia. From the Roerich Pact to Contemporary Challenges

Polina Mironenko

Abstract Nowadays Russia promotes different economic and cultural programs among which there is one of the most important sources of growth: the conservation of its vast cultural heritage. There is a particular importance in a profound study and wide accessibility to cultural heritage since the country has been experiencing fundamental social, economic and spiritual changes. Study and conservation of cultural heritage are necessary conditions for the dissemination of the national cultural wealth of Russia. Valorization of proper historical heritage helps to keep the spirituality of people so that authentic and ingrained culture doesn't get affected by false and ephemeral values. Cultural heritage basically represents the main form of existence of culture, if it is not represented to cultural heritage doesn't belong anymore to culture and at the end stops existing. In all their life, a person can valorize and internalize a small part of cultural heritage. It might be left for the other generations, so it can become a part of a heritage of the human community, but just if it's properly preserved. That's why cultural heritage protection is also a protection of the culture tout court. Issues on cultural heritage protection are common to all the societies and today Russia is facing it in a more systematic way. The history of Russian culture resembles more to that western rather than oriental: it was also followed by rather rough changes and solutions of continuity.

Keywords Conservation · Cultural heritage · Cultural landscapes

1 Background History

Its evolution was complicated by Russia's geopolitics position: being put between the west and the east, confusion was created. Russia was torn between the concept of western and oriental development, finding and affirming not without difficulty

P. Mironenko (✉)
Department of Design, Politecnico di Milano, Milan, Italy
e-mail: polinamironi@gmail.com

G. Amoruso and R. Salerno (eds.), *Cultural Landscape in Practice*,
Lecture Notes in Civil Engineering 26,
https://doi.org/10.1007/978-3-030-11422-0_10

her own identity and originality. This is why the problem of protection and conservation of the cultural heritage has always existed, at times become rather intense.

One of the most crucial moments in history was during the reign of Peter the Great (1672–1725) who transformed Russia in a radical way, thanks to the reforms he turned the country toward the West, drastically complicating the problem of relationship with her past. Despite the whole radicalism of the transformations, Peter I didn't want to abandon completely the past of Russia, its cultural heritage. Instead, during his reign, for the first time, the protection and conservation of cultural heritage appeared as a concept of extreme importance and the first steps were taken to preserve it.

For instance, at the end of the XVII century according to the decree of Peter I, there were promoted several documentation campaigns of the ancient Buddhists temples in Siberia so they would produce a notable collection of illustrations and documents. There was also one noticeable decree that said that during the years when the constructions in stone were forbidden in the whole Russia, except St. Petersburg, Peter I issued a special permit authorizing the construction in stone in the city of Tobol'sk. It was specified that Kremlin of Tobol'sk should not be built for defensive reasons or for representation of the military strength, but it had to represent the greatness and the beauty of Russian architectural art instead. As well as the construction of the road which led to China through Tobol'sk, meant building a road for a nation that had always been and it would always remain a friend of Russia [1].

Peter's efforts were continued during the reign of Catherine II who promoted decrees on measurements, the collection of drawings, researches, and documentation of buildings of historical and artistic value through their layout and plans and the descriptions of ancient cities and the protection of main archaeological monuments.

The attention to the protection of natural and cultural heritage grew significantly in the XIX century when was established the law that forbade the demolition of the buildings made in the XVIII century or the reconstruction of unity that could alter the native image.

An important role in the protection of the natural and cultural patrimony has historically been developed by public scientific organizations: *the Archaeological Society in Moscow* (1864), *the Russian Historical Society* (1866), *the Society for the protection and the conservation of the art and the antiquity in Russia* (1909) and others. During their conventions, those institutions discussed problems related to the protection of the historical and cultural heritage, they dealt with the development of the legislation on protection of the monuments, and they encouraged the creation of government institutions for the protection of the cultural and historical values. Among these organizations, there is one in particular that has to be highlighted is the *Archaeological Society in Moscow*, that included archaeologists, architects, artists, writers, and art historians. The main purposes of the *Society* were not the only study of ancient monuments of the Russian antiquity and their protection from destruction, but also from wrong restorations during the works of remaking or rebuilding.

2 The Roerich Pact

On April 15, 1935, in Washington representative members of the USA and other twenty nations of the American continent signed an agreement "*On the protection of the artistic and scientific institutions and the historical monuments*" that later became famous in the international legal practice as Pact Roerich. From 1999, this day has been considered the *Universal Day of Culture under the Banner of Peace*, known also as the *World Day of Culture*, an observance held annually around the World to promote the protection of culture, the *Roerich Pact and the Banner of Peace* [1].

The idea to settle an organized preservation of the artistic and scientific world treasures belonged to the famous artist and Russian public character Nikolaj Konstantinovich Roerich, and appeared during the studies of the Russian antiquities at the beginning of the XX century. Roerich conceived the idea of the protection of artistic and scientific achievements of humanity, which he proposed to the *Society of Architects in Russia*. The artist traveled a lot in Russia, visited the excavations, studied the origins of the Russian culture. "The man who is not able to understand the past cannot think about the future", Roerich wrote [2].

The Russian-Japanese war of 1904 made the artist think seriously about the threat that was hiding in the technical improvement of the means of destruction. In 1914, Nikolaj Roerich turned himself to the Russian government and the governments of other belligerent countries with a proposal to provide the conservation of the cultural treasures with the conclusion of an appropriate international agreement, but his plea remained unanswered. In 1929 Roerich prepared and published a draft of an essay on the protection of the cultural heritage, in various languages followed by a plea to the governments and the people of all the countries.

The signing of the pact was an important step in the history of humanity. The document contains the antimilitarist idea of Roerich called "*Peace through the culture*", that implicates not only the protection of properties and cultural treasures for the future generations but also international responsibilities based on the principle of the inviolability of goods and cultural values belonging to the humanity. Roerich believed that all the cultural values bring a spiritual power and by destroying them humanity destroys the base on which it can spiritually develop.

The project of an agreement received a world recognition and had a huge impact on the international community. The idea of Nikolaj Roerich was defended by the most illustrious intellectuals of the epoch such as Romain Rolland, Bernard Shaw, Albert Einstein, Herbert Wells, Maurice Maeterlinck, Thomas Mann, Rabindranath Tagore. The *Roerich Pact* was the first international act dedicated to the protection of the cultural heritage and was the only agreement in that field that was adopted by a part of the international community before the Second World war.

In the picture of the *Pact*, there is an approved a distinctive mark proposed by Roerich, that would mark cultural objects. That mark was the "*banner of Peace*"—a white cloth with three red dots placed in a red-magenta circle on a white background which represented the past, present and future achievements of the humanity, surrounded by the ring of the eternity.

On the *banner of Peace*, Nicholas Roerich wrote: "*This sign of the triad which is to be found all over the world may have several meanings. Some interpret it as a symbol of past, present and future, enclosed in the ring of Eternity; others consider that it refers to religion, science and art, held together in the circle of culture, but whatever be the interpretation the sign itself is of the most universal character. The oldest of Indian symbols, Chintamani, the sign of happiness, is composed of this symbol and one can find it in the Temple of Heaven in Peking.*"

The Pact contains the general principles of protection of cultural values and on the respect that had to be shown. The protection of the cultural goods in the Pact exists in an unconditional way and is not weakened by the clauses of the military necessity that reduces the effectiveness of the protection of the cultural goods in case of armed conflicts.

At the Article 1 the Pact stated: "*The historic monuments, museums, scientific, artistic, educational and cultural institutions shall be considered as neutral and as such respected and protected by belligerents.*

The same respect and protection shall be due to the personnel of the institutions mentioned above.

The same respect and protection shall be accorded to the historic monuments, museums, scientific, artistic, educational and cultural institutions in time of peace as well as in war."

The Roerich pact has developed an important role in the further formation of the international juridical norms and the public activities in the field of protection of cultural heritage. In 1949 on the 4th session of the general conference of UNESCO, it was decided to begin the works on the international legal regulation in the field of the protection of the cultural property in case of armed conflict.

On May 14, 1954, the UN "*Convention for the Protection of Cultural Property in the event of armed conflict*" was adopted in The Hague on May 14, 1954 (it became effective on August 7, 1956). The Roerich Pact was used as a foundation for this *Convention*. In March of 1999 on an initiative and with the direct participation of UNESCO, the second protocol of the convention in The Hague of 1954 was adopted. The text of the Convention of The Hague openly states the principles of protection of the cultural goods during the war, established by the Conference in The Hague in 1899, 1907 and in *Roerich Pact*.

3 Cultural Landscapes and Properties Inscribed on the UNESCO World Heritage List in Russian Federation

On October 12, 1988, Russian Federation accepted the *World Heritage Convention* from UNESCO for the conservation and promotion of the heritage of humanity. Currently in Russian Federation, there are 26 properties that are inscribed on the

World Heritage List and considered as a universal outstanding value: 16 cultural sites and 10 natural sites [3].

Unlike the European Countries, where the formation of the juridical framework for the national cultural heritage protection had started in middle of the XX century, in Russia such legislative action was launched only in the 1978. The particular nature of Russia is the condition of historical-cultural stress that the society had lived in the centuries XX–XXI, followed by the destruction of an enormous layer of properties and cultural and historical (material, spiritual, mental) values, which had deprived Russia of its enormous potential in the field of the touristic and economic development.

The modern Russian system of protection of the monuments, either from a legal point of view or in terms of financial approach, has maintained the characteristics keys of the Soviet system, even if in comparison to the past times, state's ability of preserving and managing restorations of about ten thousand of historical and cultural sites, has decreased. According to experts, currently, the amount of the government funds spent on preservation and restoration of the monuments of federal importance doesn't go over 15% of the necessary requirement. Around two-thirds of the monuments of federal importance need restoration [4].

Currently, in Russia, there are two organizations that play a key role in the protection of historical and cultural heritage. The first one is the society for the protection of the monuments of history and culture also called *VOOPIK*. The society was founded in 1966 and is a public and voluntary organization that manages several programs such as "Russian estate", "Temples and monasteries", "Russian necropolis", "The Russian emigration".

The second is the *Foundation of the Russian Culture*, founded in 1991, finances a series of programs and projects, among which "Small cities of Russia". In 1992, to enhance the scientific side of protection, the *Russian Institute of research for the cultural and natural heritage was founded.* Its assignments include identification, study, preserving, use and popularization of the cultural and natural heritage.

In 1992 was founded a *Committee for the restitution of cultural heritage* with the purpose of resolving mutual pretensions between Russia and other foreign Countries.

Many cultural goods of ancient Russia are related to religion that was a center of the daily interest of Russian people. Currently, the *Russian Orthodox Church* is contributing to the reconstruction, restoration and renovation of churches and monasteries. Despite the destruction and the misfortunes that took place in the past, there have been preserved over 1200 religious sites.

Outstanding places, rich in cultural heritage, belong to noble Russians who took form on the latter half of XVIII–XIX century. The estates belonged to the family and were called "the noble nests". There were thousands of them, but only about ten that have remained; some were destroyed during the revolution and the civil war and others disappeared with the time and desolation. Among those which survived, Arkhangelskoe, Kuskovo, Marfino, Ostafyevo, Ostankino, Shakhmatovo; they have been turned into museums, reserves and nursing homes while other properties were damaged and need to be repaired under a special programme of emergency assistance.

The role of Russian estates in the development of Russian culture was enormous. In the XVIII century, they were the base of the Russian Enlightenment; thanks to them the XIX century became the golden age of Russian culture.

The relative restoration and protection are essential steps for the conservation of cultural heritage. These problems were faced by the *Society for the study of the Russian estate* that was running in the 20s (1923–1928).

Another assignment which is strongly related to the mission of preserving Russian estates of a lesser importance is the rebirth and the development of the small cities of Russia. They are currently over 3 thousand small cities with a total population of around 40 million people [5]. Just like estates, they represent true Russian style of life, express the soul and the beauty of Russia. Each of them had a unique aspect and a proper style of life. There have been developed complete programs for rebirth of the historical and cultural environment of ancient Russian cities as Zarajsk, Podol'sk, Rybinsk and Staraya Russa and so many others.

The federal law "On the objects of the cultural heritage", approved in 2002, allows, together with the government ownership, a private ownership of architectural monuments, but privatization of cultural heritage sites is not widely adopted as a best practice. The main difficulty is the lack of separation among federal and municipal ownership of monuments, absence of a clear definition on the subject of protection in the law since it is not entirely obvious to what particular element of the monument the regime of protection is applied. Generally, the protection of the cultural heritage in the modern Russian society is becoming more complex and intense and, for this purpose, it requires a constant attention and the development of specific heritage skills. The level of culture development of every nation should be evaluated according to the actions dedicated to their own cultural heritage because by preserving the past, it is possible to imagine a future of wealth and prosperity.

References

1. The Grandars (2016) Preservation of cultural heritage. Retrieved from http://www.grandars.ru/college/sociologiya/sohranenie-kulturnogo-naslediya.html
2. Anastasian S (2015) Literaturnaya Gaseta. Retrieved from http://www.lgz.ru/article/-15-6505-15-04-2015/pakt-rerikha-predtecha-yunesko/
3. Commission of the Russian Federation for UNESCO. Retrieved from http://www.unesco.ru/en/?module=objects&action=list&page=1
4. Sivolap T (2012) To the question of preserving the cultural heritage in Russia. Retrieved from https://cyberleninka.ru/article/v/k-voprosu-sohraneniya-kulturnogo-naslediya-v-rossii-nekotorye-aspekty-resheniya-problemy
5. Alexandrov A (2009) International cooperation in the field of cultural heritage. MGIMO, Moscow

Placemaking Workshops: Application of the PPS Method

Tomasz Jeleński

Abstract This article describes three cases of placemaking workshops conducted by the author in three different countries: Ukraine, Tunisia, and Poland, and against different cultural and political backgrounds. In each case, the application of placemaking methods encouraged public participation, showed the potential to facilitate the decision-making process, and helped resolve potential or existing conflicts while building confidence in democratic procedures and institutions. This research highlights the importance of the PPS method which helped to build a team of stakeholders sharing similar views, ones convinced that a positive change is possible and are ready to cooperate. Such attitudes are especially valuable in places where local democracy and participatory urban management is undeveloped.

Keywords Participation · Placemaking · PPS · Public space · Workshop

1 Introduction

Public space is a priceless environment for human communication, interaction, and local economies. It is a space for culture and a medium for symbolic content. A good public space is fundamental for a sense of community and thus an indispensable component of sustainable urban structures.

Democratization of space remains a challenge for many municipalities. Spatial quality and public functions may no longer be imposed top-down upon a place, but it is possible to improve the quality of a public space by using adequate methods of participatory planning and design workshops. Even if the first attempt at such actions do not bring spectacular results at once, it is necessary to continue refining the methods of participation, for which there is no effective alternative in a democratic system. To develop a culture of participation, one needs to raise the standards of debate on space, respect the diversity of stakeholders, and stay

T. Jeleński (✉)
Cracow University of Technology, Kraków, Poland
e-mail: tjelenski@pk.edu.pl

G. Amoruso and R. Salerno (eds.), *Cultural Landscape in Practice*,
Lecture Notes in Civil Engineering 26,
https://doi.org/10.1007/978-3-030-11422-0_11

open-minded while looking for a consensus. The municipal administration may find valuable partners to work hand in hand on the above issues in non-governmental organizations, which do the educational work in the field of space culture, and prepare the public for wider participation.

2 Methods

2.1 Participation Through Workshops

The roles of the planner and the architect are shifting from the positions of leaders and artists who impose their concepts, to the roles of advisers and experts who offer their services to space users. Workshop methods are a means of this shift. The flow of knowledge on the nature, identity, and problems of particular places (and the clash of diverse opinions, behaviors and motivations that occur during a workshop), change the urbanist's perspective from subjective, external, and aesthetic to one that is more objective and sensitive to the local traditions and complex historical, social, cultural, and psychological contexts of the place [3].

The work of the architect and urban planner has always required decisions of a psychological and social nature [7]. The development of workshop design methods is a practical consequence of a large interest in various trends of environmental psychology and the appreciation of surveys (especially qualitative ones). These methods are intended to help work out a consensus or at least a compromise in conflict situations. A fundamental problem is the efficient organization of the dialogue environment and selection of representative partners, as well as establishing effective communication channels between professionals and the community [9]. What helps in that approach are local grassroots initiatives, led by NGOs, based, to a considerable degree, on a voluntary work. In addition to space quality and economic vitality improvement, the purpose of workshops is to activate and integrate local communities.

2.2 Placemaking

The word *placemaking* means a certain type of creation of characteristic places, ones that are particularly valuable and focus on the key functions of a public space, primarily for local communities. Such a process of creation may only take place with the involvement of various stakeholders, crystallizing the sense of community while it happens. Walljasper [12] argues that even the most struggling communities can be revived, not by top-down actions and infusions of money, but by the people who live there. The challenges which might be addressed include crime, comfort and safety, traffic control, image, and economic vitality. Real-life examples prove the effectiveness of placemaking processes in which active individuals—sometimes supported by urban planners, architects, developers, and policymakers—change their environments in a sustainable manner, taking small steps while motivating others to make change.

Placemaking should be a continuous process with the ensuing feedback: a common vision attracts partners, financial resources, and new initiatives. Solutions are dynamic and implemented in stages; they are based on previous, initially small successes. More daring experiments are introduced judiciously and with caution. The introduction of potential formal novelties should be accompanied by the careful monitoring of the place users' response. Changes are based on constant appraisal and improvements. Involvement increases alongside the feeling of public control and ownership.

One of the most important Polish organizations that disseminates placemaking and public space design and management workshop methods is the Foundation of Active People and Places (MiLA).[1] Three selected projects carried out by this organization in 2013 will be discussed further in this paper. The methodology used in MiLA's projects is based mostly on the abundant experience of the organization, Project for Public Spaces (PPS), a non-profit established in 1975 which endeavored to assist local communities in reclaiming spaces that were socially degraded or car-dominated, and to create or strengthen the more traditional characteristics of public places [11].

PPS's activities are based on the methods of observation and evaluation of space dynamics described by Jacobs [2], Whyte [13] and Gehl [1]. PPS combined theory with dozens of years of practical experience and developed an alternative approach to complex public space problems: a set of patterns and effective evaluation methods were hammered out to assist in the understanding of urban space functioning and potential place value. The PPS method helps various stakeholders define and express their aspirations, needs, and priorities. Participation appears at the stage of developing initial visions. Experts play a supporting role. Solutions result from an in-depth knowledge of local reality and multi-aspect place evaluation. A common vision attracts partners, resources, and new initiatives. Actions become dynamic and are built on previous small successes [4].

The fast and affordable improvement of poorly functioning public spaces releases social energy that is indispensable for continuous and productive space operation. Placemaking methodology consists of the creation of local partnerships that are responsible for day-to-day management, facility improvement, and maintenance of public space. Local governments and administrations obtain support from private entities interested in high space standards in their surroundings.

In Poland, the PPS placemaking method was originally disseminated by the Polish Environmental Partnership in cooperation with the International Centre of

[1]The foundation with its seat in Kraków, deals with development based on natural and cultural values of places, as well as civic activity and co-responsibility in communities (http://www.mila.org.pl/). In placemaking projects, it collaborates on a permanent basis with the International Centre of Education of the Cracow University of Technology, and INTBAU Poland.

Education of the Cracow University of Technology (ICE CUT). After the first European PPS conference in Kraków, the Polish version of the fundamental PPS manual, *How to Turn a Place Around* [10], was prepared and published. It was supplemented with a description of several case studies from CEE countries. The book [6] describes basic rules and tools for observation, analysis, dialogue, evaluation of places, and the organisation of placemaking processes.[2] Additionally, a series of lectures and workshops were given to a number of communities across Poland in 2009. The method became better known in the country and sought after in numerous places as an element of municipal governance.

The place-oriented approach to spatial management has revealed substantial potential for public involvement, collaborative decision making, and community building. It appeared to be an effective instrument for participatory democracy enhancement. Today, the experiences from the programme carried out in 2009 help to further disseminate placemaking methodology in Poland and abroad.

3 Results

3.1 *Case Study: Industrial Heritage of Krivyi Rih and the Saksagan River: Cooperation–Dialogue–Democracy*

Place: Krivyi Rih, Ukraine

Project Leader: MiLA Foundation of Active People and Places

Project Partners:

- The Krivyi Rih Society for Defending the Rule of Law (Ukraine)
- Główny Instytut Górnictwa (GIG) Katowice, Poland

Project value: 206,900 PLN

Source of financing:

- Development Cooperation Programme of the Polish Ministry of Foreign Affairs
- Democracy Support Programme, Solidarity Fund PL.

Project Description

The project, addressed to local non-governmental organizations and all active residents of the city, is a continuation of the Polish-Ukrainian programme initiated in 2012, which resulted in the creation of an intersectoral partnership in Krivyi Rih for the revitalization of the Saksagan River. The main objective of the project was

[2]Its publication was accompanied by a conference, *Turning Great Ideas into Great Public Spaces* (Kraków, 1–2.10.2009), co-organized by PPS, Environmental Partnership, and Cracow University of Technology.

the development of local democracy (and in particular strengthening the position of non-governmental organizations as partners in social dialogue with local authorities through integration and partnership development), as well as the genuine involvement of city inhabitants in the decision-making process concerning local matters.

Social dialogue was initiated within the framework of the project and concerned the following three topics:

- Revitalization of a selected section of the Saksagan River (with the application of the participatory method of public spatial planning),
- Development of the municipal park based on the designs selected by the Krivyi Rih residents,
- Commencement of the local industrial heritage conservation and promotion (with the application of the heritage interpretation methodology named eco-museum).

The project included meetings with the partners, workshops, field trips, and a study visit in Poland. Moreover, extensive consultations were organized with the residents of Krivyi Rih concerning the design plans for developing the Saksagan River banks and the municipal park. The first elements of the development have already been put in place (e.g. small items of street and park furniture, information boards, etc.).

The work on the riverfront recreational park began at the end of 2013. The park's concept, as well as its most basic functions, were defined in the placemaking workshop, which took place in July 2013. During the two-day workshop, its participants had a chance to learn about the theoretical foundations of sustainable urban design and the benefits that are to be drawn from respecting the tradition of a place and encouraging local stakeholders' participation. The relevant experience *inter alia* of the PPS and INTBAU organizations was presented. In the second, practical part of the workshop, the Place Game formula was implemented, with the use of evaluation forms developed and used by the PPS. The Place Game is an organized way of brainstorming potential improvements to public spaces in a dialogue between the people who use them. The result was a prioritized list of ideas and postulates forming a consistent programme for the new riverfront park.

The results of the workshop were subsequently shared with the Krivyi Rih residents (more than 300 questionnaires were collected), which formed the foundations of the civic design plan for the development of this area. In autumn 2013, the city district authorities issued the necessary permit, thus the implementation of the civic design began. The time schedule of the construction of particular elements of the park reflects the priority list developed during the workshop and subsequent consultations [8].

3.2 Case Study: Young Leaders of the Local Communities of Sousse and Its Surroundings (Jeunes Leaders de La Communauté Locale de Sousse et Environs)

Place: Monastir, Tunisia
Project Leader: MiLA Foundation of Active People and Places
Project Partners:

- L'Association des Jeunes Méditerranéens pour les Echanges Culturels (AJMEC)
- Association 'Eureka'
- Ecole Nationale d'Ingénieurs de Sousse

Project value: 278,455 PLN
Source of financing:

- Development Cooperation Programme of the Polish Ministry of Foreign Affairs
- Democracy Support Programme, Solidarity Fund PL.

Project Description
The main objective of the project was the civic education of the young people of Sousse and its neighboring towns: Kalaa Kebira, Hammam Sousse, Monastir, and M'saken, who were to learn and apply in practice the innovative methods of participatory planning and active approach.

The aim was achieved by:

- providing training for a group of 25 youth leaders to strengthen the mechanisms of local democracy and the participatory approach to planning and implementation of local development (training and a travelling workshop)
- the formation and support of Youth Councils in five towns
- the implementation of five youth initiatives selected by way of a contest
- starting and maintaining a youth portal and newspaper
- presenting good practices in involving young people in the development of local communities in the Małopolska region in Poland.

A practical guidebook on developing sustainable communities, one addressed to local the administration and non-governmental organizations, was also prepared within the framework of the project.

As a result, young people not only were able to improve their knowledge and skills, but also to get involved in some specific actions, thus becoming partners in the dialogue with local the administration on matters they considered important [8].

The key component of the training programme was the five-day travelling workshop[3] for the Youth Councils' leaders from Sousse and four neighboring

[3]The workshops were run by: Barbara Kazior, Małgorzata Łuszczek, and Anna Jarzębska from the MiLA Foundation of Active People and Places, Tomasz Jeleński from ICE CUT and INTBAU Poland, and Justin Hyatt, the Tunisian coordinator of the project.

towns. The aim was to present them with various methods of working with local communities, the joint planning of changes, and carrying on the dialogue. Each town was the host of one workshop session:

- M'saken—24th July 2013—Future City Game
- Kalaa Kebira—25th July 2013—Civic Journalism
- Sousse—26th July 2013—Environmental Protection Projects
- Monastir—27th July 2013—Placemaking
- Hammam Sousse—28th July 2013—Oxford-Style Debate and Art of Dialogue.

The subject of the placemaking workshop was part of the harbour in the historic town of Monastir. Contrary to the town centre (*the medina*), where the traditional development has been preserved, the prevailing architecture of the harbor area is of modern, suburban character and contains the dominant transportation infrastructure and hotel function. The beach and dunes have been cut off from the town by the road called the Route de la Falaise and cater almost exclusively to the needs of tourists; most of the time they remain deserted. Access from the town centre is difficult, unattractive, and dangerous at places.

The Monastir Youth Council chose the harbor as the site of the placemaking workshop so that they could work out some options to make the place more attractive, give it a more distinctive character, and tie it back to the town. The workshop was preceded by two presentations:

- *Why places are important to cities and towns*—prepared specially for the local context, i.e. taking into account the tradition of creating public spaces in an Arab historic town (*the medina*);
- *PPS and Placemaking*: an alternative approach to planning and design, which presented *inter alia* the rules of the Place Game—the tool for place performance evaluation used by the PPS.

After the break came the practical part of the workshop, in which the Place Game formula was used. The group was divided into four teams, which went to the harbor with the task of completing the evaluation forms on site. Each team had to fill in one evaluation form. In order to do so, they had to negotiate all the answers within the team. After returning to the town hall, representatives of each team presented the results of their work. The suggestions they put forward were written down and a vote was taken to determine the priority list. During the final discussion that completed the process, participants expressed their positive surprise at the fact that tools for activating communities around places were so friendly, and that such activity did not necessarily require specialist knowledge.

At the end of the workshop a working team was formed, which was to be responsible for further actions leading to the implementation of the jointly created vision.

3.3 Case Study: Let's Talk About Mariacka

Place: Katowice, Poland

Project Leader: MiLA Foundation of Active People and Places

Project Partner: International Centre of Education, Cracow University of Technology (ICE CUT)

Project value: 44,000 PLN

Source of financing: The Katowice City Hall.

Project Description

In 2012, a group of Katowice city councilors, responding to the conflict growing around Mariacka Street, approached the author of this paper asking about the possibility of organizing an urban planning workshop focusing on the place and implementing the methods described in the Polish version of the textbook, *How to Turn a Place Around* [6].

Workshop preparation, preceded by an information campaign and a field study, started at the beginning of 2013. The project was under the umbrella of the MiLA Foundation of Active People and Places, which specialises in the activation of local communities.

The main objective of the project was to analyses the conflicts and work out a programme of necessary changes in the functioning of the Mariacka Street area in Katowice. A few years before, the street underwent a serious transformation: together with a larger area of the city centre, it was pedestrianized and renovated. The street surface was replaced, new street furniture was installed, and most importantly, a new programme of incentives for gastronomic business investors was created to encourage them to open restaurants, bars, and clubs in the area. The street was to become a vibrant promenade lined with pavement cafés, eateries, and music clubs. Residents who did not accept the new character of the street were offered places in other locations. This idea pertained especially to the council flats' tenants and others who rent their premises from the city.

The plan to liven up the street was successfully completed within two years. The city organized a number of open-air concerts and music festivals, as well as literary and visual arts events. The place soon attracted private cultural events animators, and new clubs won regular patrons and clientele interested in entertainment and artistic activities.

At the same time, a conflict between the new users and the old tenants and residents grew. Not all of them were interested in moving to other locations. Local media trumpeted controversies around some loud music events and the behavior of people abusing alcohol. Complaints about noise, dirt, deficient sanitary infrastructure, hampered access to houses, or even about security issues were becoming increasingly more resonant. City authorities responded nervously and introduced various restrictions on events, which were in turn received rather badly by entrepreneurs and regulars, who had already got used to the intense night life of the street.

The initiative of the group of councilors who suggested dialogue based on the placemaking method was approved by the city mayor. The project called *Let's Talk about Mariacka* was commissioned and financed by the Katowice City Hall, which also provided logistic support, considerably reducing the cost of the field study and workshop. The project was carried out in the period between May and October 2013 by the MiLA and ICE CUT team.

The implementation of the project

The subject of the study was the area of Mariacka Street in Katowice, alongside a few streets directly adjacent to it (in particular Stanisława Street and Mielęckiego Street, which are of similar character). The users of this area were invited to take part in surveys, carried out in the period between 17 June and 30 August 2013 (preliminary survey); and between 3 and 24 September 2013 (the post-workshop survey).

The survey questionnaires were distributed through several channels:

- given out at the information session on 17 June 2013 to meeting participants and passers-by
- delivered to flats and shops at Mariacka, Stanisława, and Mielęckiego Streets—with the support of the Katowice City Hall Press Office
- in municipal institutions (the City of Gardens Gallery and others), where printed questionnaires were laid out for distribution
- on the Internet: both online questionnaires were put up at the www.mariacka-ankieta.mila.org.pl website.

Additionally, information on the ongoing survey was displayed inter alia:

- at the Katowice City Hall website (www.katowice.eu)
- at the MiLA Foundation website (www.mila.org.pl)
- on Facebook, on the profile dedicated to the project (www.facebook.com/events/1402162120004692/)
- at www.Mariacka.eu and www.gazeta.pl portals and in other media.

The preliminary questionnaire was completed by 418 respondents, and the post-workshop questionnaire by 307. The vast majority of them used the online version.

The next stage in the pre-workshop survey was the focus group interviews with three groups of stakeholders: the residents, the entrepreneurs, and the regulars.

The aim of the pre-workshop survey was to collect information on problems, potential conflicts, and needs related to the functioning of Mariacka Street and its environs. The survey was of ancillary character and was not done on a statistically representative sample. Nevertheless, it allowed one to formulate preliminary hypotheses and to spot the most important tendencies, problems, and needs.

The preliminary questionnaire

The survey provided a systematic way of collecting information on how Mariacka Street was perceived by various groups of users: 1. residents 2. owners and

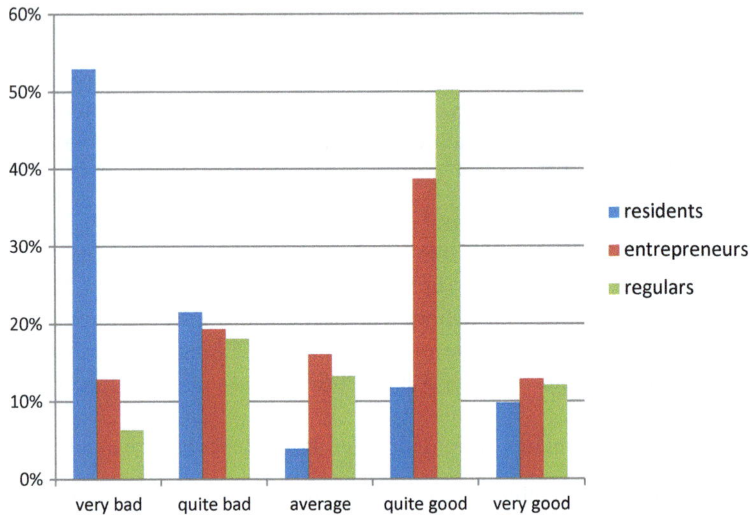

Fig. 1 How do you assess Mariacka Street? (Stakeholder responses)

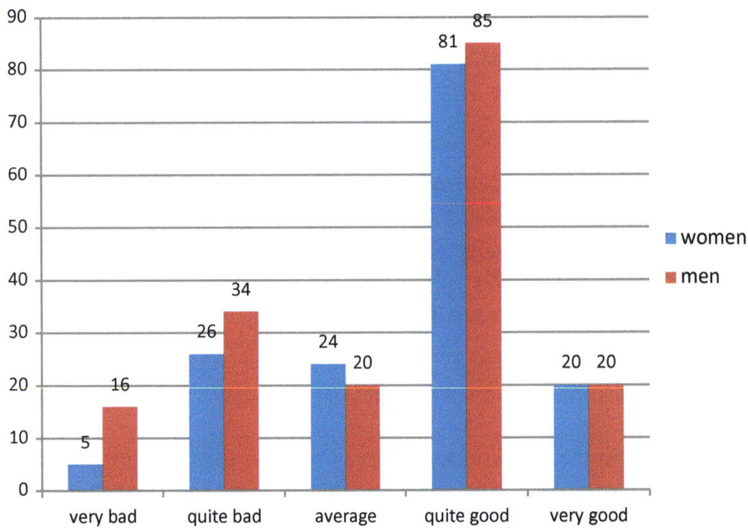

Fig. 2 How do you assess Mariacka Street? (Gender-based responses)

employees of businesses located in the neighborhood and 3. regulars (customers of the restaurants, bars, clubs, shops etc.) (Figs. 1, 2, 3, 4 and 5) [5].

Some of the questions were open; i.e. they allowed the respondent to formulate their own answers freely (e.g. on their associations with Mariacka Street, its

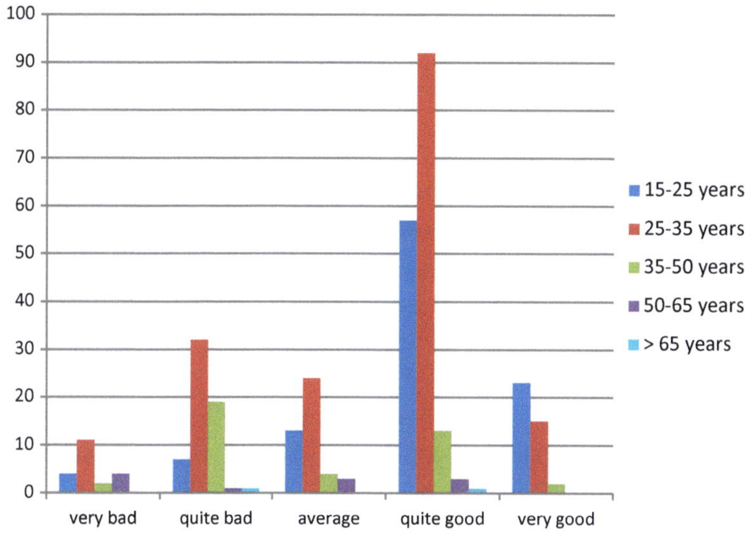

Fig. 3 How do you assess Mariacka Street? (Age-based responses)

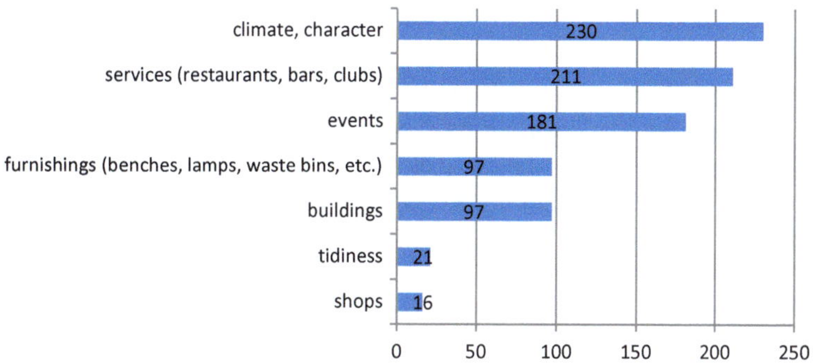

Fig. 4 What do you like about the place?

characteristic elements, source of problems, or their own ideas for developing the space).

Although answering an open question requires certain effort, more than a half of the respondents presented their own suggestions for changes. In order to facilitate the analysis, they were subsequently grouped into three categories: diversification of the programme, changes in the space, and changes concerning the organization or management of the place.

The second part of the questionnaire was in the form of closed questions, yet each one of them offered the possibility of adding a comment or suggestion other than the ones provided on the list. At the end of the questionnaire, respondents had

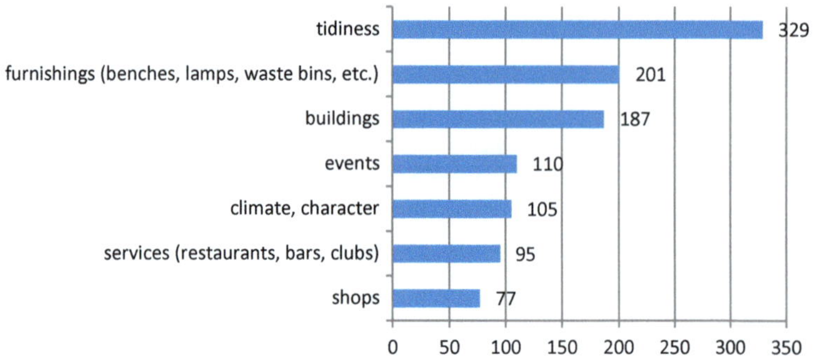

Fig. 5 What would you like to be improved?

to answer some questions about themselves, which allowed one to analyse answers in the context of the respondent's age, gender, and education level.

The answers exhibited a considerable similarity of opinions in certain aspects, particularly with regard to cleanliness, security, and what the street has to offer. Of particular significance to the survey results is the fact that the two groups that were most conflicted with each other—the residents and the entrepreneurs—put forward very similar suggestions.

Group interviews

There were several objectives to achieve by carrying out the interviews: first, to obtain opinions on Mariacka Street (both positive and negative features) with particular emphasis on the deficiencies of the place; secondly, to identify the conflict areas—define the relations between the different groups of users; thirdly, to identify the restrictions and factors hindering the amelioration of the situation; and finally, to work out acceptable proposals for the alleviation or elimination of the conflicts.

The composition of the sample was intentional—the key criterion was the character of the respondent's relation to Mariacka street. In order to ensure the adequate diversity of respondents, they were put into three groups:

A. Residents of Mariacka Street and persons residing in its direct vicinity
B. Entrepreneurs running their businesses in the area of Mariacka Street
C. Regulars—persons visiting Mariacka Street.

The method used was the *Focus Group Interview* (FGI). The advantages of this technique are the mutual stimulation of the respondents, creation of certain discussion dynamics, and confrontation of opinions. It is also important that the interviewers learn the opinions of several people simultaneously and in a relatively short time.

The project team was of the opinion that the most valuable field studies—apart from their strictly informative values—also yield a certain interpretational concept or idea, which may lead to the crystallization of new ideas and positive suggestions.

In the *Let's Talk about Mariacka* project, one implemented the *action research* formula, in which the research was a stage in the preparation for the workshop rather than a piece of strictly academic work. The assumption was that the survey should not only provide an objective examination and the best possible qualitative analysis of the existing situation, but also, by the mere fact of being carried out, should introduce some change—for example, because the respondents participating in it begin to understand the problems better, they start to realise their own points of view and the ones of other stakeholders, they learn about applicable solutions, and finally create new suggestions in dialogue with other research participants.

The first day of the workshop (31 August 2013)

The most important element of the workshop's first-day programme was an extensive presentation of the preliminary survey results. Comparing the data from several hundred questionnaires and three focus groups revealed surprisingly numerous similarities of opinions, which created quite a stir among some of the participants. The tone of the discussion became more positive and more matter-of-fact.

The last point of the program's first day was a short introduction into the Place Game (PPS), used as a tool for public evaluation of a place performance, which was planned for the second day.

The second day of the workshop (1 September 2013)

After another, more detailed, explanation of the Place Game rules, the participants formed three teams (each team composed of representatives from different stakeholder groups), which, separately, went for a walk along Mariacka and the neighboring streets to fill in their questionnaires.

The essence of the Place Game is cooperation between a team's participants in the collective evaluation of the quality and potential of a given space. An important rule of the Place Game is that the answers to all the questions on the questionnaire should be developed collectively within each team, which requires negotiating and reaching an agreement on the fundamental issues between people who normally represent different interest groups.

After returning to the workshop room, representatives from the three teams reported on the results of their work, and the main postulates of all three teams were written down on the boards previously prepared for that purpose.

After a complete list of postulates had been compiled, voting was organised to establish priorities. All participants were given an equal number of voting points in the form of sticky dots. They could use them in any way they liked, staking all of them on one (in their opinion, the most important postulate), or distributing their vote among a larger number of postulates.

Summing up the workshop, the project team offered help in establishing an action group to monitor and supervise or animate further action for the improvement of the place quality. The workshop participants expressed their willingness to appoint an interim Mariacka Street Council.

The post-workshop survey

The post-workshop survey enabled the verification of opinions on the most urgent things to be done. The order in which suggestions were listed in the questionnaire resulted from the list of priorities selected by the workshop's participants. The results of the survey slightly revised the order of priorities (Figs. 6, 7, 8 and 9) [5].

The aim of the post-workshop survey was also to verify certain ideas and postulates in individual *In-Depth Interviews* (IDI) with experts known for their profound knowledge of the subject and place, representing stakeholder groups which had not been hitherto surveyed: administration officials, non-governmental organisations, and academics (from the fields of urban studies and anthropology). The transcript of the interviews was included in the final report.

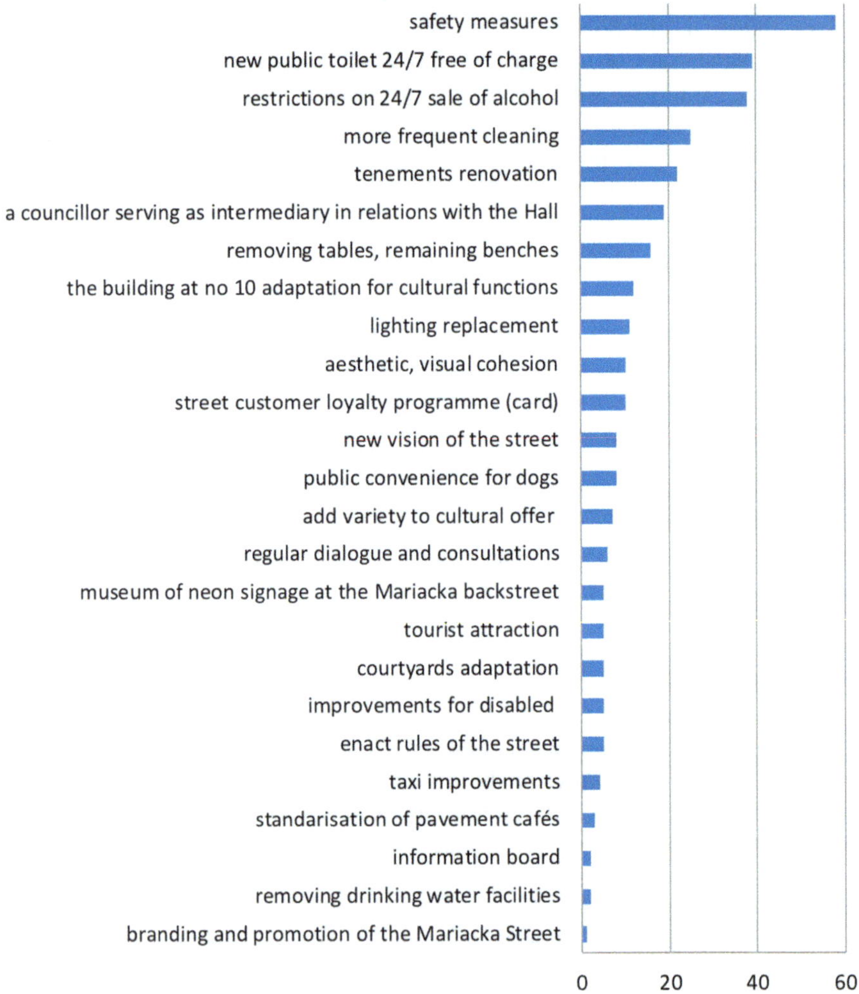

Fig. 6 Proposals—things to be done according to residents

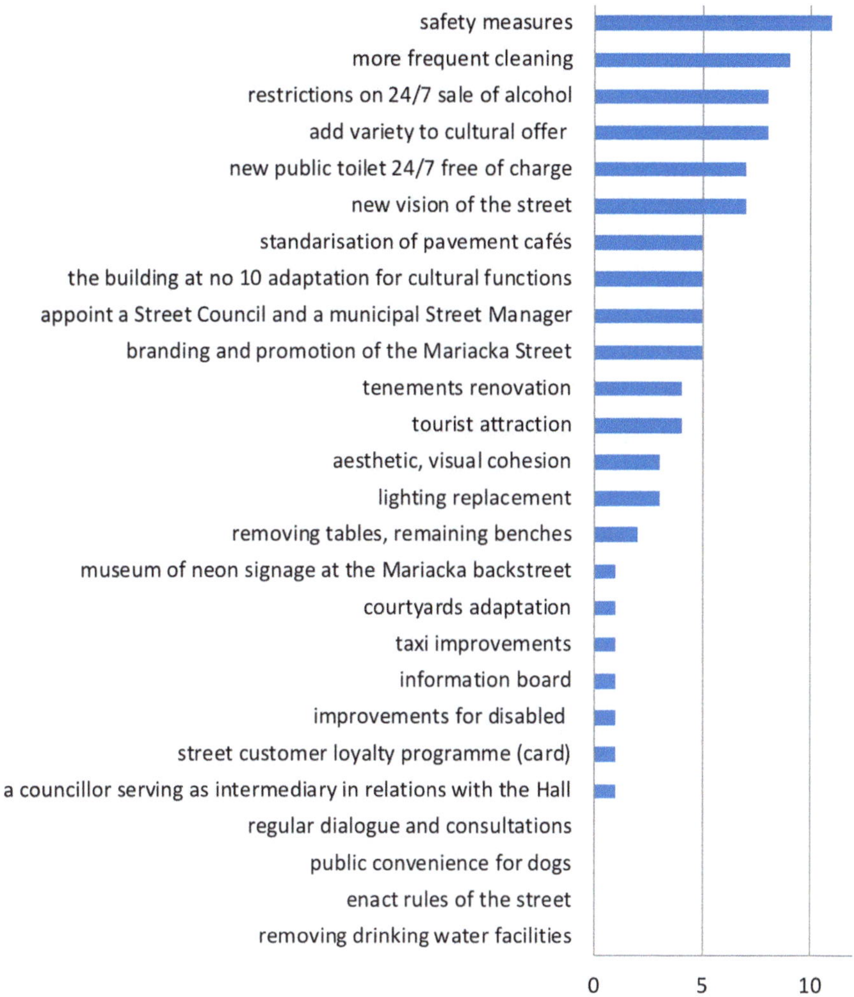

Fig. 7 Proposals—things to be done according to entrepreneurs

Summary of the project

The project confirmed the effectiveness of the method promoted by the PPS. The project team supplemented it with the preceding and complementary qualitative studies in the form of action research. Preliminary questionnaire and focus group interviews not only yielded interesting and valuable research material, but also—which is perhaps even more important from the point of view of the project objectives—made the project more credible for the group of the most active stakeholders, who subsequently came and participated in the workshop.

The preliminary survey brought quite surprising results. The initial impression—created on the basis of prior publications in the local media—that the fundamental

Fig. 8 Proposals – things to be done according to regulars

Fig. 9 Participants of the survey according to groups

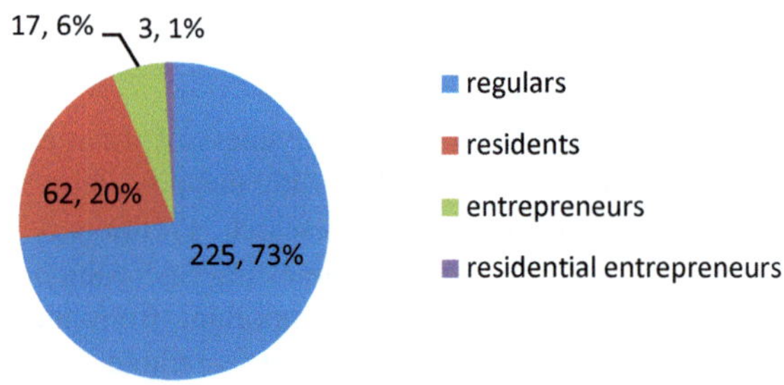

problem of the Mariacka Street area was the conflict between the residents and entrepreneurs, proved inaccurate. On the basis of several hundred completed questionnaires and three group interviews, one was able to put forward a hypothesis that the perspectives of these two groups were convergent on many points.

The residents accept the specific role of Mariacka Street and generally understand the character of gastro-business as well as the accompanying events. However, they expect equal understanding and respect from the entrepreneurs and primarily from the city administration. They demanded that they be consulted regarding potential slight departures from the generally accepted rules of social co-existence, e.g. the ones concerning quiet hours.

The stakeholders confirmed that the complaints in the media about the forms and inconveniences of the new street furniture were justified. Nevertheless, they much more often pointed out problems concerning organization or management, seeing the street furniture quality as a secondary, easy-to-fix problem.

The next survey and workshop revealed that most of the frustration related to Mariacka Street stems from the inconsistency and incoherence of the city's policy on the area. On the other hand, stakeholders representing different interest groups expressed willingness to cooperate and show flexibility in the approach to the most controversial issues as long as all the other parties of the dialogue declared equal flexibility and a conciliatory approach.

Residents and entrepreneurs, no matter how conflicting their interests seemed at first, have in fact similar needs when it comes to cleanliness, security, and comfort in their street. However, there also appeared proposals of controversial actions, e.g. 'the creation of standardized winter pavement cafes' which would increase the entrepreneurs' profits, and possibly street aesthetics, but at the same time prolong the period in which Mariacka tends to be bothersome for its residents. Since the Mariacka area residents and entrepreneurs exceeded all expectations regarding their openness to dialogue, an initiative was raised to appoint a Council which would represent the whole community in its relations with City Hall. Following the preliminary agreement reached at the workshop, the Interim Council, representing the key groups of stakeholders, commenced its work in October 2013. The Council is soon to be transformed into a fully representative body, which would be a credible partner for City Hall.

Since the administrative competences are split between different units of the municipal administration, the organizational framework in which city officials work may still prove an obstacle in effective actions. So the project participants expressed their desire for the Hall to appoint a manager or a representative for the Mariacka Street area, a person or institution that would deal with all the problems of this key part of the city centre in an integrated way. Such person would be a partner for the Street Council (now in the process of being formed), and would engage in continuous dialogue with it, facilitating contact with individual specialized City Hall units and officials, as well as with other institutions, such as those responsible for security, cleanliness, or financing integrated revitalization programs [5].

4 Conclusions

The basic assumption of this study was that local stakeholders should have actual influence on the shape of their environment. Sustainable, useful, and beautiful places cannot be built by architects only; it is more of a task for all the residents. Workshop methods require the involvement of stakeholders in the determination of the idea, vision and action programmes at an early stage of the planning or design processes. In this way the sense of democratic control is promoted, which in turn helps build consensus around places and strengthens the community spirit.

Mutual trust is the key requirement for any participatory action to succeed. In order to win and reinforce it, one needs to be active in searching for common goals and cooperation possibilities. Special obligations rest on the local government and administration, who should see it as their duty to support residents and entrepreneurs in their initiatives to improve the quality of living in the city.

The role of architects and urbanists should be seen as of auxiliary character in relation to local communities. Experts, similar to administration and nongovernmental organizations, should try to facilitate dialogue and engage in the process of grassroots implementation, a bottom-up vision based on constant monitoring, responding to change, and agreement between different groups of place users. It means in most cases that architects, planners, local councilors, and officials should accept the change in their status in relation to the community—from a leading to supporting role.

In all three cases described above, the place evaluation game was performed. Each time the context, the goal, and the way the workshop proceeded were different, but the effects were similar: the 'game' helped to build a team of stakeholders who felt that they were sharing similar views and were much more convinced that a positive change is possible, and that change mainly depends on their willingness to cooperate. Such attitude is especially valuable in places where local democracy and participatory management style is undeveloped.

The residents' participation in the planning process is often the starting point for the development of a genuine and mature community. People who see themselves as co-hosts value their environment and care for it more. A well developed and useful space, in turn, offers its users the possibility of daily encounters, thus strengthening community bonds.

Placemaking—the bottom-up ability to create valuable and popular places—requires a new, open, and more integrated approach to space development. In contrast to the traditional design or planning process, the approach resulting from focusing attention on individual features of a place, and on interrelated problems of its various users, is necessarily broader and more in-depth than the one which results from giving priority to individual aesthetics and typical functions. Due to the complex nature of the problems that need to be solved, creating a place is, to a much greater degree than just designing it, conditional on effective management. It requires the involvement of many different stakeholders on a regular basis.

References

1. Gehl J (1987) Life between buildings: using public space (trans. Koch J). Van Nostrand Reinhold, New York
2. Jacobs J (1961) The death and life of great American cities. Random House, New York
3. Jeleński T (2010a) Chapter 10: Urbanism. In: Bergier T, Kronenberg J (eds) Challenges of sustainable development in Poland—textbook. Sendzimir Foundation, Kraków, pp. 222–251
4. Jeleński T (2010b) Workshop methods of public space programming and design in the housing environment. Czasopismo Techniczne 2-A/2010 z.5, r.107. Cracow University of Technology, Kraków, pp. 181–188
5. Jeleński T, Jarzebska A, Wozniak-Szpakiewicz E (2013) Porozmawiajmy o Mariackiej: raport z badań i warsztatów. MiLA & ICE CUT, Kraków. Retrieved 26 April 2017 from http://www.mila.org.pl/attachments/article/226/MILA+MCK-PK(2013)Mariacka-RAPORT.pdf
6. Jeleński T, Kosiński W, PPS (2009) Jak przetworzyć miejsce: Podręcznik kreowania udanych przestrzeni publicznych. FPdS & PPS, Kraków
7. Lenartowicz JK (2004) Słownik psychologii architektury. CUT Press, Kraków
8. MiLA (n.d.) Fundacja Miejsc i Ludzi Aktywnych | Home. Retrieved 26 April 2017 from http://www.mila.org.pl
9. Pawłowska K (2008) Przeciwdziałanie konfliktom wokół ochrony i kształtowania krajobrazu. CUT Press, Kraków
10. PPS (2000) How to turn a place around. PPS, New York
11. PPS (n.d.) Project for public spaces | Home. Retrieved 26 April 2017 from https://www.pps.org/
12. Walljasper J (2007) The great neighborhood book: a do-it-yourself guide to placemaking. New Society Publishers, Gabriola, BC
13. Whyte WH (1980) The social life of small urban spaces. PPS, New York

Part II
Historic Settlements and Environmental Design

Place Identity Graphic Assessment and Post-disaster Reconstruction

Giuseppe Amoruso

Abstract Natural disasters and degradation phenomena interrupt the organic evolution of a place; this phenomenon invests, therefore, a vulnerable and weakened system, which inevitably becomes exposed to risk, within a territorial context already characterized by high seismicity or reduced resilience. The research identifies the tools to assess values and meanings of a territory, that have to be appropriately documented and communicated in order to inform decision-making process for conservation or reconstruction. Landscape documentation and its graphic transcription, according to new technology applications, provide a cognitive framework but also an operative vision to regenerate places and buildings according to local traditions. The research proposes the integration of models, representations and visualizations based on repertoires, high-iconic databases and predictive simulations. Promotion of local identity and psychological and environmental wellbeing requires the definition of tools for collecting and documenting local characters: analysis of urban patterns, construction techniques and tonal analysis of the urban environment, classification of architectural and landscape vocabulary.

Keywords Decision-making process · Resilience · Place-making · Landscape · Survey · Local character assessment · Representation

1 Putting Tradition in Practice

Natural disasters and degradation processes interrupt the organic and natural evolution of a place that is the material and immaterial expression of a vast heritage; this phenomenon of degeneration generally involves a highly inhomogeneous system in terms of types and construction technologies, vulnerable and weakened in its settlement infrastructure. Inevitably, exposure to the calamitous phenomenon, in this context of reduced resilience, produces greater damages than those reasonably

G. Amoruso (✉)
Dipartimento di Design, Politecnico di Milano, Milan, Italy
e-mail: giuseppe.amoruso@polimi.it

G. Amoruso and R. Salerno (eds.), *Cultural Landscape in Practice*,
Lecture Notes in Civil Engineering 26,
https://doi.org/10.1007/978-3-030-11422-0_12

acceptable and that in the perception of the population, actress, author and victim, are incomprehensible.

The integral knowledge of values and meanings of a territory, appropriately documented and shared within a community through digital systems, is the cornerstone for informing every decision-making process that intends to repair it, transform it or rebuild it. From the assessment of the post-emergency phase of recent earthquakes, for example that of 2009 in L'Aquila, it emerges that it is urgent to deal not only with reconstructing the architectural character but also with the intangible substance of which the community is constituted.

Italy was the first country in the world to establish that the protection of the landscape and the artistic-historical heritage of the nation should be included among the fundamental principles of the State as stated in the art. 9 of the Constitution (1947). In a country like Italy where populations historically coexist with the "earthquake's damn" it is paradoxical how any culture or, at the very least, awareness of the minimum precautions related to the maintenance and adaptation of traditional construction technologies for avoid serious damage to things and people has been lost.

In *Cerreto Sannita: un modello di ricostruzione post-sismica* (Cerreto Sannita: a model of post-seismic reconstruction, NT) Billy Nuzzolillo, describing the case of Cerreto, hit by the 1688 earthquake of Sannio, writes "… we want to have all the knowledge available, we want it to be permanently pursued and generated new knowledge, we want this knowledge to be transformed into competence, professionalism, rooted earthquake culture and widespread in those who govern, in citizens, in those who work in security, in those who defend our monuments and our history, in those who plan cities, in those who build roads, bridges, schools…NT" [19].

This study identifies the operational and cognitive tools for the benefit of assessing the social and environmental resilience of a place; this process generates new skills, culture and a practical thought to be shared within a community.

The research proposes the integration, in decision-making processes, of territorial models and visualizations based on repertoires, high-iconic databases and predictive simulations. Promotion of local identity and psychological and environmental well-being requires the definition of tools for local characters documentation: analysis of urban patterns, construction techniques and materials, the tonal characteristic of the environment in relation to a sequence of units of landscape that constitute the urban habitat. The documentation of the local identity and the creation of a renewed common awareness, after the emergency phase, must then deal with pragmatically some issues that seem inevitable and no longer postponed.

What are the principles on which a resilient community is based? How assess the vulnerability or resilience of a cultural landscape, a territory and a community? What can be the resilient communities to be taken as reference if they exist? What are actions, processes and constructive systems that can influence the maintenance or creation of resilient capacities within a community?

These issues were addressed by the international meeting, *Heritage, Place, Design: Putting Tradition into Practice*, hosted by the Milan Polytechnic in 2017 and promoted by the *International Network for Traditional Building, Architecture and Urbanism*. After the visit of the Prince of Wales to Amatrice (April 2017), a symbol of the devastation of the recent Italian earthquake, the meeting presented practical solutions for giving back to places and communities cultural and technical tools to reconstruct the traditional landscapes [2], in a global society that has recently suffered serious losses: the destruction of the triumph arch of Palmira (October 5, 2015) and the earthquake in Italy (in the period 24 August 2016–January 2017). This last "natural" event, in its dramatic and human implications, involves very close the topics presented in this essay: in the next years and decades Italy will have to face a challenge of considerable economic and social commitment and the it will only be able to overcome if it adopts appropriate practices at the local level, innovative in the cultural approach and in the application process.

2 Landscape as an Expression of Culture, Memory and Local Character

According to the *European Landscape Convention*, signed in Florence in 2000, "Landscape" is defined as an area perceived by people whose character is the result of the action and interaction of natural and/or human factors.

The *Convention* applies to the whole territory and concerns natural, rural, urban and suburban spaces, including land landscapes, inland waters and marine waters; it is addressed both to landscapes that can be considered exceptional, to those of everyday life and degraded landscapes. This last statement is of great importance since in the past, for example in Italy, already Benedetto Croce and other European intellectuals such as Georg Simmel in Germany and Charles Lalo in France, observed that the landscape is not nature but history and therefore we perceive it through the filter of literature and art.

In fact, the first Italian law for the protection of the landscape is due to Croce (Law n.778, 11 May 1922), while Italy was the first nation in the world to place it among the foundations of the Republican State (Article 9, paragraph 2: *La Repubblica tutela il paesaggio e il patrimonio storico e artistico della Nazione* -The Republic protects the landscape and the historical and artistic heritage of the Nation, NT) and to make constitutional the principles of the two Bottai laws, one on the artistic heritage and the other on the landscape, both approved in June 1939, during fascism.

With Croce there was established a model of beauty "connected with civil and literary and pictorial history", "panoramic natural beauties" and "landscapes of great interest", that is landscapes and cities that resembled a postcard or a picture as also addressed from the Law of protection of '39 (known as Bottai law), created to protect only "those natural beauties considered as paintings". This concept had a long tradition, rooted in Roman law and also with some precedents in the *Rescripts of the King of Naples*, which protected the views of Mergellina for example; certainly an aristocratic law that operated in legal continuity with the other previous laws. As Alfonsino Pisicchio reminds us, in the *Manifesto* for the regional law on beauty of Puglia, "this beauty had to do with the allure of the past and the "ruins" but not with the living people and their ability to build, with the ingenuity and work, other beauty. Therefore, to overcome this idea of beauty as "ornament", the Constituents, during the discussion of what was then the art. 9 of the *Constitution*, they put in the field an idea of beauty based on human work and on a so-called "minor" art, handed down for generations by master craftsmen, teachers who worked together with brain and hands. They had forged, always on the long duration, a diffused beauty that, as Concetto Marchesi said, was the very sign of identity, of memory and of national cohesion" [20, 6].

In 1992 the *World Heritage Convention* became the first international legal instrument to recognize and protect cultural landscapes. The *Convention* acknowledged that cultural landscapes "are cultural properties and represent the combined works of nature and of man", as designated in Article 1 of the *Convention* and they are illustrative of the evolution of human society and settlement over time, under the influence of the physical constraints and/or opportunities presented by their natural environment and of successive social, economic and cultural forces, both external and internal.

The Committee in 1992 adopted guidelines concerning their inclusion in the *World Heritage List*. The World Heritage Committee has identified and defined several specific types of cultural and natural properties and has adopted specific guidelines to facilitate the evaluation of such properties when nominated for inscription on the *World Heritage List*: (a) Cultural Landscapes; (b) Historic Towns and Town Centres; (c) Heritage Canals; (d) Heritage Routes.

Cultural landscapes belong to three main categories (*Operational Guidelines 2008, Annex3*), namely: (a) The most easily identifiable is the clearly defined landscape designed and created intentionally by man. This embraces garden and parkland landscapes constructed for aesthetic reasons which are often (but not always) associated with religious or other monumental buildings and ensembles; (b) The second category is the organically evolved landscape. This results from an initial social, economic, administrative, and/or religious imperative and has developed its present form by association with and in response to its natural environment. Such landscapes reflect that process of evolution in their form and component features; (c) The final category is the associative cultural landscape. The inclusion of such landscapes on the *World Heritage List* is justifiable by virtue of the powerful religious, artistic or cultural associations of the natural element rather than material cultural evidence, which may be insignificant or even absent [28].

The *European Landscape Convention* defines "landscape protection" as actions to conserve and maintain the significant or characteristic features of a landscape, justified by its heritage value derived from its natural configuration and/or from human activity, and "Landscape management" as an action, from a perspective of sustainable development, to ensure the regular upkeep of a landscape, so as to guide and harmonize changes which are brought about by social, economic and environmental processes. Finally "landscape planning" is defined as a strong forward-looking action to enhance, restore or create landscapes.

In 2008, the *Committee of Ministers* awares that landscape contributes to the formation of local cultures and that it is a basic component of the European natural and cultural heritage, contributing to human well-being and consolidation of the European identity, released a *Recommendation of the Committee of Ministers to member States on the guidelines for the implementation of the European Landscape Convention*. The fundamental principle considers the territory as a whole, meaning that the "convention applies to the entire territory and covers natural, rural, urban and peri-urban areas. It includes land, inland water and marine areas. It concerns landscapes that may be considered outstanding as well as every day and degraded landscapes". Another important topic is the role of knowledge that has to be recognized and characterized: "The identification, description and assessment of landscapes constitute the preliminary phase of any landscape policy. This involves an analysis of morphological, archaeological, historical, cultural and natural characteristics and their interrelations, as well as an analysis of changes. The perception of landscape by the public should also be analyzed from the viewpoint of both its historical development and its recent significance" [21].

The *European Landscape Convention* gives a relevant address through the definition of fundamental actions to be taken on landscape: protection, management and planning actions. The process is based on four steps and has to start from the accurate knowledge of the landscapes in terms of identification, description and assessment.

In 2011 UNESCO released a further document on landscape approach, the *Recommendations on the Historic Urban Landscapes*, the first instrument on the historic environment issued by UNESCO in 35 years.

The historic urban landscape (HUL) is the urban area understood as the result of a historic layering of cultural and natural values and attributes, extending the "historic center" or "ensemble" concept to include the broader urban context and its geographical setting.

Historical urban landscapes are complex systems that are generally perceived as environment with high-level presence of diffuse heritage; the whole concept goes beyond the specific materialization of the local culture but also highlights subjects who through their image and identity require special strategies of valorization and fruition. The 2011 *Recommendation* says also that "The historic urban landscape approach is aimed at preserving the quality of the human environment, enhancing the productive and sustainable use of urban spaces, while recognizing their dynamic character, and promoting social and functional diversity. It integrates the goals of urban heritage conservation and those of social and economic development. It is rooted in a balanced and sustainable relationship between the urban and natural environment, between the needs of present and future generations and the legacy from the past."

The *Recommendation on the Historic Urban Landscape* don't replace existing regulations or conservation approaches; rather, it is an additional tool to integrate policies and practices of conservation of the built environment into the wider goals of urban development in respect of the inherited values and traditions of different cultural contexts. This tool, which is a "soft-law" to be implemented by Member States on a voluntary basis.

The historical urban landscape is therefore the material representation of regional cultures and traditions that have developed in relation to the geographical and climatic conditions of the territory and the availability of local resources and materials. A habitat matured over the centuries, rich in history, heritage, social significance, crafts and traditions.

This was what Johann Wolfgang von Goethe wanted to summarize when he talked about the Italian built environment, a territory rich in beauty, the result of the incessant work of man down through the centuries: "I went up to Spoleto and I was also on the aqueduct.... ...The architectural art of the ancients is truly second nature, which works according to the uses and the civil purposes .... This is how the amphitheater, the temple the aqueduct arise...". In his extraordinary journey in Italy he describes our landscapes as a kind of widespread miracle, able to add to the "beauty of the first nature" the beauty of a practical art and a landscape of "everyday life" that has become a true "second nature" [11].

Also in the same territory, between 1292 and 1296, a pictorial cycle of 28 views called the *Stories of St. Francis* was painted in the lower part of the Basilica of Assisi. Going beyond the artistic question and the attribution of the work to Giotto, it is important to underline how it is a narration of the Italian medieval landscape imbued with devotion to the Saint Francis, of sacredness and of a Christian universe in perfect harmony with every earthly reality. The cycle is characterized by an innovative layout where the pictorial space interprets and suggests a three-dimensional volume where the characters of the *Stories* are naturally inserted into faded architectures and perspective wings that create practicable spaces. It can be affirmed that they represent the landscape and the contemporary environment with all its peculiarities and at different scales; a Middle Ages characterized by specific typological and environmental, civic and sacred elements: loggias, walls, porticoes, towers, city and church, individualism and communities of different scale and proportion, colored plasters, stone churches and wooden houses. The frescoes of Assisi represent a landscape not only rural and in opposition to the city but as an expression of a mental form, shared perception and figurative memory to share that reflects the society and culture of that historical era. According to Salvatore Settis, "the landscape is the historical form of the territory, the sedimentation of human presences in their interaction with nature", a balanced, integrated and organic expression of the relationship between city and countryside [24].

Umberto Eco reminds us that memory is "a faculty by which both individuals and communities base their identity (amnesiac no longer knows who he is)" [10].

In *The Invisible Cities*, Italo Calvino faces an exploration between imagination and imaginable through the description of some cities; it is a dialogue, in the form of a story, between Marco Polo and the Emperor of Tartars Kublai Khan. Calvino builds literary places, real landscapes with a conceptual structure, known as combinatorial literature, which manifests itself according to the "local" characterizations: to be precise, 55 cities, divided into 11 groups of 5 cities each where the first group it is dedicated to "city and memory": Diomira, Isidoro, Zaira, Zora and

Maurilia. In Diomira, the first invisible city, memory is the foundation of knowledge and the basic condition of the recognition of places and their peculiarities. According to the story, the traveler recognizes on his arrival what is known to him. "All these beauties the traveler already knows for having already seen them in other cities". The characteristic of Diomira is that "… he is envious of those who now think they have already lived an evening like this and have been happy that time". The bitter disappointment lies in the fact that it envies those who feel or those who can still feel sensations, in the face of an absolutely normal situation. The memory is therefore between the expression of an emotional mode and the memory itself, which becomes the object of suffering. Marco Polo writes: "Even cities believe they are the work of the mind or chance, but neither one nor the other is enough to keep up their walls. Of a city you do not enjoy seven or seventy wonders, but the answer you give to your question" [7].

In this work the influence of semiotics is evident, a discipline that analyzes signs and the way in which they make sense, and therefore deals with studying the phenomena of signification. By signification, in fact, we mean every relationship that links something materially present to something else that is absent. It is well known that our brain, on the other hand, needs to exercise and process memories through material expressions, rituals and symbol recognition.

Amnesia, which in medical literature is described as a disturbance of long-term memory, is linked to the inability to remember even ts. Johannes Spangerberg in his *Libellus Artificiosae Memoriae*, among the causes that lead to forgetfulness, indicated the corruption or diminution of faculties through old age and illnesses. Techniques for memorization cannot counteract this decline, but could offer practical expedients to prevent or reduce corruption, or so-called "forgetfulness of past species."

The art of memory is a practice that, following a precise system of rules, is aimed at the preservation and the fruition of information for the benefit of civilization and citizens [10]. This practice seems to be no longer useful whereas in the past there was nothing to rely on but an ability to store a huge amount of data (names, concepts, arguments) and consequently to develop memory techniques. Latin sources, including the *De oratore* by Cicerone and the anonymous *Rhetorica ad Herennium*, report that memory was one of five parts in which rhetoric was divided - formed as a whole by *inventio, dispositio, elocutio, memoria, pronuntiatio*—and allowed the orator to remember his own speech by associating the various parts with a series of "places" and "images" imprinted in the mind. Frances Yates writes that the solution to remembering was to organize memory spatially: a large house or city divided into a series of architectural environments—palaces, rooms, gardens, vestibules—where to place, with imagination, what had to be recalled. The exercise consisted in retracing such rooms mentally and then finding what needed to be recalled in the different places where memory had been subdivided. This practice allowed to call a *locus* and to associate and place *imagines*, easy-to-remember pictures that had to address the memory of "things" (*res*) and "words" (*verba*).

The unknown extensor of *Rhetorica ad Herennium* took care of distinguishing between two kinds of memory, a natural one, to be considered as an innate faculty and an artificial one, to be strengthened and consolidated through education. Images allowed us to remember arguments and concepts, the real subject of the discourse, while *imagines verborum* were used to remember every single word: that is, the language with which to communicate the subject. These were both a kind of artificial memory, or *memoria artificiosa*: *memoria rerum* and *memoria verborum*. A rule, often used to support speech, was to leave an imprint in the mind through an image of extraordinary beauty or ugliness (*imagines agentes*), a way to cause indelible emotion in memory [30].

These brief concepts give us back the importance of knowledge to understand the complexity of the built environment; developing knowledge is the only way to subtract the memory of places, in its tangible and intangible characters, to oblivion. The concept of heritage cannot remain only a special episode in the passive exercise of memory, since part of the cultural identity of a society is constituted by the places in which it resides.

The historical urban landscape can also be considered as a cultural landscape [23], a material expression of the models of adaptation to different sites and of environmental resilience, in which each element can be identified and documented; this landscape can be described graphically as a catalog with multiple levels of information developing appropriate semantic models [1]. Direct analysis, digital technologies and transcription methods offer tools for the innovation of the process of knowledge and conservation of this heritage, already applied in numerous international contexts.

The documentation of landscape meanings and semantic characters makes it possible to apply tools and processes for the characterization of the so-called "proximity landscape": an innovative concept that considers the landscape as an evolving organism strongly linked to its traditions that highlight the widespread and immaterial value of a community through its constructive characteristics and its relationship with environmental and natural aspects.

3 Reconstruction Practice. Building for the Future and Resilient Communities

The reconstruction of territories and cities as a result of epochal and rare events, is more often a constant practice due to the multiplication of wars, earthquakes, natural disasters and the impossibility of the inhabitants affected to manage independently the phases following the destruction. Reconstruction practices, in addition to representing perhaps the only moments of reflection on the future of cities, have provided an exceptional wealth of technical and cultural experiences related to new ways of living, the preservation of monuments, the preservation of urban identities and territorial and, more recently, adaptation to climate change climatic conditions.

In the current context historic urban landscape beauty or sites listed by Unesco as *Outstanding Universal Value*, unfortunately coexist with sprawl and degradation of territory, with environmental resources waste, and the ignorance of traditions that constitute the tangible and intangible heritage: signs, memories and projects that do not belong to the individual but represent the whole community. According to Settis, the beauty of our landscapes cannot be of any benefit if actions are not carried out to save beauty itself: first of all defining it, defending it, creating it, as the architect is called to do by the very nature of his craft [24]. In a post-emergency phase, among the many solutions, which ones can guarantee the durability, resilience and popularity requirements that an urban settlement must have? The concept of resilience has become increasingly important to the process of sustainable planning policy and practice according to recent and past reconstructions. Resilience means the overall goal of decreasing human vulnerability: an ability to remain in existence, to sustain a period of hardship or difficulty, and the capacity to recover quickly from unavoidable consequences of disasters. The main message is that built heritage is one of the main assets that communities recovering from disastrous events wish to rebuild. The ability to reconstruction planning is a key to becoming a more resilient city. Social networks as well as place attachment and place satisfaction could make people more resilient and more able or willing to rebuild their previous lives and homes. The first decision made by survivors following a disaster is whether to stay and rebuild or relocate. The mental costs to remaining in a damaged community and trying to rebuild can be very high. That is why individuals with more social capital are more likely to stay in damaged areas and work with neighbors to rebuild. Those with fewer connections, who feel less of a sense of place, are more likely to exit [12]. Looking at the diverse examples and emerging theories of urban resilience, it is possible to review a range of urban heritage conservation practice which provided effective and sustainable answer to disasters.

In the Italian scenario, the reconstruction after the earthquake of Friuli in 1976 (magnitude 6.5 of the Richter scale) is considered a virtuous case of response to one of the worst Italian disasters. In 2017 Venzone, a village in the province of Udine rebuilt after the earthquake, has been voted the most beautiful village (*Il Borgo dei Borghi* competition) because it represented "one of the most extraordinary examples of architectural and artistic post-earthquake recovery". The historic town centre reconstruction was finished in 1990 according to its original style and reusing almost 10,000 stones from the demolished buildings.

Resilience measurement is a fundamental prerequisite for systematic development of communities over their lifecycle. Resilience assessment systems measure resilience as performance (past incident and the urban system's reaction in that) or resilience as competence (city's perceived capability to adapt, recover and benefit of shocks). Such knowledge is the starting core of each reconstruction practice, addressing the principle of building back better and safer in carrying out sustainable reconstruction and recovery interventions of historic areas. Targets are to practice according to a holistic approach, socially, economically and environmentally resilient communities and also including proactively residents so that they will better cope with future disasters. Any action that improve the sense of 'preparedness', in case of loss, is important to preserve their identity and economic, social and environmental practicality and to seamlessly transmit their heritage to new generations. Knowledge- and evidence-based approaches to improvement of resilience and reconstruction principles square measure required to extend the cost-effectiveness of those activities from the total life cycle perspective. The conventional approach to natural disaster recovery is based mostly on post-impact reaction instead of risk reduction policies: prevention, preparedness, and mitigation, cultural resilience have to be embodied in the place-making process. For several decades, UNESCO's World Heritage Committee (WHC) generally promoted the opposition to reconstruction of town after a disaster. The first exception was made in 1980 for the historic centre of Warsaw, whose massive rebuilding was accepted by UNESCO as a symbol of "the inner strength and determination of the nation, which brought about the reconstruction of the heritage on a unique scale in the history of the world" [29]. Other exceptions included the listing of the Old Bridge Area of the Old City of Mostar which was justified on the basis of the restoration of cultural value, an intangible dimension of the property, and the Tombs of Buganda Kings at Kasubi, Uganda, which were destroyed by fire in 2010 and were gained provisional approval for reconstruction on condition that the new structure was based on sound documentation, traditional forms and techniques, and continuing use [8].

The current version of the *World Heritage Committee's Operational Guidelines for the Implementation of the World Heritage Convention* still echoes the *Venice Charter* when it states: "In relation to authenticity, the reconstruction of archaeological remains or historic buildings or districts is justifiable only in exceptional circumstances. Reconstruction is acceptable only on the basis of complete and detailed documentation and to no extent on conjecture" [8]. However, in light of recent extremists' attacks on heritage places, like the willful destruction in 2012 of the Sufi mausoleums at the Timbuktu World Heritage Site in Mali, decisions of the *World Heritage Committee* reflect a shifting attitude towards reconstruction which is now cautiously accepted if it seeks to reflect a pattern of use or cultural practice that sustains cultural value. The justification for this shift is based on previously mentioned exceptions and ideas published in the *Nara Document on Authenticity* in which the broadened use of intangible attributes makes a stronger case for reconstruction [27].

The professional historical investigation based on scientific grounds helps to establish a hierarchy of values and valorizes space, while social participation—in both spheres of identifying and conserving built heritage—enhances the sustainability of the process negotiated within the community. In a dynamic environment of today, more importantly than ever before, historical and traditional forms of architecture may reflect the meanings and values that people hold from the past and want to preserve, adapt, restore, reconstruct or even reinvent. Recreating cultural complexity, adequacy, and character of a townscape is possible if one builds on a historic layering of cultural attributes and forms. People and communities differ in their capacity to combat disasters because of certain pre-existing factors such as having the proper institutional and financial mechanisms in place to execute preparedness and recovery plans. Despite these differences, societies around the world are on almost a level-playing field with regards to building and maintaining at least some degree of architectural resilience. Successful community-driven historical reconstructions of European cities prove that reconstruction of architectural assets —especially when smartly combined with technical improvements of living standards—helps people to reconnect with their past, or, if those places are new to them, to understand places they are going to live in [8]. The assessment of previous experiences shows that practice has to enhance resilience and reduce vulnerability of historic areas through holistic approach reconstruction and economic and social recovery of historic landscape by local authorities and communities taking advantage of the use of new knowledge and tools.

4 The "Observational Technique" for Place-identity Assessment

The urban landscape represents a complex relational system that manifests itself mainly through the material expression of the meanings and of the spatial and social relations. In *The Architectural Tuning of Settlements*, Leon Krier, architect and piano player, explains concisely through a set of small drawings that good architecture and good settlements, like good music, can be made when the component parts are understood. A matrix of nine possible character-types of cities and towns, formal (classical), informal (vernacular) and a combination of both architecture and urban context presents the three right "tunings" that all of us can easily recognize in our living environment. Krier also addresses other characteristics of buildings and urban design; they play a specific role and, like music, there is more to a note than its pitch because duration, intensity and feeling matter. Buildings have uses (civic, private, and commercial), scale, and proportion and these qualities need to be tuned correctly within the environment [13]. The documentation of the different identities and morphological relationships, and their graphic transcription, also thanks to new technologies, provides a cognitive framework necessary to plan the regeneration of living and building according to local tradition. The cultural system should be understood as a cognitive, social and symbolic area of a territory and a community: from traditional, artistic, demographic, anthropological, monumental and environmental heritage to the territorial and urban systems.

In the UNESCO *Recommendation* on *Historic Urban Landscape, at* article IV, a series of innovative tools are presented: "The approach based on the historic urban landscape implies the application of a range of traditional and innovative tools adapted to local contexts. Some of these tools, which need to be developed as part of the process involving the different stakeholders, might include:

(a) Civic engagement tools should involve a diverse cross-section of stakeholders, and empower them to identify key values in their urban areas, develop visions that reflect their diversity, set goals, and agree on actions to safeguard their heritage and promote sustainable development. These tools, which constitute an integral part of urban governance dynamics, should facilitate intercultural dialogue by learning from communities about their histories, traditions, values, needs and aspirations, and by facilitating mediation and negotiation between groups with conflicting interests.

(b) Knowledge and planning tools should help protect the integrity and authenticity of the attributes of urban heritage. They should also allow for the recognition of cultural significance and diversity, and provide for the monitoring and management of change to improve the quality of life and of urban space. These tools would include documentation and mapping of cultural and natural characteristics. Heritage, social and environmental impact assessments should be used to support and facilitate decision-making processes within a framework of sustainable development. The process of analysis and mapping of the places is the necessary prelude to a design action oriented to their modification, and also includes the emotional and perceptive dimension of involvement of the senses aimed at the representation of space through visual thought and the production of graphic materials.

"The eye does not see things but figures of things that mean other things" Calvino writes in *The Invisible Cities*. The mapping process consists in the formation of "images", first of all mental concepts that are then interpreted and transcribed in relation to a more or less sophisticated code. The images "build" a shape, reproduce itself through their geometric, typological, aesthetic structure and give evidence of the structural and functional relations between the components.

For example, Attilio Marcolli considered perceptual phenomena related to color as manifestations of a color-space topology; the interaction between the built and the unbuilt environment creates the "color-city", where conditions of color and shade are integrated in the spatial and topological conditions, forming a city that can be defined as a *tonos–topos* combination [16]. Still on the subject of spatial relations and the chromatic environmental component, the color designer Jean-Philippe Lenclos, investigated on regional color atlas as well as studies on traditional colors of different French cities and landscapes. One of his most famous projects was his concept of *Geography of Color*. Lenclos and his wife Dominique published their first book *Couleurs de la France, Maisons et Paysages* in 1982 [14, 15]. Cities are systems that express multiple chromatic identities. Each urban area has its own "chromatic spatiality" as well as that morphological one that depends on many factors: brightness, materials, colors and contrasts, type of space, size of buildings; color therefore represents an attribute that explicitly represent the spirit of place.

The interventions on the historical urban landscape include methodologies, tools and project techniques [17]. To develop new knowledge on the settlement, typological and morphological characteristics of the widespread heritage, the tools and techniques for landscape representation that document the landscape units and the single minimum units are used; moreover, it is necessary to direct the meta-planning area to the recognition of the value of places. Other identifying characteristics are recurrent and can be found in different landscape units where the strong relationship between urban space and its chromatic component and the processing of local materials is evident. It is therefore necessary to promote operational methods to rehabilitate and reconstruct the physical and human condition, deeply investigating the representation and enhancement of cultural heritage and historical urban landscapes, according to the definition of UNESCO, but linking the cognitive action to the subsequent methodologies of placemaking and strategic design [3].

The British society was the first to experience rapid and problematic industrial growth leading to the emergence of neighborhoods for workers and immigrants and creating a machine for the production and distribution of products. In this context the first philanthropic movements were born, which dealt with creating harmonic communities and social services such as schools, hospitals and health centers. Among all the movement formed around the figure of Ebenezer Howard who with his manifesto *Tomorrow* proposed a practical model for investors and landowners aimed at the construction of industrial urban villages based on mixed and integrated functions. Patrick Geddes, a contemporary and also a supporter of Howard, advocated the civic survey as indispensable to urban planning: his motto was "diagnosis before treatment" addressing practical techniques for regional survey, analysis and planning. Such a survey should include, at a minimum, the geology, the geography, the climate, the economic life, and the social institutions of the city and region. His early work surveying the Old Town of Edinburgh became a model for later surveys. He was particularly critical of that form of planning which relied overmuch on design and effect, neglecting to consider "the surrounding quarter and constructed without reference to local needs or potentialities". Geddes encouraged instead exploration and consideration of the "whole set of existing conditions", studying the "place as it stands, seeking out how it has grown to be what it is, and recognizing alike its advantages, its difficulties and its defects". Geddes believed that cities should be seen as continuously evolving organisms, setting great worth on the continuity of tradition and physical characteristics of a place. Once the essence of a place was understood, he believed, it could be given a new lease of life through good design and by targeting detrimental elements [22]. Drawing on the scientific method, Geddes encouraged observation as the way to discover and survey with the relationships among place, work and folk (or family). In 1892, to allow the general public an opportunity to observe these relationships, Geddes settled a laboratory called the *Outlook Tower* that documented and visualized the regional landscape according to a sociological approach. In keeping with scientific process and using new technologies, Geddes developed an *Index Museum* to categorize his physical observations and maintained *Encyclopedia Graphicato*, which utilized a *camera obscura* to provide an opportunity for the general public to observe their own landscape to witness the relationships among units of society. Geddes's lesson teaches that a widespread cultural asset, like a historical center, is the main element of the identity of a territory and an expression of its community; through the peculiarities of a place, its specific traditions and the historical memory that resides in its physical resources, it is possible to design a planning matrix that links the diffused environmental values to the natural scenery, to the traditions of building and use of local resources; in this paradigm there are the meanings linked to the habitation and the form of the agrarian landscapes and of the urban structure that influences the sociality expressed by traditions, gastronomy and craft.

In the characterization and representation of a landscape of proximity it is important to proceed by highlighting the characteristics of stability and continuity and also making the parameters of unitarity and difference recognizable. The concept of place is inextricably linked to the concept of limit and boundary, spatial relationship and connotation, which intertwine with a physical-perceptive delimitation and give a representation of its soul. The place is a set of identities with boundaries, in which there is always a link between the subjects and the space. It is therefore something specific, with its own character, which identifies it and, at the same time, makes it unique. Therefore, operational actions on the territory should: develop, deploy and validate tools, information models, strategies and plans for enhancing the resilience of historic areas to cope with disaster events, vulnerability assessment and integrated reconstruction.

Also according to the aforementioned UNESCO recommendation on Historic Urban Landscape, It further recommended to identify the critical steps to implement the Historic Urban Landscape approach, which may include the following: to undertake comprehensive surveys and mapping of the city's natural, cultural and human resources and to reach consensus using participatory planning and stakeholder consultations on what values to protect for transmission to future generations and to determine the attributes that carry these values. A further step is to assess vulnerability of these attributes to socio-economic stresses and impacts of climate change and to integrate urban heritage values and their vulnerability status into a wider framework of city development, which shall provide indications of areas of heritage sensitivity that require careful attention to planning, design and implementation of development projects.

The research defines a graphic-design standard for the promotion of architectural and urban quality. The urban and morphological code establishes the graphic conventions to support the urban design process: a set of principles, rules and expectations to represent a building concept.

This is possible because codes constitute the vehicle for translating design issues into a built form but at the same time they are also instruments of representation and communication of types, materials and lexicon. The design information is collected through synthetic graphical tools organized in the form of graphic parallel of templates, so it is possible to define of a graphic vocabulary and the drafting of appropriate glossaries called *Typological Code-Pattern Book* [12]. This document addresses the understanding of the local code and identity through: the documentation of its urban patterns and building techniques, increased knowledge of the value of its vernacular architecture and urbanism.

Ray Gindroz, author of numerous books on the use of pattern books, wrote: "They were the direct descendants of the books used since Roman times, the means by which architects have passed along their knowledge of design to builders in remote places. From Vitruvius, to Palladio, to Asher Benjamin, to the American Vignola, architects provided helpful guides for the building industry. In the second half of the nineteenth century, Pattern Books became part of builders' marketing programs. These attractively designed books were easy to understand. Their pages combined realistic drawings of houses along with floor plans and important details. There were many choices of floor plans and arrangements of architectural elements, but all using details and proportions to the style. Pattern Books set the rules, but each builder found ways of interpreting them, elaborating them, or even bending them. The result is the balance between individual expression and unity found in traditional neighborhoods. The patterns and elements of style were expressed differently in each region and often elements were "cross-bred" across different styles. They represented a consensus among architects, builders, realtors and home buyers on the way to design buildings and communities" [26]. The pattern book is associated with typological representations for landscape sections, both urban and extra-urban, which describe the change in the image of the territory as it passes from the rural to the urban context, describing the structuring and characterizing elements of the different parts of the territory [9]. According to the *European Landscape Convention*, "Landscape" designates a certain part of the territory, as perceived by the populations, whose character derives from the action of natural and/or human factors and their interrelation. Therefore it is necessary to identify their own landscapes and their articulations and units, on the whole of their territory, analyze their characteristics and evaluate them, taking into account the specific values attributed to them by the subjects and the populations concerned.

The knowledge of a territory considers a vast field because the single landscape units coexist organically, in proximity and integrating according to a system of associations of urban and natural elements. Familiar to the discipline of ecology, the transect is an ordering system that arranges a sequence of natural habitats. It has proved a particularly useful way to detect transitions and distribution patterns. The transect can be extended to the human habitat as a means of creating a coherent rural-to-urban gradient. In addition to providing a system of classification, the transect is an instrument of design upon which the usually specialized urban components can be correlated. The idealized geographic continuum of the transect cab ne divided into locally-calibrated tiers that are distributed from natural to urban core; in actual application is often manifested as a mosaic of areas [18].

5 Conclusions

If buildings and cities are a portrait of human condition, scholars and citizens, according to their skills and their expectations, have to invest resources to let this heritage to remain alive. Cultural heritage in general is made up of the products and processes of a culture that are preserved and passed on through generations through the regeneration of memory.

Moreover, the *Convention for the Safeguarding of the Intangible Cultural Heritage* (2003) defines the *Intangible Cultural Heritage* (ICH)—or the living heritage - as the protagonist of the cultural diversity of humanity and its maintenance a guarantee for continuous creativity. Intangible cultural heritage indicates practices, representations, expressions, knowledge, skills - as well as tools, objects, artifacts and associated cultural spaces—that communities, groups and, in some cases, individuals they recognize as part of their cultural heritage. This immaterial cultural heritage, transmitted from generation to generation, is constantly regenerated by communities and groups in response to their environment, their interaction with nature and their history and provides them with a sense of identity and continuity, thus promoting respect for cultural diversity. and of human creativity.

Natural disasters periodically invest landscapes or historical centers, putting their organic nature in crisis and making them fragile. Such places, so beautiful as to be defined as "city-nativity scene" had first been depopulated, in this case we speak of abandoned villages especially in the internal areas of Italy, and then they became comfortable aggregates for second homes and therefore usually non-recipients of investments structural changes by the institutions. The economic balance based on the farm and on the passive maintenance of the distinctive, typological, figurative and constructive characteristics of the architecture has actually become a superficial, banal and compromised by alien constructive systems and by the vague alibis of cultural or gastronomic tourism. The maintenance of memory and cultural identity must now be supported no longer in a mono-functional but organic, conscious, economically articulated and respectful manner of architectural traditions.

The analysis of the current planning processes has led to the evaluation of alternative solutions to the conventional cultural approach that has caused the inevitable expansion of the suburban fabric around the inhabited centers and the erosion of the natural territory and of the traditional built heritage with a strong loss of quality urban and environmental. It is essential to direct research to review, map and systematically characterize existing experiences and good practices in Europe and globally, through evidence and common metrics to evaluate and establish their replicability conditions, and recommend how historic areas can be rendered more resilient and better prepared to face future disaster events.

A "practical" research on the landscape must also deal with the legal systems and overcome "the absurd laws that ruin cities" as recalled in a recent Settis writing [25] reaching a structured knowledge of the historical and environmental context. The survey of urban and environmental characteristics is also aimed at highlighting the indicators of the quality or degradation of public spaces and of the fabric built to allow the recovery and enhancement of inhabited areas and places of historical and artistic interest, the improvement of quality architectural and the recovery of the landscape value of the territory also through the elimination of incongruous works and regulatory instruments [4, 5]. This process of integrated knowledge provide science- and evidence-based guidelines and models to local authorities for carrying out sustainable reconstruction within a participatory and community–based context.

References

1. Amoruso G (2017) Developing semantic models for the historic urban landscape regeneration. In: Amoruso G (ed) Putting tradition into practice: heritage, place and design. Springer International Publishing AG, Cham, pp 769–777
2. Amoruso G (2017) Putting tradition into practice: heritage, place and design. Springer International Publishing AG, Cham
3. Amoruso G (2015) The image of historic urban landscapes. representation codes. In: Brusaporci S (ed) Handbook of research on emerging digital tools for architectural surveying, modeling, and representation. Hershey, IGI Global, pp 566–595
4. Amoruso, G, Tarchini, ML (2004) La proyectación de la calidad urbana: un código gráfico digital para la representación y el recupero de la ciudad histórica. In: IX Seminario Internacional Forum UNESCO—Universidad y Patrimonio, Buenos Aires
5. Amoruso G (2005) La qualità urbana attraverso la rappresentazione dell'immagine degli abitati storici. In: Volta V (ed) Immagine della città europea. Convegno Internazionale, Università degli Studi di Brescia, Brescia
6. Bonesio L (2007) Paesaggio, identità e comunità tra locale e globale. Reggio Emilia, Diabasi
7. Calvino I (1994) Le città invisibili. Mondadori, Milano
8. Cameron C (2017) Reconstruction: changing attitudes. The UNESCO courier. july-september 2017(2):56–59
9. Duany A, DPZ (2011) Garden cities: theory & practice of Agrarian Urbanism. The prince's foundation for the built environment, London
10. Eco U (1998) Fondo Young sulla memoria e la mnemotecnica. Presentazione del Fondo. http://web.unirsm.sm/young/eco.htm (accessed on Dec 1, 2018)
11. Goethe JW (1959) Viaggio in Italia 1816-1817. Firenze, Sansoni editore
12. Jeleński T (2018) Practices of built heritage post-disaster reconstruction for resilient cities. Buildings 2018(8):53
13. Krier L (2008) The architectural tuning of settlements. The Prince's Foundation for the Built Environment, London
14. Lenclos J, Lenclos D (1982) Les couleurs de la France. Maisons et Paysages, Le Moniteur, Paris
15. Lenclos J, Lenclos D (1999) Les Couleurs du monde: geographie de la couleur. Le Moniteur, Paris
16. Marcolli A (2006) Colore - Città. Il caso della città di Venezia. In: Bisson M, Boeri C (eds) Variazioni sul colore. Contributi per una riflessione sul progetto e sulla cultura del colore. Franco Angeli, Milano: p 17–26
17. Marconi P (1997) Introduzione. In: Radi L, Radi L (eds) Foligno in particolare: elementi tipologici dell'edificazione storica. Comune di Foligno, Area gestione del territorio, Foligno, pp 13–20
18. Neal P (2003) Urban Villages and the Making of Communities. Spon Press, London
19. Nuzzolillo B (2002) Cerreto Sannita: un modello di ricostruzione post-sismica. Edizioni Sanniopress: Benevento: p 7
20. Pisicchio A (2018) Manifesto per la legge regionale sulla bellezza. Bari, Regione Puglia. Assessorato alla Pianificazione territoriale - Urbanistica, Assetto del territorio, Paesaggio, Politiche abitative
21. Recommendation CM/Rec(2008)3 of the Committee of Ministers to member states on the guidelines for the implementation of the European Landscape Convention (Adopted by the Committee of Ministers on 6 February 2008 at the 1017th meeting of the Ministers' Deputies)
22. Robb S (2017) Conservative surgery in Edinburgh. In: Context. Institute of historic building conservation, 150, p 19–22
23. Salerno R (2017) Enhancing not-outstanding cultural landscapes in a European perspective: a challenge for digitization. In: Amoruso G (ed) Putting tradition into practice: heritage, place and design. Springer International Publishing AG, Cham, pp 3–8

24. Settis S (2017) Architettura e democrazia. Paesaggio, città, diritti civili. Einaudi, Torino
25. Settis S (2017) Le leggi assurde che rovinano le città. Fatto Quotidiano, 29 Agosto 2017
26. The Prince's Foundation (2003) Civitas. Traditional urbanism in contemporary practice. The Prince's Foundation for the Built Environment, London
27. UNESCO (1994) Nara Document on Authenticity
28. UNESCO (2013) Managing world cultural heritage. World Heritage Resources Manual, World Heritage Centre UNESCO, Paris
29. UNESCO World Heritage Historic Centre of Warsaw. Available online: http://whc.unesco.org/en/list/30/ Accessed 1 Dec 2018
30. Yates FA (1966) The art of memory. Routledge and Kegan Paul, London

L'Aquila Model. Strategies and Restoration Processes for Historic Centre

Mario Centofanti, Stefano Brusaporci and Pamela Maiezza

Abstract Nine years after the earthquake of 6 April 2009, it is possible to analyze strategies and current processes the reconstruction of L'Aquila and the centres of its territory, highlighting positivity and criticality, in order to identify the possible optimization of methodologies and procedures, useful for present and future emergencies. An undoubted positivity is the financing system for interventions of protection of buildings of historical/architectural and landscape values. A first criticality is the "non synchronic reconstruction", both in terms of financial resources and in terms of validation and implementation procedures. A second problem is the "parcelization" of interventions. A third problem is denotable as "divided knowledge", that is the process of knowledge is discretized, according to the individual building or block project's validation procedure. In the case of the historic centre of L'Aquila, the paradox is that we have an extraordinary knowledge, unique and without precedents, with thousands of surveys, tests and analysis on materials.

Keywords Urban heritage · Conservation · Restoration · Urban design · L'Aquila

Centofanti is the author of paragraphs 1 and 2, Brusaporci of 3 and 4, Maiezza of 5.

M. Centofanti · S. Brusaporci (✉) · P. Maiezza
Department of Civil, Construction-Architectural and Environmental Engineering, University of L'Aquila, Rome, Italy
e-mail: stefano.brusaporci@univaq.it

M. Centofanti
e-mail: mario.centofanti@univaq.it

P. Maiezza
e-mail: pamela.maiezza@graduate.univaq.it

G. Amoruso and R. Salerno (eds.), *Cultural Landscape in Practice*, Lecture Notes in Civil Engineering 26,
https://doi.org/10.1007/978-3-030-11422-0_13

1 Introduction

Nine years after the earthquake of 6 April 2009, the so-called "L'Aquila-Model" has emerged to international attention (Fig. 1). It is interesting for the strategies and processes put into place in the reconstruction of L'Aquila and of historical centres in the interlard. The aim is to highlight positivity and criticality, in order to identify the possible optimization of methodologies and procedures useful for present and future similar emergencies. The research field is the physical reality of the "historicized city" and the "settlement system" of the territory.

The issues complexity has to be underlined, considering a whole historical centre damaged—of significant dimension of about 170 ha—and a territory scattered with small historical centres (Fig. 2). This settlement grows from the pre-roman and classical historical urban system based on "pagi" and "vici", according to a medieval complex phenomenon of settlement centralization called "incastelment" that characterized the inner Apennines areas of central Italy between the 10th and 11th centuries [15, 34].

Therefore, it is fundamental the conservation of places' identity characteristics, through a process of urban re-composition, in line with the historical events of the building (at the half of the 13th century) and transformation of L'Aquila, intended as "city-territory" that is a city historically integrated with its territory [16].

Fig. 1 Piazza Duomo in L'Aquila: the ongoing reconstruction process

Fig. 2 The restoration of the city: construction sites along the streets of the historic centre

2 Critical issues in the reconstruction process

2.1 Added Value. The Financing System and the Protection of Historical/Architectural and Landscape Values

In L'Aquila reconstruction, an undoubted positivity is related to the financing system of interventions, which complements the objectives of protection of historical/architectural and landscape values, assigning a remarkable increase for buildings "of cultural interest", of "prestige" and of "landscape interest" (Fig. 3). This is an important and unprecedented novelty in the Italian regulatory landscape [19].

2.2 The Asynchronous Reconstruction

In L'Aquila reconstruction process, the first critical issues is "The asynchronous reconstruction".

Both in terms of financial resources and validation and building procedures, the interventions planning and the realization management are structured according to an organization with parallel and independent lines (Fig. 4).

It causes different process speeds, also with very different rapidity.

The cause rises from the distinction between publics, privates, ecclesiastic implementing bodies, according to building's types and kind of owners; but also

PROTECTION / LEVELS OF RECONSTRUCION COST	
PROTECTION TYPE	**INCREASE ON BASIC COMPENSATION**
Buildings with declaration of interest (restriction) former DL 42/04	Up to 100%
Buildings with declaration of particular landscape interest (OPCM n.3996 of 17 January 2012)	Up to 100%
Buildings of particular historical-artistic value: presence of valuable elements or typological and constructive complexity (OPCM n. 3917 del 30 December 2010)	Up to 60%

Fig. 3 The financing system of interventions, which assigns a remarkable increase for listed buildings

IMPLEMENTING BODIES / CONTRACTING AUTHORITIES	TYPE OF BUILDINGS
Private, single owners	Single Homes
condominiums	multi-family buildings
Obligatory consortia between owners for building aggregates of historic centres	Building aggregates including convents and religious belongings
MIBACT Segretariato regionale per l'Abruzzo	Worship Buildings, Min. Int. and Diocesan Fund
Provveditorato Interregionale OO.PP Lazio-Abruzzo-Sardegna	Public buildings
Azienda Gran Sasso Acqua S.p.a.	Buried subservices

Fig. 4 The asynchronous reconstruction planning and management of the implementation

from different kind of procedures and a diversification of verifiers and managers of technical-administrative procedures (Fig. 5).

Therefore, there are differentiated speeds: between L'Aquila, the minor centers of the municipal territory, and the minor centers of the seismic crater; between public and private works; between private works on buildings subjected to the restrictions of historical-artistic protection and private buildings without restrictions; between the realization of network infrastructures and buildings.

Procedure	Types of procedure	Reviewers and managers of technical-administrative processes
A	Individual private buildings or aggregates buildings or portions of aggregate buildings subject to declaration of interest	MIBACT Soprintendenza Unica del Cratere
B	Individual buildings or aggregates buildings or portions of aggregates not subjected to	USRA e Comune dell'Aquila
C	Worship Buildings Fondo Min. Int. e Diocesi - Fund Min. Int. and Diocese	MIBACT Soprintendenza Unica del Cratere
D	Public Works	Provveditorato Interregionale OO.PP.
E	Buried subservices	Region

Fig. 5 The asynchronous reconstruction related speed differential in the validation process

2.3 The Parcelization of the Interventions

The second critical issue is the "intervention parcelization". The "Reconstruction plan" of the historical centre of L'Aquila is a mosaic of the individual parcellized proposals for intervention, related to the procedural process (Fig. 6).

Fig. 6 Municipality of L'Aquila: Reconstruction plan of the historic centre

There is not an overall strategic vision aimed at the contextualized urban-building re-composition and at the conservation of "urban identity" and its spatial and figurative qualities.

The Reconstruction Plan is intended as a mosaic of intervention intentions and not as a coherence and compatibility framework and as an urban restoration project.

2.4 The Divided Knowledge

The third critical issue is the "divided knowledge".

The process of knowledge is discretized, similarly to the validation procedure of individual building or block projects. In the case of the historic center of L'Aquila, the paradox is that we are faced with an extraordinary cognitive heritage, unique and without precedents, with thousands of surveys and tests on materials.

However, this knowledge is parcelled for single building/block and it is scattered between the archives of the various bodies in charge of project validation and control of procedures. It is not expected to form a structured database, useful for a critical and related knowledge, to constitute a framework of reference and coherence for the individual parcelled interventions and a support for the strategic plans at the urban scale.

2.5 Investing in Knowledge

Therefore, investing in knowledge appears as a primary objective to be pursued, always, in the immediate and in parallel to the emergency and first intervention activities. The aim is to recognize the identity values that are the foundation of the architectural/urban restoration project and of the reconstruction processes [5].

According to the specificity of each single architectural signifier, this project aims to define a perfect balance between the seismic risk mitigation needs (*firmitas*) and re-functionalization (*utilitas*), and the reintegration of the formal and figurative values of architecture (*venustas*). Together with the safeguard of the authenticity of the historical presence (as signifier and meaning), in full compliance with the "requirements" of the pre-existence.

3 The Stone-Town and the Values of Identity

L'Aquila retains in itself—in the sense that it contains them—its historical memory and its own identity, and yet this latter has been repeatedly threatened in its signifiers by the seismic events, because the city has been obliged several times to rebuild itself, to reinvent itself, to re-propose itself according to new formal and figurative registers.

Anyway, the understanding of the sedimentation and historical stratification process represent the pivotal basis for reconstruction rules, both on the urban scale and on the building scale. These are the two levels in which the antinomy between conservation and innovation appears, both according to the critical identification of the pre-existence to be protected, and to the innovative culture of the project.

The historical centre is a monument, for memory and as document of community life: "The whole ancient city […] can be defined as the figurative and living image of an historic reality, expressed in the ways of the architectural language, diversified and changing over time, in which every building or element constitutes a formal moment of life […]. Consequently, the restoration, as critical operation aiming at the understanding and conservation, involves the entire urban environment and the whole ancient city, becoming 'urban restoration'" [3] (Fig. 7).

However, the concept of "unity of the work of art" can not be used in a strict sense for the historical centre. In the work of art the aesthetic value is a priority in restoration design, while the historical centre is more like a sylloge, that is a structured set of fine and value-rich parts connected according to precise formal, figurative, spatial, fruition and use structuring: "Structural relationships of urban kind, between volumetric system and open space system (streets and squares); spatial relationships between built and open spaces; architectural relationships between nodal elements and serial elements of the fabric; the different building types, the system of volumes, the system of interconnections between volumes and finishes (roofs, windows…), the relationships between the different building typologies".

Follows: "A wise use of the historical value leads to the fact that in every new intervention in the Historical Centre primarily we must aim to preserve the

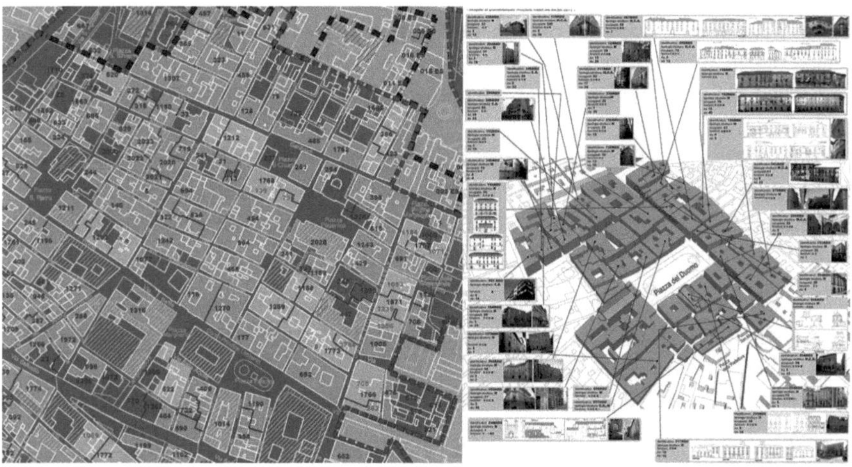

Fig. 7 Plan of reconstruction of the historic centre and buildings of artistic and historical interest

structural laws, the connective relationships that characterize it and that it offers as its quality" [2].

Moreover, in the case of L'Aquila, the historical building fabric is characterized by a complex formal, functional and figurative stratification (for example buildings with facades of the 18th or 19th century, realized after the earthquake of 1709, which have valuable inner courts of the 16th century) outcome of the different historical phases of post-earthquake destruction and reconstruction.

4 Recontextualizing the Restoration Project

The city already contains in its historical conforming identity the reasons and rules for its reconstruction. In this sense, one of the peculiar constituent components of the urban morphology of the city of L'Aquila are the spatial and figurative units of the squares [23]. Different for consistency and importance, the squares are widespread throughout the historic centre and are present almost entirely from the city foundation. Over the centuries, they have been transformed and stratified, also due to earthquakes, until they reach the current configuration. The multi-secular dialectic between the churches, fountains and palaces that rest on the squares entrust the architectural characterization.

There are one of the starting points for the urban restoration coordination project. Moreover, according to the varied damage situations that have occurred, new formal and figurative balances in relation values will be deduced, harmonizing the individual components (buildings) with the urban space environment but above all the space/square. Similarly for "urban scenes" and "urban gaps", because of damage and earthquake collapse, with the interruption of the historical building fabric.

The pre-eminent technical and scientific orientation aimed at pure structural consolidation that focuses on the building's material physical-chemical components, has to be integrated with proper approach of critical-conservative restoration, rooted on the architectural value's judgment, according to the historical and artistic expression. [4].

Before that technical procedure, restoration is above all an act of culture, of historical-critical comprehension; moreover, it is also a creative design act [9].

Therefore, the Reconstruction Plan requires the need to delineate in parallel and in addition the following items:

- an additional cognitive approach based on the historical-critical analysis of the ancient city, aiming at the recognition of its form, its values, its invariants [1, 11, 13, 16, 22, 24, 26, 32];
- a historical-critical analysis for the foundation and the presupposition of concepts and planning tools for the resolution of coordination and correlation issues among the fragmented interventions on blocks, moving from the relational,

formal and figurative reconfiguration of urban scenes and urban spaces (streets and squares) and compensation of lacunas [2, 11, 28, 30, 31];
- a conscious design orientation [6, 7, 12, 14, 17, 18, 25, 29], able to combine the full respect of pre-existence, even if residual, with the creative reinterpretation of the relationship between old and new [7, 10, 21, 27].

5 Territory and Landscape: The Minor Historical Centres

As explained in the Introduction, there is a stagnation in the post-earthquake intervention processes in the smaller towns of the inter-land. They are usually characterized by the presence of historical centres of early medieval foundation (Fig. 8).

From a methodological point of view, the cognitive approach to the identity characteristics of the places aims to the identification and "recognition" of values and to the orientation of protection and enhancement actions, through the understanding of the territory as an intersection of historical settlement systems with the naturalistic-environmental system [8].

Follows a "reading" of the territory as "complex" and "integrated" system, to highlight correlated systems, that is "anthropic elements related to the landscape" and "anthropized natural elements". They are:

- The historical centres of the territory;
- The territorial historical-artistic buildings;

Fig. 8 Photographic view of the plain of Navelli in L'Aquila territory (2005)

- The water and production facilities system: springs; waterways; hydraulic locks; mills; paper mills; branches; hydraulic power factories;
- The system of routes and of the historical viability including the railways;
- The system of pastoralism/transhumance: “tratturi”, churches, “grange”, chapels; fountains; taverns; cave and tholos;
- The archaeological sites: fortified enclosures; necropolis; tombs; urban settlements; isolated monuments; archaeological sites.

5.1 Criticalities

In parallel with the drafting of the “Heritage Card”, in the context of the overall urban planning of the territory, it is useful to develop a careful “critical analysis”, aimed at the preparation of directives and regulatory equipment, such as heritage protection actions.

Particular attention must be paid to the analysis of the “transformation processes of historical settlement systems”, necessary for:

- reconstruction interventions, in progress or planned (Reconstruction plans, Urban-building renovation projects, Reconstruction projects for building aggregates);
- interventions for anthropogenic degradation, degradation due to abandonment, degradation due to damage and/or earthquake collapse; degradation resulting from interventions carried out in the emergency phase;
- study of the residual architectural and urban values.

With reference to the analysis of transformation processes, in relation to the requests for heritage protection, the historical centre of Arischia is an exemplariness between all the other minor historical centres of L’Aquila Municipality.

From the urban master plan PRG of 1978, there is not any norm of protection of the historical building fabric. Therefore, building renovation interventions are possible, even of demolition and reconstruction.

In particular, the historic centre of Arischia, rather large dimensionally, expressed and still expresses historical-typological and significant architectural language values, such as the court-house buildings (perhaps unique in the Apennine context for their particular characteristics), the “vignale”-houses, the “profferlo”-houses (that is houses with a particular system of entrance with stairs on the road).

It is evident that the tragic current condition of the post-earthquake requires particular caution.

Nevertheless, where the residual typological, formal and figurative values still allow an adequate reading, it would be desirable to direct the reconstruction plan to combine the needs of consolidation and re-functionalization [20].

References

1. Antonini O (2004) Chiese dell'Aquila. Architettura religiosa e struttura urbana, Carsa, Pescara
2. Benedetti S (1982) La cultura del Restauro nel "recupero" dei centri storici. Storia Architettura 1:89–104
3. Bonelli R (1963) Il Restauro architettonico, voce Restauro in Enciclopedia Universale dell'Arte, Istituto per la Collaborazione Culturale Venezia-Roma, vol XI, pp 344–351
4. Brandi C (1963) Teoria del restauro. Einaudi, Torino
5. Brusaporci S (2014) Issues of historic town surveying: visualizing urban values. In: SCIRES-IT, vol 4, pp 55–62
6. Brusaporci S (2017) Digital innovations in architectural heritage conservation: emerging research and opportunities. IGI Global, Hershey
7. Brusaporci S, Amoruso G (eds) (2018) Built heritage: conservation vs. emergencies. Buildings (Special Issue, MDPI, Basel)
8. Brusaporci S, Lucchese V, Maiezza P (2018) Leggere il paesaggio. In: Bianconi F, Filippucci M (eds) Il Prossimo paesaggio. Realtà, rappresentazione, progetto. Gangemi Editore, Roma, pp 129–134
9. Carbonara G (1997) Avvicinamento al restauro. Liguori, Napoli
10. Carbonara G (2011) Architettura d'oggi e restauro. Un confronto antico-nuovo, UTET Scienze Tecniche, Milano
11. Casiello S (ed) (2011) I ruderi e la guerra. Memoria, ricostruzioni, restauri, Nardini, Firenze
12. Centofanti M (1984) L'Aquila 1753–1983. Il restauro della città, Colacchi, L'Aquila
13. Centofanti M, Colapietra R (1997) L'Aquila i palazzi. Ediarte, Roma
14. Centofanti M, Brusaporci S (2009) Per il restauro del centro storico dell'Aquila. Arkos - Scienza e Restauro. 20:21–29
15. Centofanti M, Colapietra R, Antonini O, Pasqualetti T (2010) Aquila dalla fondazione alla renovatio urbis, Textus, L'Aquila
16. Centofanti M, Brusaporci S (2011) Il disegno della città e le sue trasformazioni. Citta' e Storia. 1- 2011:151–187
17. Centofanti M, Brusaporci S (2011) L'Aquila invisible city: surveying, preservation and restoration of the city. In: SAVE Heritage: sefeguard of architectural, visual, environmental heritage, La scuola di Pitagora, Napoli
18. Centofanti M, Brusaporci S (2012) Architectural 3D modeling in historical buildings knowledge and restoration processes. In: Less More architecture design landscape, La Scuola di Pitagora, Napoli 2012
19. Centofanti M, Brusaporci S (2013) "Aquila monumentale": temi e problemi del progetto di restauro post sisma. In: Conforti C, Gusella V (eds) Aid Monuments—Conoscere progettare Ricostruire. Aracne, ROMA, pp 80–94
20. Centofanti M, Brusaporci S, Lucchese V, Maiezza P (2016) The representation of the landscape as construction of knowledge. Landscape & Archeology. I QUADERNI DI CAREGGI. Uniscape, pp 318–324
21. Centroni A (ed) (2004) Manutenzione recupero della città storica. 'L'inserzione del nuovo nel vecchio' a trent'anni da Cesare Brandi, ARCo, Gangemi, Roma
22. Ciranna S, Vaquero Pineiro M, (eds) (2011) L'Aquila oltre i terremoti. Costruzioni e ricostruzioni della città. Città e Storia, 1–2011
23. Conforti C (ed) (1992) L'Aquila. Città delle Piazze, Carsa, Pescara
24. D'Antonio M (2010) Il Convento domenicano dell'Aquila. S.P. L'Aquila, Colacchi, L'Aquila, D.A
25. D'Antonio M (2011) S. Domenico a L'Aquila. Il restauro del Complesso monumentale, Carsa, Pescara
26. dodo.coco.momo.momo., Per l'Aquila, N. 25, 2009
27. Ferlenga A (ed) (2007) Antico e Nuovo. Architetture e architettura, Il Poligrafo, Padova
28. Fiorani D (ed) (2005) Il colore dell'edilizia storica. Gangemi, Roma

29. Milani L., Morisi C., Calderini C., Donatelli A.(eds) (2011) L' Università e la ricerca per l'Abruzzo, Textus, L'Aquila
30. Ranellucci S (2003) Il restauro urbano. UTET, Torino
31. Spagnesi G (1989) Una storia per gli architetti. NIS, Roma
32. Spagnesi G, Il Centro Storico dell'Aquila. Memoria e progetto, Studium, Roma
33. Toubert P (1995) Dalla terra ai castelli: paesaggio, agricoltura e poteri nell'Italia medievale. Einaudi, Torino
34. Wickham C (1982) Studi sulla società degli Appennini nell'Alto Medioevo: contadini, signori e insediamento nel territorio di Valva (Sulmona). CLUEB, Bologna

Vulnerability of Architectural Heritage in Seismic Areas: Constructive Aspects and Effect of Interventions

Maria Rosa Valluzzi and Luca Sbrogiò

Abstract In this chapter, the effects of past interventions applied to historical city centers struck by a series of earthquakes over time are analyzed in terms of local and overall damage. Three villages in central Italy, Castelluccio di Norcia, Campi Alto di Norcia and Castelsantangelo sul Nera, containing overall about 150 buildings, are examined here. Classification of damage was based on the European Macroseismic Scale. The study provided vulnerability maps detailing the influence of interventions on buildings in historical city centers, which may contribute to better evaluation of damage scenario and maintenance plans.

Keywords Seismic vulnerability · Survey form · Masonry · Retrofit · Intervention

1 Seismic Vulnerability of Existing Masonry Buildings

Architectural heritage is particularly prone to damage in seismic areas, due to construction defects or limitations, which combine, together with the effects of lack of maintenance and deterioration exposure of materials. Common vulnerabilities associated with building construction, often detectable in existing masonry buildings, involve one or more of the following aspects: (i) the poor quality of the masonry; (ii) the scarce connections among components (walls, floors and roof); (iii) the structural irregularities; (iv) the inadequate stiffness of horizontal components (floors and roof); (v) the existence of thrusting structures (e.g., arches and vaults, but also pounding elements in floors and roofs). These deficiencies result in partial or even overall collapses ('mode 1' mechanisms), mainly due to overturning and out-of-plane bending of walls and assemblages (e.g., corners), which can occur

M. R. Valluzzi (✉) · L. Sbrogiò
Department of Cultural Heritage, University of Padova, Padua, Italy
e-mail: mariarosa.valluzzi@unipd.it

L. Sbrogiò
e-mail: luca.sbrogio@phd.unipd.it

G. Amoruso and R. Salerno (eds.), *Cultural Landscape in Practice*,
Lecture Notes in Civil Engineering 26,
https://doi.org/10.1007/978-3-030-11422-0_14

even at low-medium earthquake magnitudes [12, 3, 20, 16]. In addition, poor-quality masonry, mainly due to the lack of transverse shear elements in multi-leaf walls, irregular texture and composition, low-grade properties of the constituent materials (mortar and bricks/stones), deterioration of binding power in ancient mortars [6, 7], may even involve anticipatory brittle collapse ('mode 0') due to disaggregation of constituent materials, which may occur under even lower seismic magnitudes and before any other sort of mechanism can be activated.

Such brittle behavior (modes 0 and 1) can be inhibited by better quality of materials and more adequate construction systems and details, so that more extensive collaboration among parts can be activated and exploited ('box-like' behavior). This results in overall pseudo-ductile behavior (adequate displacement capacity without collapse), which can concentrate dissipative shear cracks in lintel elements, thus preserving as much as possible the integrity of piers ('mode 2' mechanisms), provided that their continuity in height is ensured by the regular lay-out of openings.

All these behaviors were observed as the effects of a series of seismic events in central Italy from August to October 2016 and are still active, i.e., in the highlands encompassed by Macerata (N), L'Aquila (S), Ascoli Piceno (E) and Perugia (W), known as Sibillini Mountains Park. These areas and their surroundings are some of the most active seismic areas (0.255g expected PGA according to the seismic hazard map; more than 0.7g actual PGA recorded in 2016) [15, 19]. High seismicity in this area has always meant that local population are scattered in many villages and 'castles', so that the very concept of 'cultural landscape' is embodied in them, with the result that they have been built and rebuilt over the centuries after every earthquake. They are the product of traditional building techniques, which gave rise both to their vulnerability and to attempts at avoiding it.

Figure 1 shows the location of the centers studied here, in the area struck by the 2016 sequence: Castelsantangelo sul Nera (province of Macerata: MC), Campi Alto di Norcia and Castelluccio di Norcia (Perugia: PG). They are all fortified villages built in the late Middle Age to keep control of fertile lands and roads between the borders of Norcia and Visso, the two largest, at the time of their expansion.

Figures 2 and 3 show the typical damage, detected in central Italy after the 2016 earthquakes, encompassing damage modes 0 and 1 (unfavorable conditions), and mode 2 (favorable conditions), due to constructive aspects.

Unfortunately, 'mode 0' failure is a major problem. The supply of building materials near it greatly influences the quality of masonry, which ranges from including very poor sand and clay mortar with almost spherical sandstone rocks (Fig. 2b) to better-dressed limestone with pure lime mortar (Fig. 2a).

Many houses show signs of earthquake damage and also the basic systems used to prevent major damage, ascribed to mode 1: buttressing, tie-bars and special systems, such as the detachment of barrel vaults from the façade to preserve the former when the latter collapses, as already observed in Campi Alto di Norcia [8, 4].

Lastly, the preferred mode 2 occurred in a very few cases and mostly in more recent buildings, dating to the 19th century, when some rules for seismic-proof

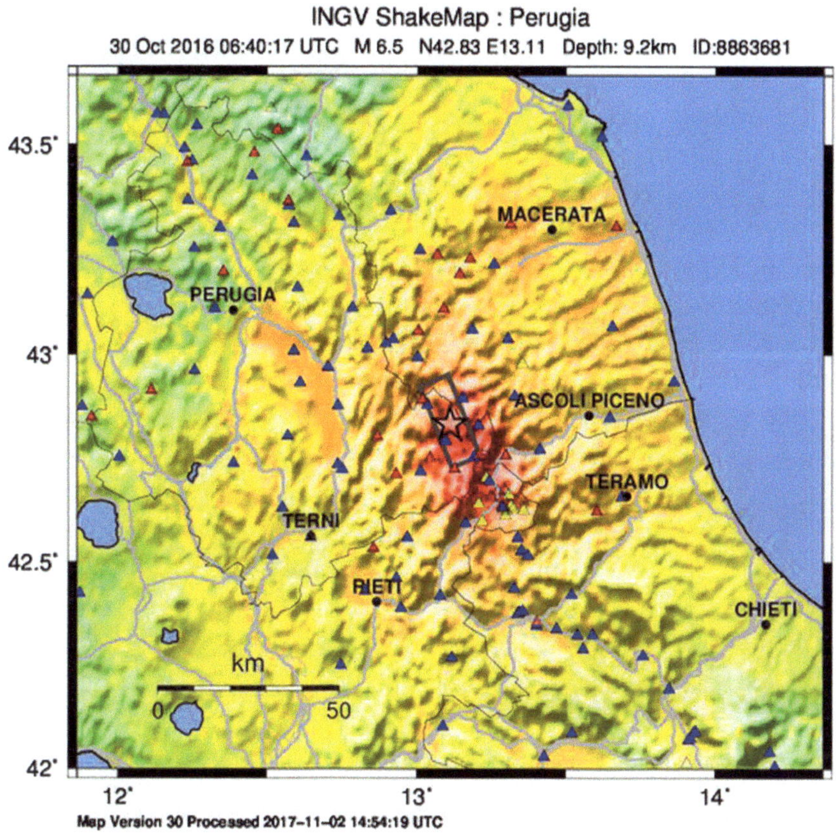

PERCEIVED SHAKING	Not felt	Weak	Light	Moderate	Strong	Very strong	Severe	Violent	Extreme
POTENTIAL DAMAGE	none	none	none	Very light	Light	Moderate	Mod./Heavy	Heavy	Very Heavy
PEAK ACC.(%g)	<0.06	0.2	0.8	2.0	4.8	12	29	70	>171
PEAK VEL.(cm/s)	<0.02	0.08	0.3	0.9	2.4	6.4	17	45	>120
INSTRUMENTAL INTENSITY	I	II–III	IV	V	VI	VII	VIII	IX	X+

Scale based upon Faenza and Michelini, 2010, 2011

Fig. 1 Shake map of October 30 2016 earthquake (not to scale). Case studies are located within the red area, which includes the Sibillini Park [15]

practices (after the 1859 earthquake that struck the region) were issued. However, these regulations apply only to large towns (Visso, Pieve Torina, Norcia), while traditional systems survived in villages.

More commonly, since 1979 (year of the Valnerina earthquake, Mw 5.9) and until the early 2000s (1997 Colfiorito earthquake, Mw 5.8), various retrofitting techniques have been applied to repair damage, mainly substitution of components (especially floors and roofs) and the addition of heavy, incompatible structures (e.g., reinforced concrete).

Fig. 2 Unfavorable behavior of masonry structures under seismic actions. Out-of-plane collapse (modes 0 and 1) due to poor-quality construction: **a** collapse due to poor-quality masonry (Gualdo, near Castelsantangelo sul Nera); **b** local collapse due to lack of connections in multi-layer masonry at corners (Castelsantangelo sul Nera); **c** pounding due to irregular interaction between house and contiguous bell-tower (Castelsantangelo sul Nera); **d** facade overturning in terraced house with original timber floors parallel to façade (Castelsantangelo sul Nera); **e** effect of thrusting roof resting on two-leaf masonry wall (Castelsantangelo sul Nera)

1.1 Procedures for Vulnerability and Damage Evaluation

The earthquakes of the early 1980s stimulated the development of simplified tools and procedures to evaluate both the damage and vulnerability of ordinary masonry buildings, mainly through survey forms. Of these, 'GNDT[1] II level' [2, 13, 11] and AeDES[2] forms [1, 10] are the most used.

The GNDT form evaluates a normalized index (Iv) with three levels of vulnerability, which can be related to verbal assessment of the overall vulnerability of a building (from 'very high' to 'very low'), as shown in Table 1 [9].

[1]Acronym for 'Italian Group of Defense against Earthquakes'.

[2]Acronym for 'Level 1 Form for Post-Earthquake Damage and Usability Assessment and Emergency Countermeasures in Ordinary Buildings'.

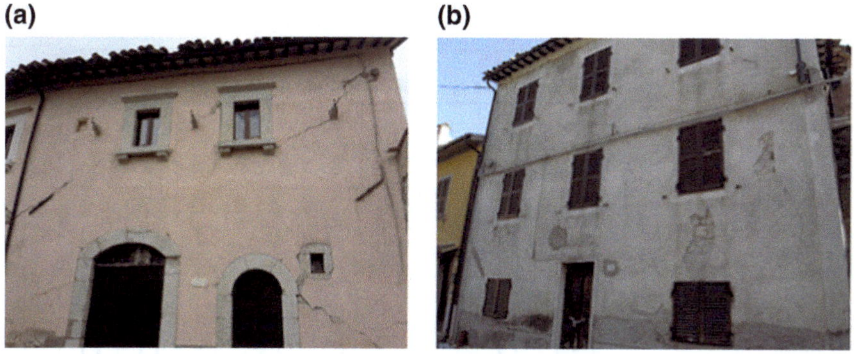

Fig. 3 Favorable behavior of masonry structures under seismic actions. In-plane damage limited to repairable elements without collapse (mode 2): shear failure mainly extensive in masonry piers **a** (Castelsantangelo sul Nera) and, more favorably, on lintels **b** (Pieve Torina, MC)

Table 1 Correlation between GNDT form scores and vulnerability classes according to EMS 98 (from [9])

GNDT form score	Iv normalized	Iv range	EMS 98 vulnerability class	Verbal assessment
0	0	0.00–0.10	C	Retrofitted
52.5	0.13	0.11–0.20		Very low
		0.21–0.40	B	Low
203.75	0.51	0.41–0.60		Medium
		0.61–0.80	A	High
393.75	1	0.81–1.00		Very high

The AeDES form dates back to the 1997–98 seismic events which struck the Umbria and Marche regions and which has been extensively validated in many post-earthquake contexts (Molise 2002, L'Aquila 2009, Emilia 2012). It relates the grading of damage levels of five main components, according to the European Macroseismic Scale 1998 (EMS 98) [14], also taking into account the extent of damage throughout the building in question (Fig. 4). Evaluation requires simpler information than the GNDT form, because it was designed for on-site use, just after seismic events.

These tools have been used in vulnerability and damage evaluations since the early 2000s for case studies throughout Italy's central regions. However, recent events have shown that they would need updating, in order to represent better the peculiar actual conditions of built heritage all over Italy, seriously affected since the adoption of r.c.-based techniques.

Level - extension / Structural component / Pre-existing damage		DAMAGE									
		D4-D5 Very heavy			D2-D3 Medium-severe			D1 Slight			Null
		> 2/3	1/3 - 2/3	< 1/3	> 2/3	1/3 - 2/3	< 1/3	> 2/3	1/3 - 2/3	< 1/3	
		A	B	C	D	E	F	G	H	I	L
1	Vertical structures	☐	☐	☐	☐	☐	☐	☐	☐	☐	○
2	Floors	☐	☐	☐	☐	☐	☐	☐	☐	☐	○
3	Stairs	☐	☐	☐	☐	☐	☐	☐	☐	☐	○
4	Roof	☐	☐	☐	☐	☐	☐	☐	☐	☐	○
5	Infills-partitions	☐	☐	☐	☐	☐	☐	☐	☐	☐	○
6	Pre-existing damage	☐	☐	☐	☐	☐	☐	☐	☐	☐	○

Fig. 4 Damage evaluation module in AeDES form [18]

2 Aim and actual effect of interventions

Interventions in masonry buildings in seismic areas mainly focus on rehabilitating collaboration among parts, at both local (cross-section, texture, connections) and more extensive building scales (strengthening of structural components and assemblages, overall working). Table 2 lists intervention techniques aimed at reducing the possible vulnerability of masonry buildings and their structural components.

The proper design of interventions according to the vulnerability in question, the adoption of compatible materials and techniques, and the suitable installation are all essential requisites, if effective results in enhancing structural behavior are to be attained [5, 17, 21].

Nonetheless, analysis of historic city centers struck by earthquakes in the last 40 years in Italy has revealed severe damage and collapse attributable to additional vulnerability triggered by heavy retrofitting techniques: although they were ideally conceived to improve mutual collaboration among the structures, in actual fact they entail more complex and hybrid behaviors (Fig. 5). It is the case of the use of r.c. to substitute floors and roofs (aimed at increasing the in-plane stiffness), to strengthen vaults (to prevent collapse), or applied as ring beams at floor and roof levels (to connect walls against overturning). This practice was commonly adopted in masonry buildings from the 1980s onwards, according to the knowledge and recommendations available then. However, even more recent techniques, thought to be more compatible with ancient fabrics, e.g., horizontal steel trusses as ties, have proved to be inadequate, mainly due to the lack of connections to masonry.

Table 2 Intervention techniques to reduce vulnerability in masonry buildings

Failure mode	Building component	Vulnerability	Damage	Intervention technique	Main improvement	Main prerequisite for effectiveness
Mode 0	Wall	Low masonry quality	Disaggregation	Grout injection	Inner compactness, homogeneity	Presence of voids (e.g., incoherent core in multi-leaf walls)
				Jacketing (r.c., composites)	Collaboration among layers	Connections applicable throughout thickness
				Repointing	Mortar quality	Regular texture
Mode 1	Wall	Slenderness	Out-of-plane deformation	Intermediate connections (ties, confinement rings)	Constraints along wall	Good-quality masonry
	Wall-to-wall assembly	No or insufficient connections	Overturning	Tie-rods	Connection of opposite walls	Good-quality masonry
				Ring beams	Force redistribution among walls	Good-quality masonry
				Confinement rings	Connection among walls	Good-quality masonry
				Stitching	Local connection at corners	Good-quality masonry
	Floor	No diaphragm effect, inadequate floor-to-walls connection	Beams sliding at supports, pounding	Overlap of timber boards, diagonal strips, thin r.c. shells	In-plane stiffening	Effective connectiosn between new and old elements
				Substitution (same concept) with steel/r.c. precast beams and hollow bricks	In-plane stiffening	Strong masonry
				Substitution (different concept) with r.c. beams, hollow bricks and thin r.c. shell	In-plane stiffening	Strong masonry

(continued)

Table 2 (continued)

Failure mode	Building component	Vulnerability	Damage	Intervention technique	Main improvement	Main prerequisite for effectiveness
	Roof	No diaphragm effect, thrust	Beams sliding at supports, overall sliding	Overlap of timber boards, diagonal strips, thin r.c. shells	In-plane stiffening	Effective connections between new and old elements
				Substitution (same concept) with steel/r.c. precast beams and hollow bricks	In-plane stiffening	Strong masonry, inhibition of thrust
				Substitution (different concept) with r.c. beams, hollow bricks and thin r.c. shell	In-plane stiffening	Strong masonry, inhibition of thrust
	Vault	Thin cross section, thrust	Overturning at abutments, deformation, collapse	Tie-rods, transverse stiffening walls	Reduction of trust and deformability	Good-quality piers
				Thin shells (r.c., composites)	Stiffness, holistic unity	Effective connection/ bonding
Mode 2	Piers and lintels	Irregular distribution of openings	Shear failure in piers or lintels	Pier reinforcement (see 'wall')	Mechanical strength	'box-like' behavior
				Lintels repair/substitution	Mechanical strength	

(a)
(b)
(c)
(d)
(e)
(f)
(g)
(h)
(i)
(j)
(k)

◂**Fig. 5** Damage caused by unfavorable interaction of existing masonry structures with retrofitting techniques: **a** corner overturning due to r.c. ring beams (Campi Alto di Norcia); **b** damage at interface between r.c. beams and masonry (Castelsantangelo sul Nera); **c** disaggregation of ancient masonry together with new masonry (Gualdo, near Castelsantangelo sul Nera); **d** unreinforced first-floor overturning under reinforced upper storeys (Campi Alto di Norcia); **e** collapse of jacketed multilayer masonry (Nocelleto, near Castelsantangelo sul Nera); **f** failure due to irregular distribution of r.c. jacketing (Castelluccio di Norcia); **g** shear failure in precast r.c. beams and consequent overall collapse due to lack of horizontal floor diaphragms (Campi Alto di Norcia); **h** partial collapse due to poor-quality masonry under heavy r.c. floors (Pretare, near Arquata del Tronto, AP; **i** overall (Castelluccio di Norcia) and **j** partial (Castelsantangelo sul Nera) collapse under heavy r.c. floors and roofs; **k** V-shaped overturning under r.c. ring beam, even in presence of anchoring rebars (Castelsantangelo sul Nera)

3 Vulnerability Maps of Historic Centers

Due to extensive observation of unfavorable effects of modern interventions after the 2016 seismic events in central Italy, a new form has been proposed to flank the assessment procedures mentioned above (i.e., GNDT II level and AeDES forms). This form collects information on interventions and modern alterations of heritage buildings as part of a purpose-designed method able to implement data collected on entire (small) historic towns in GIS [19].

Among others, the three villages studied here (Campi Alto di Norcia, Castelluccio di Norcia, Castelsantangelo sul Nera) offer an interesting overview of damage mechanisms proper to 'original' buildings, i.e. with only traditional interventions, as well as 'altered' buildings affected by invasive techniques.

Campi Alto di Norcia and Castelsantangelo sul Nera have a similar layout: an egg-shaped boundary wall built on slope with terraced houses on the lower half, and a tower and church at the top. Instead, Castelluccio di Norcia is built on a small cliff overlooking the surrounding plain with a curtain wall at its base. Over the centuries, these 'castles' have been abandoned because of the development of *extra moenia* suburbs, closer to the road systems, like for Castelluccio di Norcia and Castelsangelo sul Nera. Campi Alto di Norcia, being further from the valley road, is now almost completely abandoned (Fig. 6).

The new procedure collects data on interventions, both architectural (e.g., changes in plans' layouts of openings distribution) and structural, with reference to vulnerability assessments and damage states available over the last 40 years (i.e., from 1979 onwards), according to building types existing in the three centers (Fig. 7).

In particular, comparisons between the conditions and the damage scenarios surveyed after the 1979 and 2016 seismic events provide significant results in terms of evaluation of intervention effects. Figure 8 shows a considerable shift towards higher damage states in Campi Alto di Norcia: no cases of collapses can be found in 1979 and 'medium' damage (i.e., D3 category, according to EMS 98) involves much of the building stock. In 2016, damage state surveys show the opposite

Fig. 6 View of three towns surveyed after 2016 seismic sequences: **a** Campi Alto di Norcia (PG); **b** Castelluccio di Norcia (PG); **c** Castelsantangelo sul Nera (MC)

Fig. 7 Building types in **a** Campi Alto di Norcia (PG), **b** Castelluccio di Norcia (PG), **c** Castelsantangelo sul Nera (MC)

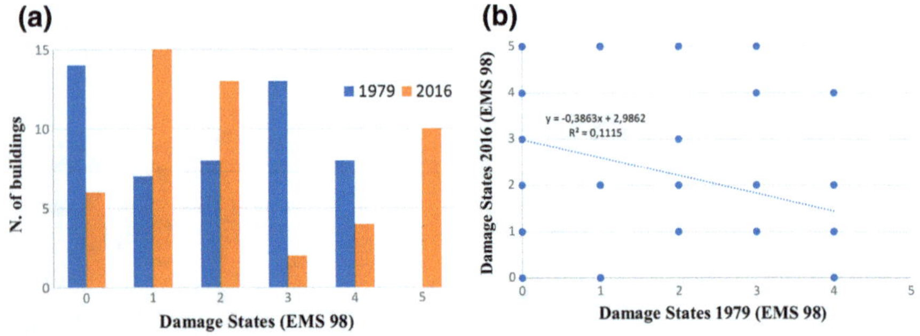

Fig. 8 Campi Alto di Norcia, distribution of EMS 98 damage states in 1979 and 2016: **a** absolute frequencies; **b** correlation between them

behavior, as most buildings are listed either in EMS 98 classes D1-D2 (low or almost no damage) or D5 (collapse).

GIS mapping clearly shows that, for instance (Fig. 9), the buildings severely damaged in Campi Alto di Norcia in 1979 are now not damaged, and vice versa. On one hand, this means that retrofit interventions have been effective, whereas buildings with generic structural updating became more vulnerable than the previous states. On the other hand, effective interventions imply the complete transformation of the original architectural qualities of a building, because all components (wall, floors, roof) are involved in the strengthening process.

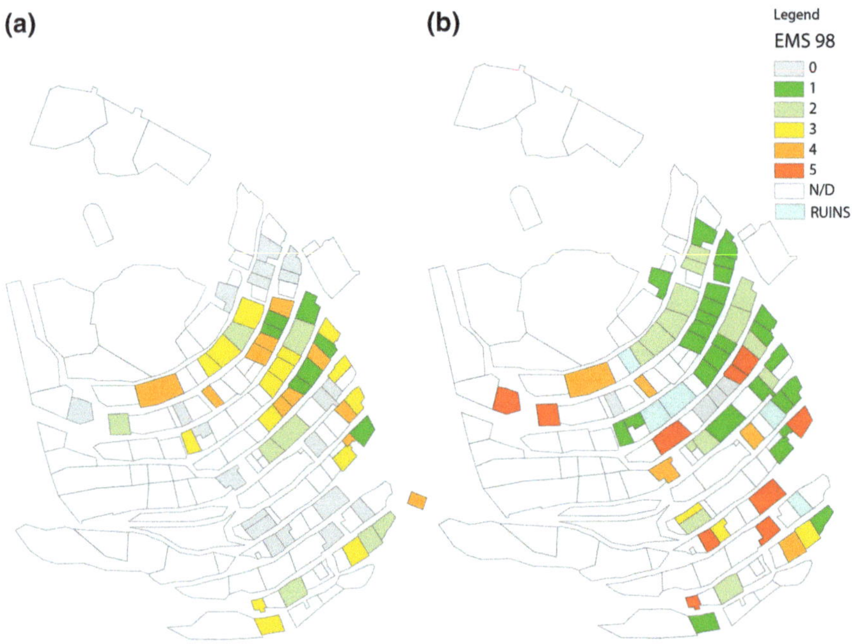

Fig. 9 Campi Alto di Norcia, comparison between EMS 98 damage level distributions after 1979 **a** [19] and 2016 **b** earthquakes in retrofitted buildings

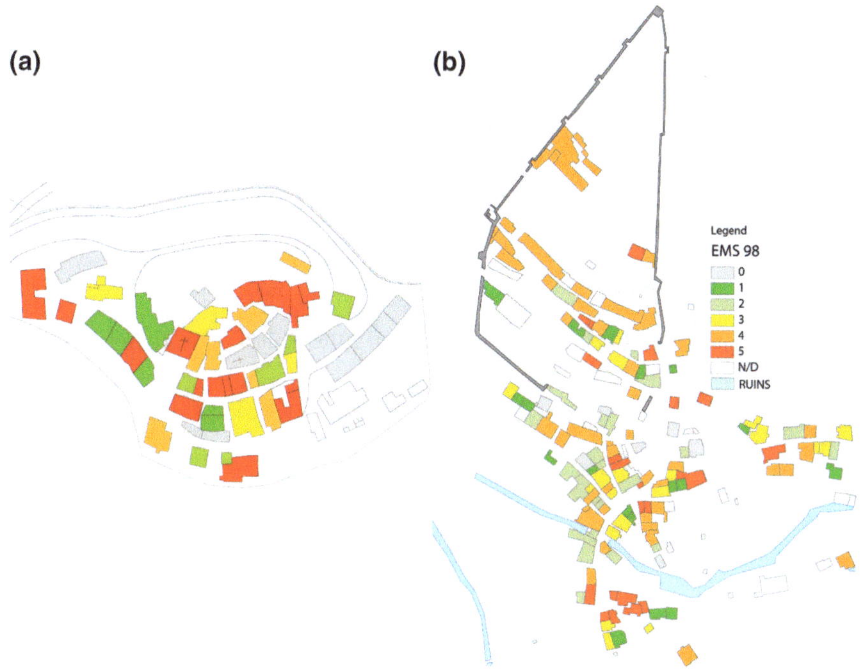

Fig. 10 EMS 98 damage level distributions in **a** Castelluccio di Norcia and **b** Castelsantangelo sul Nera after 2016 earthquakes

Table 3 Proposed values to be used in AeDES damage index evaluations

Structural component	Damage score	Weight
Vertical bearing structures	0–9	1.00
Floors	0–9	0.85
Stairs	0–9	0.50
Roofs	0–9	0.65
Internal nonstructural partitions	0–9	0.50

Also in Castelluccio di Norcia and Castelsantangelo sul Nera, for which no information about damage states after previous earthquakes is available, a certain shift towards higher EMS 98 damage states (D4-D5 compared to D2-D3) can be observed (Fig. 10). This trend may be an effect of the high number of partial 'mode 1' failures caused by interventions (mainly installations of r.c. ring beams and floors).

Lastly, a comparison between actual damage states and existing vulnerability assessment procedures (GNDT II level and AeDES forms) has been proposed. To this aim, the AeDES damage module assessment (Fig. 4) has been transformed into a normalized index (Id) with assigned scores of 9 for the worst damage state and 0 for 'no damage', and calculation of the weighted average according to the values listed in Table 3.

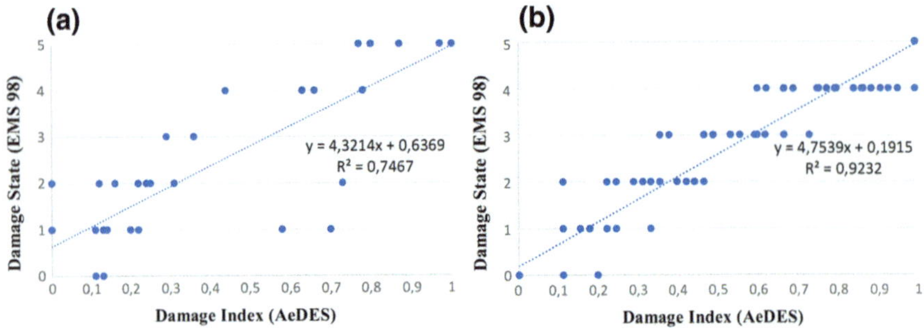

Fig. 11 Distribution of EMS 98 damage states in 2016: Campi Alto di Norcia (**a**) and Castelsantangelo sul Nera (**b**)

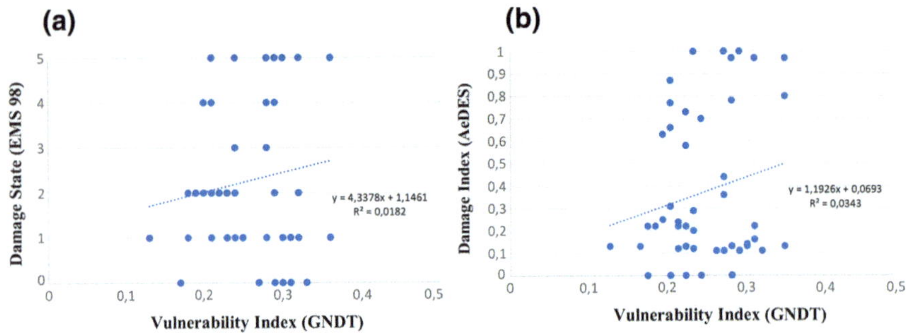

Fig. 12 Campi Alto di Norcia, correlation between vulnerability index (GNDT II level method) and EMS 98 damage state (**a**) and AeDES damage index (**b**), both showing very little correlation. Iv scale is limited to 0.5, to distinguish among points (max Iv = 1)

Comparisons among the resulting Id to damage states according to EMS 98 show a significant correlation between indexes (Fig. 11). Instead, when Id is compared with a vulnerability estimator such as the GNDT index (Iv), almost no correlation is found (Fig. 12): that is, vulnerability factors identified by the procedure were either wrongly judged or ill-defined, again presumably because of structural interventions.

4 Conclusions

The latest earthquakes in Italy in 2016 confirm the essential role played by masonry quality and connections among components for proper functioning of improvement techniques commonly proposed to strengthen horizontal structural components. This aspect is particularly important when heavy retrofitting interventions (substitution of floors and roof, or strengthening of vaults with r.c.-based slabs) are applied to existing multi-leaf masonry structures with few or no inner and/or outer connections.

It is well known that these techniques, widely used in the past after previous earthquakes, greatly influence the seismic behavior of masonry buildings. In particular, with reference to the three city centers analyzed here, research showed that retrofitted buildings suffered severe damage or even collapse. According to EMS 98 damage classification (D3: 'average' and D4-D5 'high' damage states), Castelsantangelo sul Nera, Castelluccio di Norcia and Campi Alto di Norcia, showed the following percentages of damaged buildings: 45, 41 and 10% of buildings in D4-D5 and 15, 12 and 4% of buildings in D3.

In addition, building alterations after such large-scale interventions also affected the 'representativeness' of vulnerability assessment methods. The application of common expeditious and simplified procedures (GNDT II level and AeDES forms) to the three city centers repeatedly struck by earthquakes over the past 40 years, with subsequent retrofitting measures, showed high scatter of the vulnerability judgments with respect to the actual damage observed.

Further research will focus on developing calibrated procedures for vulnerability assessment, based on the observations of the effects of intervention techniques on masonry buildings in other historical city centers in the area.

Acknowledgements Authors would like to thank D. Darù and M. Vettore for their contribution to on-site surveys. The research is framed within the DPC/ReLUIS 2014–2018 Italian project.

References

1. Baggio C, Bernardini A, Colozza R (2000) Manuale per la compilazione della scheda di primo livello di rilevamento danno, pronto intervento e agibilità per edifici ordinari nell'emergenza post-sismica (AeDES). Servizio Sismico Nazionale-Gruppo Nazionale per la Difesa dai Terremoti, Roma (in Italian)
2. Benedetti D, Petrini L (1984) Sulla vulnerabilità degli edifici in muratura: proposta di un metodo di Valutazione. L'industria delle costruzioni 19(1):66–74 (in Italian)
3. Binda L, Cardani G, Saisi A, Valluzzi MR (2006) Vulnerability analysis of the historical buildings in seismic area by a multilevel approach. Asian J Civ Eng (Building and Housing), 7(4):343–357
4. Binda L, Cardani G, Saisi A, Valluzzi MR, Munari M, Modena C (2007) Multilevel approach to the vulnerability analysis of historic buildings in seismic areas—Part 1: detection of parameters for the vulnerability analysis through on site and laboratory investigations. Int. J Restor Build Monuments, 13(6):413–426
5. Binda L, Gambarotta L, Lagomarsino S, Modena C (1999) A multilevel approach to the damage assessment and seismic improvement of masonry buildings in Italy. In: Bernardini A (ed) Seismic damage to masonry buildings. Balkema, Rotterdam
6. Binda L, Saisi A (2005) Research on historic structures in seismic areas in Italy. Prog Struct Mat Eng 7:71–85
7. Borri A, Corradi M, Castori G, De Maria A (2015) A method for the analysis and classifica-tion of historic masonry. Bull Earthq Eng 13:1–19

8. Cardani G (2003) La vulnerabilità sismica dei centri storici: il caso di Campi Alto di Norcia. Linee guida per la diagnosi finalizzata alla scelta delle tecniche di intervento per la prevenzione dei danni. Ph.D. Thesis, Politecnico di Milano, Milano, Italy (in Italian)
9. Cifani G, Lemme A, Miozzi C, Russo G (1997) Aspetti metodologici. In: Repertorio dei meccanismi di danno, delle tecniche di intervento e dei relativi costi negli edifici in muratura sisma Marche 1997, Decreto del commissario delegato per gli interventi di protezione civile n. 28 del 10 aprile 2002, pp 233–268 (in Italian)
10. DPCM (2011) Approvazione del modello per il rilevamento dei danni, pronto intervento e agibilità per edifici ordinari nell'emergenza postsismica e del relativo manuale di compilazione. Gazzetta Ufficiale della Repubblica Italiana n. 113 del 17 maggio 2011 (Supplemento ordinario n. 123) (in Italian)
11. Ferrini M, Melozzi A, Pagliazzi A, Scarparolo S (2003) Rilevamento della vulnerabilità sismica degli edifici in muratura. Manuale per la compilazione della Scheda GDNT/CNR di II livello, Regione Toscana, http://www.regione.toscana.it/documents/10180/12262198/vsm_man.pdf/095d3648-191d-43aa-ae88-ad78cff79fb3. Last access June 2018. (in Italian)
12. Giuffrè A (1991) Letture sulla meccanica delle murature storiche. Kappa (ed) (in Italian)
13. GNDT-SSN (1994) Scheda di esposizione e vulnerabilità e di rilevamento danni di primo livello e secondo livello (muratura e cemento armato). Gruppo Nazionale per la Difesa dai Terremoti - Servizio Sismico Nazionale, Roma (in Italian)
14. Grunthal G. (1998) European Macrosismic Scale 1998. Conseil de l' Europe, Cahiers du Centre Europeen de Geodynamique et de Seismologie, vol 15, Luxemburg
15. INGV http://www.ingv.it/it/. Last access June 2018
16. Modena C, Valluzzi MR, da Porto F, Casarin F (2011) Structural aspects of the conservation of historic masonry constructions in seismic areas: remedial measures and emergency actions. Int. J Architect Heritage 5(4–5):539–558. https://doi.org/10.1080/15583058.2011.569632
17. NTC (2008) Approvazione delle nuove norme tecniche per le costruzioni, Gazzetta Ufficiale della Repubblica Italiana, n. 29 del 4 febbraio 2008 (Supplemento ordinario n. 30) (in Italian)
18. Pinto AV, Taucer F (2007) Field Manual for post-earthquake damage and safety assessment and short-term countermeasures (AeDES), in JRC scientific and technical reports. Ispra, IT
19. Sbrogiò L, Cardani G, Valluzzi MR (2018) Damage survey of a historic town and comparison with past events after the 2016 central Italy earthquake. In: Proceeding of ECEE 16th European conference on earthquake engineering, Thessaloniki (Greece), 18–21 June 2018 (12 pp)
20. Valluzzi MR (2007) On the vulnerability of historical masonry structures: analysis and mitigation. Mater Struct 40:723–743. https://doi.org/10.1617/s11527-006-9188-7
21. Valluzzi MR (2016) Challenges and perspectives for the protection of masonry structures in historic centers: the role of innovative materials and techniques. Rilem Tech Lett 1:45–49. https://doi.org/10.21809/rilemtechlett.2016.10

Reducing the Loss of Built Heritage in Areas of Tourist Interest

Giuliana Cardani

Abstract Like the fruits of those seeds sown centuries ago, buildings and places are now an essential part of our cultural heritage. When their golden Age long has passed, they risk abandonment, because maintenance costs are often prohibitive, or they risk being completely transformed. So tourism could provide an important contribution to sustaining their maintenance. Tourism is an essential resource for the promotion of cultural sites and for keeping them alive. Traditional and modern technology should be employed whether for a single building or a whole village or even a small island. On the other hand, increasing exploitation of such sites may reveal itself to be a double edged sword. By attracting an uncontrolled number of visitors this could result in a more rapid decline of the resource. Some places have been radically transformed to accommodate ever increasing numbers of tourists, with detrimental results. Rather than being enriched by the authenticity of the site, mass tourism is liable to damage the authenticity that it seeks. Sometimes the final result is less authentic and the traditional spirit of the place destroyed, sacrificed to a need to comply to standard "hit and run" tourist destinations. The paper wants to present one of the many Italian cases where the need to deal with an ever growing tourist pressure has reached such a high level that intervention now becomes urgent: the small Isola Superiore of Stresa, named Fishermens' Island located in the centre of Lago Maggiore (Italy). Some suggestions are here reported in an attempt to reduce the loss of cultural heritage in the small island as well as in other similar situations.

Keywords Authenticity · Tourism · Consumption · Built heritage · Minor architecture · Fishermens' island

---

G. Cardani (✉)
Department of Civil and Environmental Engineering, Politecnico di Milano, Milan, Italy
e-mail: giuliana.cardani@polimi.it

G. Amoruso and R. Salerno (eds.), *Cultural Landscape in Practice*, Lecture Notes in Civil Engineering 26,
https://doi.org/10.1007/978-3-030-11422-0_15

1 Introduction

Like the fruits of those seeds sown centuries ago, buildings and places are now an essential part of our cultural heritage. After their maximum and best use period of greatest interest, they risk being abandoned, because maintenance costs are often prohibitive or risk being completely transformed. Tourism could therefore make an important contribution to supporting their maintenance.

The ICOMS "International Cultural Tourism Charter—Managing Tourism at Places of Heritage Significance" clearly reports that the protection, conservation, interpretation and presentation of the heritage and cultural diversity of any particular place or region is an important challenge for people everywhere. However, the management of that heritage is usually the responsibility of the particular community or custodian group. Excessive or poorly-managed tourism and tourism related development can threaten their physical nature, integrity and significant characteristics. The environment, culture and lifestyles of host communities may also be degraded, together with the visitor's experience of the place [1].

Some places have been radically transformed to accommodate an increasing number of tourists, with damaging results. Instead of being enriched by the authenticity of the site, mass tourism risks destroying the authenticity sought. After some time what remains is less authentic and the traditional spirit of the place dissolved, sacrificed to the need to comply with standard tourist destinations of the "hit and run" sort. When the use of the heritage is primarily touristic, even more so if it involves mass tourism, the reasons that attract visitors to a certain place should also be analysed together with the way in which this product is "consumed". To this end, the policies for tourism exploitation of an area should not be left to the sole interests of the stakeholders but must be programmed and controlled by qualified scientific staff.

This is what it is suggested for the Isola Superiore in Stresa, called Fishermens' Island (Isola dei Pescatori), a small historic fishing village located in the centre of Lake Maggiore (North-West Italy), which has arrived to this day with minimal changes but has recently undergone an uncontrolled invasion of commercial touristic activities, which do not take into account the traditional and historic values of the place. Nowadays this reality needs to deal with an ever growing tourist pressure but without eroding its authenticity, before it is too late.

2 Tourism as a Double Edged Sword

The cultural value of heritage is not only intrinsic, but is closely linked to the community that comprehends and receives it as a strong testimonial of the past to be passed on to posterity. Without this awareness and without any response on our behalf, all that remains is to accept these new forms of consumerism, through the commercialization of cultural heritage, considered as generic commercial products, the results of which are already before our eyes. It is not necessary to limit ourselves to citing the most famous Italian cities; there is a large number of minor places, but nonetheless famous, which are invaded for short periods; they have been completely transformed merely to satisfy the requests of tourists, paying attention only to the financial gain aspect. On the other hand, some places are abandoned without touristic development despite their history and beauty, risking to be left without resources for their maintenance. In this complex context, the debate on the sustainability of historical and artistic heritage tourism needs to be developed.

2.1 The Case of Isola dei Pescatori di Stresa

The example presented here refers to a small touristic island on Lago Maggiore, in the North West of Italy: Isola Superiore better known as Isola dei Pescatori, in Stresa (Fig. 1). The island is part of the Borromean islands and for years has been included in the international tourism circuit as one of the most visited areas in Italy [3], although it is the only island outside the property of the Borromeo family—powerful feudal lords of the area since the sixteenth century.

With its 100 m in width, 350 m in length and its fish shape, the poorest Isola dei Pescatori, despite being the smallest of the three main Borromean Islands—which are Isola Bella and Isola Madre—, is the only one to be permanently inhabited throughout the year. Today there are only about thirty people living there, a drastic reduction in numbers compared to more than a hundred of the last century. It can be accessed by the boats of the national shipping company and of local cooperatives, their frequency being greatest in the summer months, decreasing in winter. These boats are also used by dozens of people employed in restaurants, cafes and businesses on the island, along with motorboats of the public service and private boats of residents and visitors.

The history of the island is ancient: in medieval ages there was already a village of fishermen gathered around a small proto-Romanesque church of the XI century, now only one of the chapels of the seventeenth century church visible now [4], a

Fig. 1 Current aerial view of the Isola Superiore in Lake Maggiore and a drawing of 1840

small cemetery and an array of narrow streets with a “fishbone” shape, ending at the “tail” of the island, where the port and some landing stages for boats are positioned and used by fishermen as an outdoor space dedicated to the making and repair of nets. The layout of the island is still the original and a testimony of medieval town planning, which has a considerable historical and scientific interest, and still has some medieval portals hidden between the houses [5]. The island has certainly been inhabited in even more remote times, as evidenced by some rocks still visible with cup marks carved into the rock (*coppelle*).

The name Fishermens’ Island should be a name given in the last two centuries because in 19th cent. documents speak of olive groves in the area and the inhabitants of the island remembered that in ancient times it was also called the Island of Olives [2]. In the 18th century olive-growing had disappeared and the external features of the built heritage were already completely defined and similar to the current ones. The island thus hosts a small village, consisting of simple multi-storey houses, arranged in long and narrow blocks, with long balconies for drying fish, narrow streets with steps perpendicularly connected to a unique central street, one small square and the promenade on the lake border. None of the houses has a basement but they have a warehouse on the ground floor to reduce damage in case of flooding. Currently this level is almost all occupied by the modern commercial activities.

The ancient village, famous for its picturesque atmosphere, has inspired artists and musicians who decided to move there. The island has been inhabited mainly by fishermen: at difficult times, such as World War II, fishing was more than ever a fundamental resource for the island community. This work activity remained the primary one until the 80s of the 20th century. This prevalent activity has ensured the presence of people over the years and they have taken care of the built environment and shores. In order to guarantee their livelihoods, the fishermen themselves contributed to the conservation of various fish species, using traditional methods that restocked and maintained the natural lake habitats.

The life and economy of fishing has been integrated in the last two centuries by the tourism of Italians and foreigners, including various artists and some illustrious guests, present in private residences and quality hotels along the lake, which flanked some trattorias and inns and the commercial activity of souvenir kiosks, once managed by islanders.

Currently, several private houses, no longer inhabited by residents, have been taken over by the short-term rental market, while most commercial and catering activities are run by owners who reside elsewhere, as do almost all their staff.

2.2 *Erosion of the Cultural Heritage: Threatening Aspects of the Island*

Several factors are threatening the island's environment and its characteristic built heritage. This vulnerability is also strongly linked to the fluctuating dynamics of its frequency: the tourist invasion in the summer months and depopulation in winter. This strain on such a small and fragile territory puts the environmental, social and cultural identity of the island under pressure. The main risk is that this identity disappears forever to the benefit of a commercial, uneven and conflicting declination of each activity. Fishing has for centuries been the main source of income for the inhabitants of the Isola Superiore, so as to give its second name and the typical character. This island's economy has supported the interests of mass tourism more than those of its own inhabitants, fishing activity is reduced to a few people, destabilizing the territory and its fragile balance. No one looks after the coastline anymore, so no one can see that the increase in engine power of tourist boats is slowly and materially eroding the shores and the whole contour of the island (Figs. 2 and 3), damaging old and new jetties and bringing the water closer to the main external way, usually occupied by souvenir kiosks. These same kiosks, along with the *dehors* of bars and restaurants, now extended, reduce the walking path areas and obstruct the view of the lake, with an untidy occupation of public land. Even the entropized landscape risks severe deterioration due to the absence of planning strategies, maintenance plans, and coordination of different knowhow aimed at its protection (Fig. 4). The medieval footprint and the picturesque nature of the island are constantly threatened more by the lack of governance than by the needs of tourist businesses. The tourists visiting the Isola dei Pescatori always wonder at the reason for this name, as no clear or visible sign of this once dominant activity is now apparent and "hit and run" tourism of day-trippers is now prevalent. Is it too late to preserve the cultural value still existing on the island and so what can we do now?

Fig. 2 Isola Superiore or Pescatori: **a** private and **b** public trip boats to reach the island

Fig. 3 Isola Superiore or Pescatori: **a** tangible erosion of the shores and **b** consequent damage of the stone masonry piers and jetties due to the wave motion caused by the trip boats

Fig. 4 Isola Superiore or Pescatori, houses needing a restoration plan: **a** the old primary school and **b** a 18th cent. private house with inflected walls

3 Proposal to Preserve Cultural Heritage from the Slow Consumption

A growing number of visitors can be an extremely important resource for a territory, presenting many challenges and opportunities [1], but without control, it can often turn out to be a double-edged sword. The impact on the territory, in fact, can be either positive or negative.

It is necessary to improve the value of the whole territory, in order to better manage this resource, where the cultural heritage exists, recognizing the value of the testimonies of the place and the community that lives there. These can help to define the peculiarities of that context with a rich and varied set of resources that characterizes that territory, starting from the historic vernacular buildings, often badly preserved or provisionally repaired and which require guidelines for planned preservation (Fig. 4).

In order to safeguard these resources, it is necessary to promote specific actions, both on an architectural and territorial scale. The tangible and intangible heritage, as well as the customs and traditions of the places and people who live there, must be recognised and valued above all by the citizens themselves. To do this, local government must promote active integration coordinated by research institutes such as universities, schools capable of discovering and enhancing cultural resources and raising awareness of their value (like the local touristic and hotel schools). A primary objective for managing heritage is to communicate its significance and the need for its conservation to its host community and to visitors. Reasonable and well managed physical and intellectual and/or emotive access to heritage and cultural development is both a right and a privilege. It brings with it a duty to respect the heritage values of the present host community, indigenous custodians or owners of historic property, and for the landscapes and cultures from which that heritage evolved [1].

The fundamental objective of protecting historical heritage must foster interaction between the public administration and citizens, assisted by researchers with different skills (restoration, design, management, technology, etc.). This means sowing the seeds of the awareness and fundamental role that must be played by citizens, and controlling and sometimes opposing solutions that are not consistent with the joint project.

The exchange of specialist knowledge with the widespread awareness of the people who live daily in the territory, can become a valid tool in controlling abuses dictated only by economic interests.

A key role is therefore played by educating local authorities to ensure forward-looking policies and sustainable conservation strategies, in order to sow the seeds for the defence of the local heritage of each site, thereby spreading the same seeds to professionals and the local population, including the youngest.

This strategy could be effectively applied to the small fishermen's island in Italy. In this sense, architectural conservation can contribute to the future development of tourist sites of historical interest.

Icomos says: Tourism should bring benefits to host communities and provide an important means and motivation for them to care for and maintain their heritage and cultural practices. The involvement and co-operation of local and/or indigenous community representatives, conservationists, tourist operators, property owners, policy makers, those preparing national development plans and site managers is necessary to achieve a sustainable tourism industry and enhance the protection of heritage resources for future generations.

As the principles of the Icomos charter reported, the relationship between heritage sites and tourism should be managed in a sustainable way for present and future generations, in order to avoid conflict and indigenous people should be involved in establishing goals, strategies and policies for the identification, conservation and management of their heritage resources, cultural practices and contemporary cultural expressions, in the tourism context. The promotion, distribution and sale of local crafts and other products should provide reasonable social and economic returns to the host community, while ensuring that their cultural integrity is not degraded [1].

On the Island the goal is to keep the identity of the local community, focused on fishing, kept alive by supporting the activities of the few fishermen still partially active and encouraging the entry of young people in this sector. The renewal of awareness, developed in the last few years, of the genuity and integrity of food could give a kick start to the consumption of local fish as a starting point in the reconstruction of the whole fishing chain. Aiming at re-launching this specific culture, recovering the traces, even the smallest, present on the territory of the fishing culture material to the entire supply chain (as shipbuilding and repair of wooden boats, fishing tackle, nets, protection and restocking of fish heritage, conservation, processing and marketing of fish, etc.) is a need, which also involves the companies, the schools (like the hotel school in Stresa), research centres (in the area there is a national research institute of ecosystem study) and activities already present on the nearby dry land, apart from the tourist season. The rescue of this activity, without pretending to become the predominant one on the island, is able to guarantee the constant control and protection of the environment and landscape, starting with the lake and its shores. All shores can be properly repaired and constantly maintained, as well as the "*sassere*", rows of stones built in the water near the banks, to slow down the wave motion, the movement and reduction of the sand. This important link with the natural environment can again be transmitted to the younger generations.

The project could provide training courses for teachers of local schools, students, fishermen and restaurateurs depending on the theme of interest (material culture, oral knowledge of fishermen, lake environment, sustainable tourism, economic and nutritional aspects, etc.).

The setting up of a small open-air museum is suggested, a sort of documentation centre on fishing, starting from a building (in this case the old small primary school, once the residential house of the musician Ugo Ara, Fig. 4a) where all the information and documents can be obtained, and spreading along the island, to discover traditional workplaces, materials, and boats.

The didactical path can then reach, through the alleys and the ancient vaulted passageways of the fishermen's houses, the following [6]:

(a) the pier of the nets, which is the strip of land to the north of the island, without buildings, where fishermen used to roll out the nets to dry;

(b) the furnace, the place where once a month the fishermen dyed the cotton and linen nets, putting them steeping in boiling water with chestnut peel, so that the tannin makes them more resistant;
(c) the historical jetties of the fishing boats, the latter named "*burchielli*";
(d) the fishing places, in case of fishing-tourism, routes to be followed out on the lake with fishermen.

The project of preservation and safeguard of the whole island, besides improving its economy, trying to re-evaluate the historical activity of fishing and to improve the tourist offer, seeks to widen the culture on offer to the visitors, allowing to assign spaces and places [7] also to other temporary cultural activities connected to the artistic and literary personalities on the island or those who have frequented it. This activity should be managed so as to be sustainable not only thanks to the tourists' offers, but also by a growing involvement of inhabitants, traders, companies, philanthropists and people who decide to move permanently to this island, allowing to extend and better distribute visits throughout the year and better managing the different activities.

Fishing depends on water quality and biodiversity in the lake ecosystem and the museum should become an instrument of environmental protection and the growth of a culture of integration between nature and built environment. This human environmental relationship, in such a particular context, could become a new theme for didactic-educational activities to be offered to the already extensive presence of school groups and tourists.

This small improvement could not only safeguard the identity of the island, but also have an impact on all the activities already present there and on those on the mainland, in a synergic way.

4 Conclusions

The exploitation for the purpose of tourism of an historical-architectural heritage is a double-edged sword: if the resources generated by its use can provide resources for its conservation and protection, on the other hand intensive use may damage the asset. Local authorities have the responsibility to regulate and manage a balanced use of the assets, ensuring that a greater number of visitors can get to know and appreciate the heritage, without causing any deterioration.

The role of the scientific and technical experts must not be limited to providing and applying the most appropriate techniques for conservation and protection, but must be based on the reasons why visitors choose to visit a particular place and how they use it.

As well as the urban context surrounding an architectural work must not be disregarded during a conservation plan, in the same manner the analysis of tourist use cannot disregard the socio-cultural reasons that propel visitors. Only with a holistic approach that places these reasons at the centre of the analysis, will it be possible to undertake an integral protection of the cultural heritage in all its tangible and intangible aspects.

The article resumes some reflections developed for the planning of a balanced use of historical assets, affected by a growing uncontrolled tourist presence, which threatens to compromise the material preservation of the heritage. In this sense, architectural conservation can contribute to the future development of tourist sites of historical interest.

The role of recognizing the values of cultural heritage must be made active. It is possible to propose thematic itineraries, through which to make a significant contribution to the preservation and fruition of the cultural heritage represented not only by the individual architectural artefacts, but also by that set of tangible and intangible assets that are connected to them.

Acknowledgements The author wishes to acknowledge the contribution of Prof. G. Ottolini and the inhabitants of the Fishermen's Island for their initiatives and the enduring passion with which they continue to hope that the ancient traditions of the island will not be definitively erased by the growing mass tourism. The author wishes to thanks C. Zonin, V. Cozzi, M. Ugolini and also the students of the School of Architecture of the Politecnico di Milano, who in 2016, during an Architectural preservation studio, carried out surveys of the historic buildings with the colleagues C. Achille and C. Campanella and were able to recognize the values of this small island even outside the tourist season.

References

1. Icomos (1999) International cultural tourism charter: managing tourism at places of heritage significance. Retrieved from https://www.icomos.org/charters/tourism_e.pdf. 12 May 2018
2. De Vit V (1877) Il Lago Maggiore, Stresa e le Isole Borromee, vol 1. Aldina F, Alberghetti e figli, Prato, Italy
3. Borromean Islands (2018) Retrieved from http://www.stresa.com/borromeanislands. 12 May 2018
4. Burba V (2016) Le quattro sorelle Ricostruzione storica sulle origini delle Isole del Lago Maggiore. Retrieved from: http://appuntiretrodatati.blogspot.it/2016_04_28_archive.html. 12 May 2018
5. Marzi A (1991) Le Isole Borromee, matrici medievali e trasformazioni barocche. I luoghi di culto, le fortificazioni, lo stupore dell'Arte e gran teatro lacustre, Le Rive, vol 2, Editrice S.E. M.M. srl, Omegna, Italy
6. Pisoni CA, Spadoni I, Zacchera M (2010) Nelle reti del tempo—Storie parole immagini di pescatori alle Isole Borromee e sul lago Maggiore. La Compagnia dè Bindoni per il Magazzeno storico verbanese, Germignaga, Italy
7. Khakzad s, Griffith D (2016) The role of fishing material culture in communities' sense of place as an added-value in management of coastal areas. J Mar I Cult 5:95–117. https://doi.org/10.1016/j.imic.2016.09.002

Small Historic City Centers of Mediterranean Europe: Critical Points and Potentialities for Environmental Sustainability

Valentina Pica

Abstract Conserving historic urban environments with a multidisciplinary and integrated approach is currently one of the most universally urgent and challenging cultural heritage conservation issues, that also deals with environmental sustainability. More specifically, Southern Europe small towns are progressively being abandoned, while they should be preserved with integrated studies and interventions, oriented to innovative solutions. The current conditions and historic resources of the Albaicín quarter of Granada (Spain), as well as of other small historic city centers in Italy, are being pointed out. A method for a comprehensive recovery plan approach of these centers is proposed, that should start from a territorial analysis, throughout different steps, such as: identifying existing assets; historical studies of their evolution; mapping and general classification of the built heritage; study of the accessibility of the historical center's sites; a deep analysis of their critical points and opportunities and a critical valuation of the existing planning regulations. It is also important to outline the common mistakes which are being currently made, in order to apply corrective measures in the next future, based on a kind of "peer to peer" urbanism. This paper also argues in favor of numerous international projects, aimed at the renewal and resiliency of small historic Mediterranean centers, in order to foster good practices in their conservation worldwide.

Keywords Small historic centers · Recovery plans · "Peer to peer" urbanism · Integrated interventions

V. Pica (✉)
Marie Sklodowska-Curie Fellow, Roma Tre University, Rome, Italy
e-mail: valentina.pica@uniroma3.it

G. Amoruso and R. Salerno (eds.), *Cultural Landscape in Practice*, Lecture Notes in Civil Engineering 26,
https://doi.org/10.1007/978-3-030-11422-0_16

1 An Integrated Approach to Small Historical Centers Revitalization

The continuous and fast transformation of our modern cities have been carrying out relevant issues on environmental sustainability. According to the Charter of Cracow (2000), historic cities conservation is a urgent topic. The Charter states that "*Historic towns and villages, in their territorial setting, represent an essential part of our universal heritage, and should be seen as a whole with the structures, spaces and human factors, normally in the process of continuous evolution and change*" [1]. Scattered heritage should be politically considered a huge resource for European territorial revitalization and cultural tourism growth, while at the moment it almost consists of a problem to be solved, moreover in Southern Europe, the poorest part of the EU. This region presents one of the richest scattered heritage in the world, but this is very fragile and is under threat, because of its degradation, due to political and economic needs, and the seismicity of this area. The current reality is that its small centers' traditional buildings haven't been rigorously mapped and requalified and are progressively being abandoned because of their marginal locations, far from the biggest cities main services. A first mapping and management tool of these historic centers at risk has been realized recently in Italy [2], but is not completed and the lack of an accurate map of the urban fabrics of the marginal small city centers is quite known, and is not even contemplated in the Heritage Risk Chart, the GIS tool of the Italian central government administration [3].

Moreover, European rules, and more specifically the Directive 2010/31/UE, don't require that these fabrics be retrofitted in order to preserve their historical value [4]. Nevertheless, over the last three decades there has been a marked trend to the rehabilitation for residential use of historic city centers, art heritage cities and areas with a particularly high concentration of historical assets and local traditions [5]. For a few years there has been an increase of the attractiveness of the Italian small historic city centers, those with less than 20.000 inhabitants, which are the majority of the urban settlements in Italy.

In Spain and in other European Countries on the Mediterranean basin, like Greece, there are similar urban and rural historic assets. They are, like in Italy, the object of a new residential demand, partly due to their environmental value, which have been largely preserved from urban and industrial development.

A recent research carried out on the historic small city of Fermo, in the Italian region of Marche, and directed by the urbanist Luigi Piccinato at the Department of Architecture of the Roma Tre University, shows how the main characteristics of small historic centers guarantee an optimal quality of life, thanks to the social, economic and other intangible assets that they can generate [6].

Nowadays, the valorization of the historic small towns under marginal conditions, situated far from the biggest cities, and the monitoring of their attractiveness in order to preserve their architectural characters, mean a challenge shared between the workers in the sector, the administrators and the private stakeholders.

These ones are increasingly preferring to live in small peripheral towns with a healthy and well conserved natural environment. They can be persuaded to move out of the biggest cities on condition that small centers can be provided with the necessary facilities and services.

Nonetheless, these private actors who are looking for a better place to live are increasing but they aren't a large number yet. Therefore, it is time for the scientific community to pay attention to the several stakeholders in order to create projects for the small historic centers' revitalization. These actors can be external investors, small, medium and big local companies, associations, and can be involved into collectively managed and crowdfunded events, or experimental workshops and working tables, in collaboration with local administrations.

1.1 Method Issues

Adopting an integrated model of development strategies in order to preserve small historic centers of Southern Europe and to contribute to their revitalization is necessary. This model should be grounded in historic urban fabrics conservation, in the valorization of cultural environmental heritage and in the cultural identity issues, in order to contrast the marginal conditions of the majority of these small towns. In the case the environmental plans and the special projects of revitalization of these marginal areas are successful, coordinated intervention programs must be foreseen in order to protect their environmental and architectural beauty.

Contemporary technologies and research methods allow to propose efficient solutions with relatively low budgets. There are several instruments nowadays, such as energy efficiency retrofits and Smart City planning parameters, which can be tailored to small towns [7]. It is important, therefore, to contribute to the creation of a shared method based on integrated intervention strategies in a multidisciplinary framework (urbanism, architectural planning, conservative planning, historical analysis, urban codes and administrative rules, etc.). This method should start from a territorial analysis, based on digital tools like a GIS database, and should be realized throughout different steps. Firstly, it is important to identify and analyze the existing assets, in order to develop complete historical and critical studies of their diachronic evolution. The operative basis of this kind of analysis includes the typological study tools of the urban fabrics determined by architect Saverio Muratori and his school [8].

Therefore, it is basic to realize the mapping and general classification of the built heritage and existing infrastructures, with the study of the accessibility of the historical center's sites. Finally, it is important to include a deep analysis of the critical points and opportunities and a critical valuation of the existing planning regulations in these historic centers, the majority of them being located in regions of high seismic hazard, moreover in Italy [9].

This analysis has to be guided by a full recognition of the development and improvement opportunities, in order to create useful resources to the revitalization of the analyzed historic centers and their environment.

Various tools of Landscape Sustainable Design, which have been developed over the last few decades even in the Permaculture discipline, can be applied in order to create team building, problem solving, and to study systematically all the weaknesses and the strengths of each part of every project [10].

It is an holistic approach, which empowers people for historic centers requalification, and can become the central corpus of an innovative and organic method for urban restoration, carrying out projects from a broadened perspective.

Small historic centers won't be conserved without taking into exam their surrounding landscape, and this line of work will contribute to their well regulated "smart growth".

At the moment, complete recovery plans of integrated interventions grounded in a holistic management of sites and their surrounding context and oriented to this "smart growth" and regeneration of small historical centers, haven't been defined yet. Although the recent realization of several innovative scientific studies [11] and research projects [12], the most significant being the *SmartPolis project* at the BHLab of the ITABC of the CNR [13], as well as the setting up of pilot projects [14], these recovery plans aren't always comprehensive, as they are often conducted from a sectorial perspective.

Nowadays, operative proposals based on a rigorous and shared method are almost inexistent, in the framework of urban restoration, not being those which have been formulated over the past decades fully exhaustive and accepted among the various stakeholders.

Local and central policies focused on an effective multidisciplinary intervention plan development are also absent, while they are necessary for the protection of small historic centers and their landscape.

In Italy, the most universally known country bound to urban recovery, since the first legislative system, which have allowed to establish a definition of the concept "protected historic center", came into force (which is called Bottai Law, nn. 1089 y 1497 of 1939) until today, after more than 70 years, the disciplinary debate on this issue has highly evolved.

The fundamental steps of this debate have been marked by the administrative machinery in the Gubbio Chart in 1960, with which is established for the first time the necessary acknowledgement of a preliminary typological classification of the historic centers, including the urgent individualization of the areas which shall be rehabilitated. Then, the law 457 of 1978 has determined the funding by consistent sums of the recovery of historic heritage and has introduced a specific urban planning tool, the Recovery Plan, although it has not stated the clear respect of restoration practices.

In view of all this, an historic center is actually known in Italy and in Southern Europe, theoretically, as a cultural, economic and social asset to be preserved and revitalized, for its characteristics of urban area with a fragile scattered heritage, with a specific identity and high historical value, both referring to specialized buildings and domestic ones [15].

Nonetheless, the practical and full application of the statements and guidelines of this philosophic and theoretical framework are still almost absent. Besides, there are several denounces of the distance between the world of architects and planners and the real needs of the citizens, the public administration and the legislative system [16]. Furthermore, the long time that it takes to realize recognitions, to make decisions and receive administrative answers may go further than the direct control of the projects.

1.2 Sustainable Strategies and Virtuous Examples

Within this scenario, it is arguable that an effective operative method of urban recovery should have an interdisciplinary approach and be of "macro design", with interventions at several scales which can be calibrated on performance criteria and precise and flexible prescriptive rules, and adaptable each time to specific architectural "micro projects" [17].

It is a *work in progress* design procedure, which allows sustainability and effectiveness because of its cyclic closed loop, from macro to micro in a biunique way, letting the decision-making process be corrected at every step, if necessary.

The prioritization of the main measures follows this line, which is almost evolutive, and can be tailored to the reality and modified according to the local conditions, and the real needs of the stakeholders.

It is not about proposing a dogmatic approach, by which decisions are warranted, but a reflexive and dynamic approach, which can define the complex connections between the different characteristics of built heritage in order to establish various levels of intervention. These levels can consist, according to the different circumstances, in simple conservation, transformation, requalification, new construction or demolition, without unuseful restrictions.

It is also important to investigate the possibilities to increase the attractiveness of small historic centers through a virtuous combination of urban interventions grounded in the "modulation of protection" [18], historic buildings restoration and new technologies application at a urban and architectural level. Within this process, it is important to understand at every step the needs of local customers and of external or potential stakeholders, in order to optimize time and resources.

It is fundamental to propose guideline methods that can be related to technicians and to the administrators, who can interpret them critically in order to realise virtuous projects.

A valid line of interventions that could be followed or integrated within this framework is that of the "scattered museums" criteria, following the example realized by Giovanni Manieri Elia and Alessandra Centroni in the small historic center of Arsoli (RM, Italy) [19].

The experimental projects managed by Scalora and Monti following these guidelines on the historic centers of San Giacomo, Ascoli Piceno, S. Erasmo en Gaeta and Paganica, after the heart quake of Aquila on April 2009, deserve to be quoted [20]. In all these projects the morphological analysis of the historic city is useful for creating good diachronic (it means of the evolution of the multi-layered city) and synchronic observation points, which allow to analyze the several constructive phases that are present at the same time, as well as for carrying out comparative methods, which correlate the historic urban fabric of similar small centers located in different countries.

The general aim of these experimental projects is that to critically elaborate reconstructive hypothesis of the shape of the historic centers in its principal evolutive phases, from the study of static configurations to dynamic ones.

These hypothesis can provide a useful database for the formulation of the shared guidelines in order to foster sustainable processes, through a comprehensive operative framework valid for technicians, urbanists, stakeholders and administrators.

These guidelines can also be largely distributed among the customers and towards a larger public, in order to attract investors.

There can be created different documents that can be shared worldwide, in order to elaborate web forums, online laboratories and opinion-making blogs. An important example of webpage with open source documents has been created by the international project HISTCAPE, of the programme INTERREG IVC, financed by the European Regional Development Fund. It promotes sustainable ideas and methods for environmental and cultural revitalization of small historic centers worldwide; moreover, it fosters the development of local economy and of small businesses targeted at local customers, as well as social participation in the historic fabrics reuse, through horizontal governance, involvement and empowerment of the community on local cultural heritage management. These issues are disseminated through specific open source publications [21].

Nowadays social re-use is not diffusely contemplated yet in Southern Europe, while it can be an effective device to fulfill the administrative and political gap of funding recovery interventions, by activating new forms of financing, like crowdfunding, based on platforms for fundraising. Crowdsourcing platforms as well could optimize economical resources, because specific interventions could be tailored to the real needs of the stakeholders, and this would be the basis of a "smart growth" of this kind of settlements, which should be regulated by specific guidelines for the historic housing protection.

It is an experimental approach, and it fosters a new model of heritage management based on a bottom-up process [22], a social sustainability and participative model of planning, also called "peer to peer" urbanism. A successful example of "peer to peer urbanism" application is the Artena Project, in central Italy [23]. The entire initiative works through participation and public engagement and it is fully bottom-up. It achieves its goal through self-financed sub-projects, which also financially support the whole initiative. The project aims to design the physical, virtual, and socio-economic space of the city, believing that every space has "bio-political" value. Whether it is impossible to go back to a rural civilization, it cannot be denied the need for recovery of environmental as well as urban and socio-economic characteristics, still preserved in Italian villages, with the aim to improve the quality of our lives. A new socio-economic network around heritage could be activated by this line of interventions, based on considering historic centers as a resource and an opportunity for territorial and social development.

This kind of approach raised within the framework of heritage reuse and energy building retrofitting, both promoted by the International Community after the first crisis of oil in 1973 (see the Brundtland Report of 1987, World Commission on Environment and Development, WCED). From then on, important conferences and meetings have been organized by international institutions to deal with this issue, such as the Brazilian meeting at Rio de Janeiro of the United Nations for Environment and Development in 1992, which put the basis for the "Agenda 21 for culture", which is the first document with worldwide mission that advocates establishing the groundwork of an undertaking by cities and local governments for cultural development.

The Agenda 21 for culture was agreed by cities and local governments from all over the world to enshrine their commitment to human rights, cultural diversity, sustainability, participatory democracy and creating conditions for peace. It was approved by European Community by the 4th Forum of Local Authorities for Social Inclusion of Porto Alegre, held in Barcelona on 8 May 2004 as part of the first Universal Forum of Cultures. This agreement fosters cultural participation as vital elements of citizenship and has been followed by the Agenda 2030 for the Sustainable Development, an action program signed in September 2015 by the governments of the 193 member countries of the UN. It includes 17 Sustainable Development Goals—SDGs—in a large action program for a total of 169 'targets'.

2 Case Studies

Current conditions of historic small city centers of Southern and Central Spain were analyzed during my Ph.D. study, when I could deeply analyses ten emblematic mansions belonging to the XVI Century Granada, in Spain [24], the majority of them being situated in the historic quarter of the Albaicin. Also, this study has taken into account the main architectural expressions of greater contemporary urban areas of the Kingdom of Castile.

The historic houses of the Albaicin have been studied and surveyed in relation to the entire city modern transformation, bringing to light the delicate and complex transition of the town, from the Nasrid period to Christian times. The drawings of the original conditions of the sites, previously to the most recent renovation works many of these buildings have undergone, have been presented.

The method of this work was based on, first of all, the drawing of a precise and thorough planimetry of the ten surveyed houses, together with in situ tasks based on physical data; subsequently, a study of archival data was made, also analyzing the

Fig. 1 General Map of XVI Century Granada, with the medieval walls (grey line), the urban fabric (yellow color), the Albaicin quarter (red color) and the productive gardens (green color). With letters a, b, c, d, e, the most important mansions of that period have been marked

specialized existing bibliography on the subject. The integrated analysis of the buildings and their surroundings has been fruitful and could be registered on a data map created with the Geographical Information System, where also the Medieval settings, and those of later constructions, have been individualized and dated (Fig. 1).

The surveyed domestic buildings have also been compared with other similar outstanding constructions of the same period that are located in the city, which have either been preserved or disappeared. These houses, although have been demolished, were well documented and studied during the XIX and XX Centuries [25], and their study has confirmed most of the preliminary hypotheses, and have guided the reconstruction of the spatial distribution of the original architecture in the historic period discussed (Fig. 2).

The Albaicin appears under conditions of marginality, up on a hill in front of the Alhambra, with low accessibility and inadequate infrastructure facilities.

For this reason, since the 60’s of the last Century it has been progressively abandoned by local inhabitants, who moved to the part of the city downhill, and its degradation is still visible, even if the local government has tried to finance recovery plans in the last decades, opening the door to a process of gentrification.

This analysis has documented and outlined the huge loss of the character and of the original materials of housing in the historic center nowadays, after several rehabilitation works and the construction of new buildings (Fig. 3).

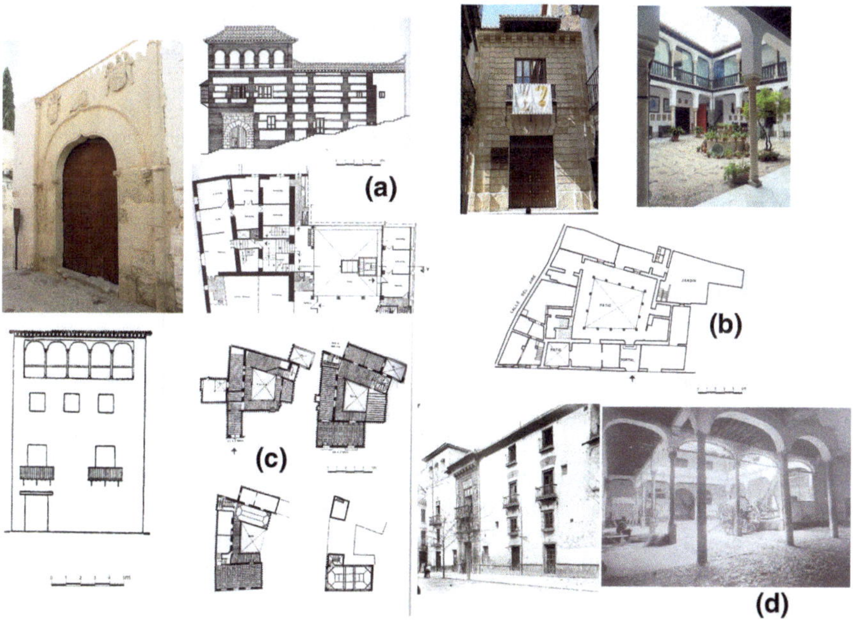

Fig. 2 Important mansions of XVI Century, Granada. **a** Zafra House, Calle Zafra, 5; **b** Pisas House, Calle Convalecencia, 1; **c** House located in Calle S. Escolastica and demolished in the early XX Century; **d** Cordova Palace, in Calle Sierpe Alta, demolished in 1919

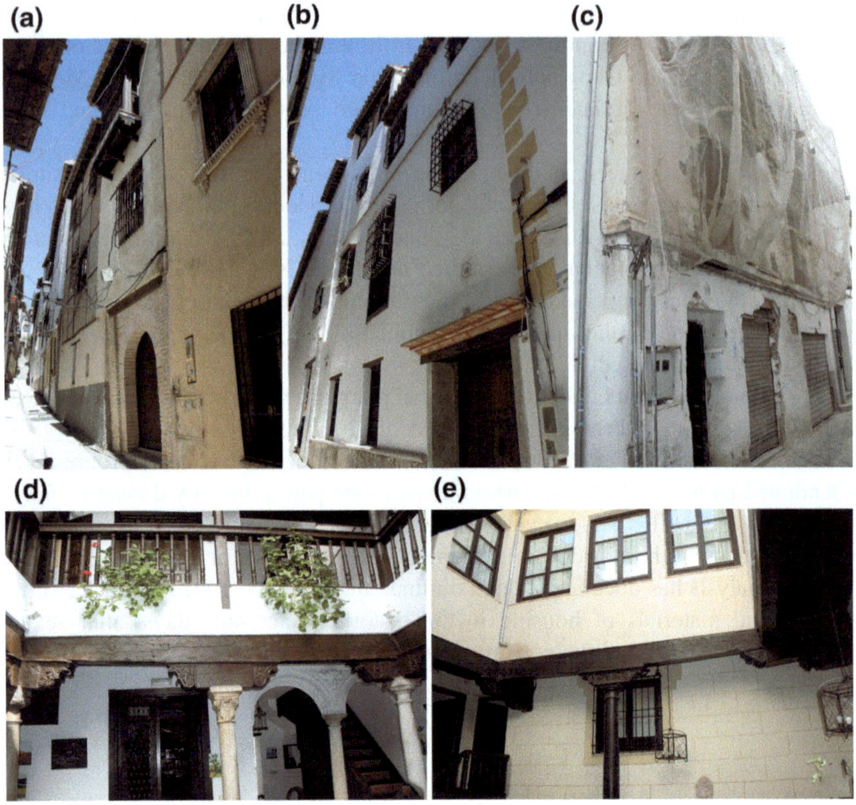

Fig. 3 Examples of rehabilitation works and new constructions in XVI Century mansions in the quarter Albaicin of Granada. **a** Calle Horno de Oro, 8, which still conserves a kind of traditional wooden balcony, although the existent has been realized recently, similar to the original one. **b** Calle Horno del Oro, 12, a typical transformed facade. **c** Calle S. Juan de los Reyes, 113. Deteriorated facade of a medieval house. **d** Internal patio of a house in Cuesta de los Aceituneros, 6, in a good state of conservation, although quite modified. **e** Internal patio of the mansion located in Carrera del Darrio, 9, very modified, without the upper wooden gallery

The morphological analysis has been focused as a typological study of the existing historic buildings, with a special attention on Al-Andalus remains over which the new householders constructed or adapted their mansions. Thus, the research also dealt with some special multicultural characteristics of the constructive techniques, as well as of the society of that period. The basic type of Castilian and Andalusian houses was that organized around a central patio [26]. The hypotheses brought forward have been supported by the archaeological documentation that has been elaborated during the excavation works realized during the last three decades in the houses concerned in this study (Fig. 4).

Moreover, during this work, the more recent transformations that these historic buildings suffered over the contemporary period have been pointed out.

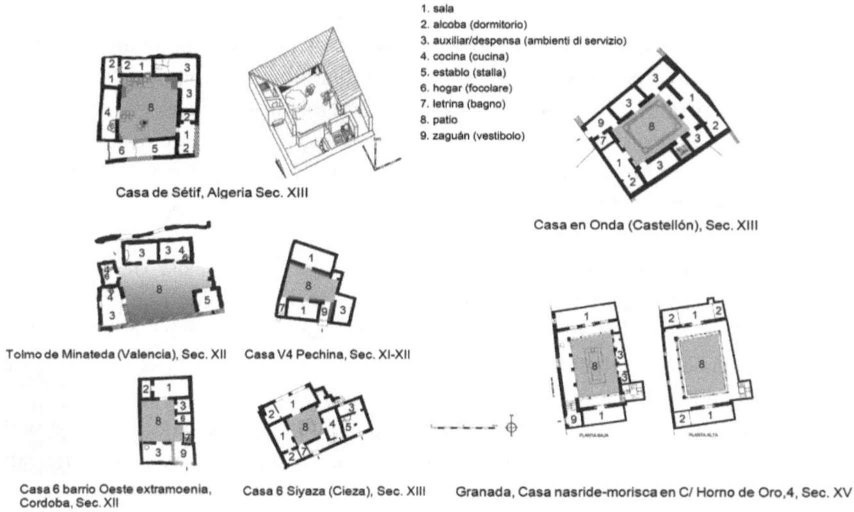

Fig. 4 Analysis of the functional specializations of the medieval patio house in Andalucía and Maghreb. The pictures on the left have been extracted from: Gutierrez Lloret, S. (2012) Gramática de la casa: perspectivas de análisis arqueológico de los espacios domésticos medievales en la península Ibérica (siglos VII–XIII). *Arqueología de la Arquitectura* (9), pp. 139–164

Even nowadays its treasures are so exposed to severe deterioration that their cataloguing, documentation and appreciation as heritage are becoming crucial. The inscription of the Albaicin into the List of World Cultural Heritage by the UNESCO in 1994, was of none guarantee of its rigorous protection. The main line of interventions followed has been that of "facadism", keeping only the front of the house, and rebuilding entirely the rest.

This kind of intervention has been denounced and criticized several times by the experts worldwide, but it hasn't been strictly banned by the local government, and it is still being adopted [27].

Despite the fact that the ten surveyed houses are listed in the special protection plans ("*Planes Especiales de Protección*", Albaicín Plan of 1990 and the Town Centre Plan of 2002) their formal characteristics, as well as structural and decorative historical value, have suffered and are still suffering at the moment, gradual plunders or destruction. These buildings are not classified as *BIC* (the administrative symbol for Spanish "Bien de Interés Cultural", a category which defines national monuments), so aren't rigorously protected and it is vital to understand the restorations errors carried out on them.

This kind of analysis has outlined the needs of the city of Granada to formulate more correct and efficient criteria for its better conservation. It has been also possible to deduce the most important priorities and guidelines, following environmental and social sustainability, for the "green" conservation of the Albaicín, under a new socioeconomic paradigm framed into "pro-common and common good", "co-living" and "social joint responsibility" criteria.

The housing developed by ancient and Castilian society in XVI century Granada still have elements consisting of historic resources, on which an innovative strategy can be grounded, especially in relation to sustainability and bioclimatic features. The internal "patios" of the houses allow to develop open spaces, which can be used for social and productive activities (exhibitions, common use spaces, events, urban kitchen gardens, etc.). This areas can be also a guarantee for a good and sustainable quality of life, more sociable, cooperative and based on the sharing economy (Fig. 5).

Therefore, the Andalusian patio houses typology can consist of a powerful resource to develop collaborative projects of urban recovery, based on participative planning. This kind of projects can be fomented and financed by crowdfunding through open source platforms, which can be based on a webgis, open online laboratories and free sharing crowdsourcing pages [28].

The typology of the medieval Andalusian patio house is quite similar to that with a courtyard, present in contemporary Southern and Central Italy, and partly still nowadays. This courtyard can be lateral, on one side of the ground floor, and the domestic units are organized in a line towards the main streets, following a scheme, which is very common in Central and Northern Italy, or it can be central, like in Granada and in other historic city centers and small towns centers of Andalusia (Fig. 6).

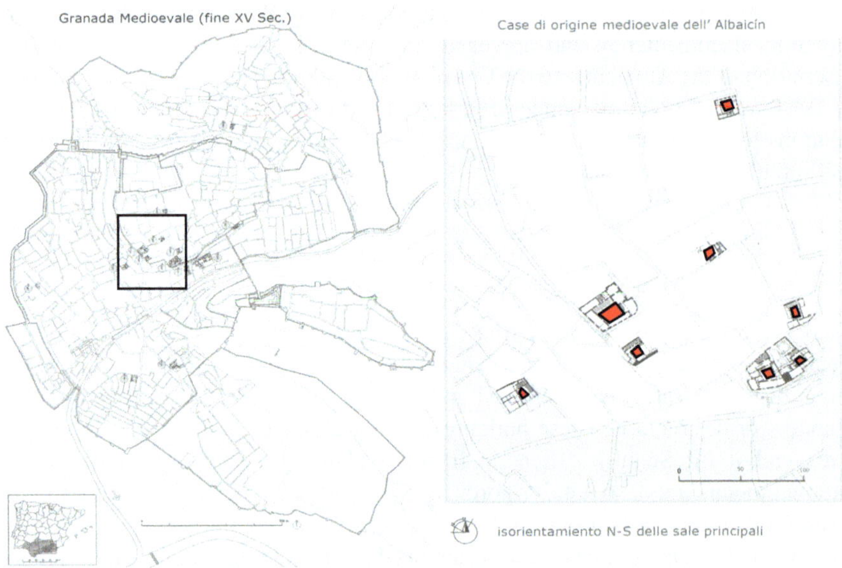

Fig. 5 Orientation N-S and internal patios (red color) of the main mansions in XVI Century Granada

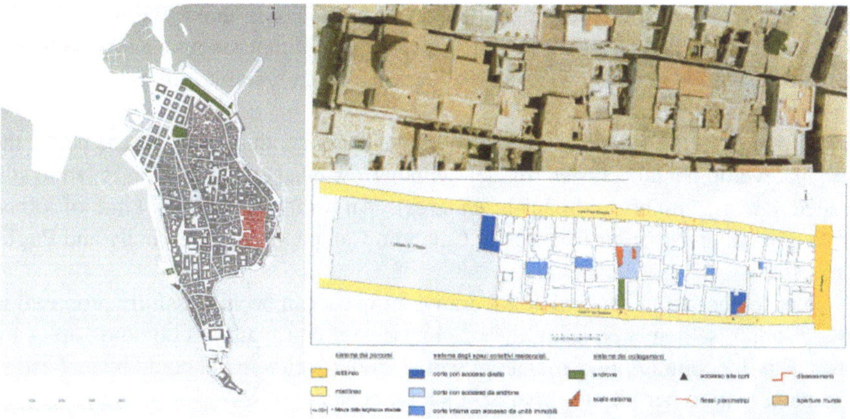

Fig. 6 Scalora, G., Monti, G. Siracusa, Sicily. Ortigia, Giudecca quarter. Individualization of medieval courtyards (blue and cyan hues), In: Scalora, G.; Monti, G. (2010) Op. cit., pp. 76–78

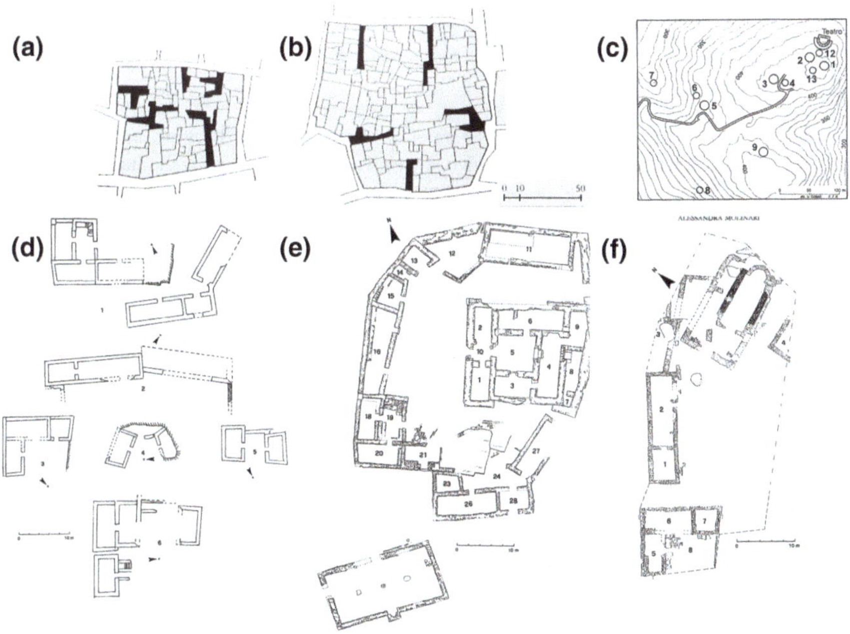

Fig. 7 Sicily. Urban pattern in Trapani (**a**) and Mazara (**b**), with narrow dead end streets (black color), from: E. Guidoni (1979) La componente islámica nella formazione delle cittá italiane. In: *Gli arabi in Italia. Cultura, contatti e tradizioni*, Milano, pp. 575–597, espec. p. 578. Segestas' rests of medieval mansions: Monte Barbaro quarter (**c**); 5th zone (**d**); castle area (**f**), from: MolinarI, A. (2000) Edilizia pubblica e privata nella Segesta medieval. In: *Castrum 6: Maisons et espaces domestiques dans le monde méditerranéen au Moyen Age, Collection de l'École Française de Rome*, 105–6, Roma, pp. 177–197, espec. pp. 180–184

This last type is more developed in Southern italy, or in that Central small cities where the kingdom of Frederick II, Holy Roman Emperor and King of Sicily, which was very connected with the Northern Africa and Arabic cultures, was settled during the XIII Century.

In this case the home units are distributed around this open space, being the urban development a "cluster" model, with narrow dead end streets (in Palermo still called *darbi*, similar to Spanish *adarves*). An example of this kind of urban structure still exists in the regions of Calabria, Campania, Molise, Sicily and Puglia (Fig. 7).

Peer to peer urbanism and collaborative projects can be successfully proposed in these Italian settlements, using the open spaces of courtyards as common areas for the co-living community, for example, in a similar way to what could be realized in Granada or in other similar Spanish historic centers.

Acknowledgements The author would like to recognize the contribution to the research on the Albaicin of Granada of the "Laboratory of Archeology and Architecture of the City" (LAAC) of the School of Arabic Studies of the Consejo Superior de Investigaciones Cientificas (CSIC) in Granada.

References

1. Charter of Cracow. In: International conference on conservation. Kraków, art. 8 (2000)
2. Atlante dei centri storici a rischio sismico, a webgis database created by the Italian Protezione Civile, in collaboration with the Department of Architecture of Roma Tre, under the direction of Prof. Arch. Paolo Marconi. Retrieved from: http://www.protezionecivile.gov.it/jcms/it/censimento_centri_storici.wp (2013)
3. Cacace C, Fiorani D (2014) Centri storici, vulnerabilità, rischio e gestione della conservazione. Una proposta d'implementazione dello strumento Carta del Rischio. In: Della Torre S (ed) Protezione dal rischio sismico. In: Proceedings of the conference "preventive and planned conservation", Monza, Mantova, 5–9 May 2014. Milan, Polytechnic and Nardini Ed, pp 107–117
4. De Fino M et al. (2013) Energy assessment and retrofitting of Mediterranean historical towns. In: Proceedings of the conference sustainable retrofitting, Sep, p 977
5. Giovannetti F (2011) Il futuro dei borghi storici è nell'arte di costruire premoderna. ARCO. (Associazione per il recupero del costruito). Progetti d'eccellenza per il restauro italiano. Cured by Alessandra Centroni, Maria Grazia Filetici. Roma: Gangemi ed., pp 143–154, espec. p 145
6. Piccinato G (2008) Fermoimmagine. Studio sulla felicità urbana. Quodlibet studio. Città e paesaggio. Album, Macerata
7. Cerasoli M, Rizzo, B. (2015) Il futuro tecnologico dei centri storici. Cerasoli M (ed), Libro degli Atti del 9° Congresso Città eTerritorio Virtuale - Città Memoria Gente. Roma, 2–4 ottobre 2013. Roma: Roma TrEpress, Università Roma Tre
8. Cataldi G (2003) Da Muratori a Caniggia: matrici, derivazioni e fondamenti disciplinari della scuola italiana di tipologia progettuale. In: D'Amato Guerrieri C, Strappa G (eds) Gianfranco Caniggia: dalla lettura di Como all'interpretazione tipologica della città. Proceedings of the Conference (Cernobbio, 5 July 2002), Mario Adda, Bari

9. Scalora G, Monti G (2010) La conservazione dei centri storici in zona sismica. Un metodo operativo di restauro urbano. Accademia Universa, Varese
10. Permaculture design, a step by step guide. Aranya (2015)
11. Espejo Fernández J (2012) Tratamiento de la movilidad urbana en centros históricos en rehabilitación 722-TMF-289. Unpublished degree thesis, School of Civil Engineering of the Polytechnic University of Catalonia (UPC) Barcelona
12. Ibarlucea Bustamante E (2001) Cascos Históricos: regeneración urbana. El caso de Bilbao, in: Azkoaga, Cuadernos de ciencias sociales y económicas, (8), Bilbao
13. Gigliarelli E et al. (2013) Cultural heritage, urban planning and smart city. In: Proceedings of the 6th international congress "science and technology for the safeguard of cultural heritage in the Mediterranean Basin", Athens, Greece
14. Meloni C (2012) Prove di Smart City per l'Aquila. In: Proceedings of the workshop-smart, sostenible e sicuro: il futuro dei centri storici. Lucca, 18 Octobre 2012, ENEA event at International Congress Lu.Be.C. 2012
15. Cerasoli M (2010) Il recupero dei centri storici: la modulazione della tutela. Planum, pp 1–10
16. Filetici M G (2011) L'arte del costruire al servizio della conservazione: ma a quali condizioni? ARCO. (Associazione per il recupero del costruito). Progetti d'eccellenza per il restauro italiano. Centroni A, Filetici MG (eds) Gangemi ed, Roma, pp 29–34
17. Carbonara G, *Introduzione*, in Scalora G, Monti G (2010) La conservazione dei centri storici in zona sismica. Un metodo operativo di restauro urbano. Accademia Universa, Varese, p. VI
18. Cerasoli M, BiereArenas, RM (2016) The sustainable future of the smaller historical centers, between "modulation of the protection" and new technologies. In: Virtual city and territory. "Back to the sense of the city: international monograph book", Centre de Política de Sòl i Valoracions, Barcelona, Spain, pp 297–312
19. Centroni A, Manieri Elia G (2011) Il Museo Diffuso e la riqualificazione del centro storico di Arsoli (RM). In: ARCO. (Associazione per il recupero del costruito). Progetti d'eccellenza per il restauro italiano. A cura di Alessandra Centroni, Maria Grazia Filetici. Roma, Gangemi ed, pp 119–128
20. Scalora G, Monti G (2010) La conservazione dei centri storici in zona sismica. Un metodo operativo di restauro urbano. Accademia Universa, Varese
21. EPA (United States Environmental Protection Agency) (2015) How small towns and cities can use local assets to rebuild their economies. Retrieved from: https://www.epa.gov/sites/production/files/2015-05/documents/competitive_advantage_051215_508_final.pdf
22. Frieman Y (2012) Bottom up. [online]. Press floder of the 7th edition of the International Festival of Architecture, EME3 2012. Barcelona, 27 June 2012. [consulted 20 July 2013]. Retrieved from: http://www.eme3.org/wp-content/uploads/2012/03/DP-eme3_2012_es_26-06-2012-BD.pdf
23. Abbate A, Gentili A, Minervino G, Serafini S (2013) Gli spazi comuni post-bellici e l'identità biourbana di Artena. In: Walter Fabietti, Carmela Giannino, Marichela Sepe (eds) La ricostruzione dopo una catastrofe: da spazio in attesa a spazio pubblico. Proceedings of the Workshop Biennal Public Space 2013. Urbanistica Online Dossier (5), pp 113–116
24. Pica V (2016) Houses of the Castilian Ruling Class in XVIth Century Granada. Typologies, adaptation and urban context. Conservation criteria. Ph.D. Thesis directed by Ph.D. architect A. Orihuela Uzal and codirected by architect E. Nuere Matauco, and lectured on 2 February 2016 at the Polytechnic University of Madrid, ETSA, unpublished
25. Torres Balbás L (1923) Granada, la ciudad que desaparece. Arquitectura: órgano de la Sociedad Central de Arquitectos (53), pp 305–318
26. Orihuela Uzal A, Almagro Gorbea A (2001) De la casa andalusí a la casa morisca: la evolución de un tipo arquitectónico. La ciudad medieval: de la casa al tejido urbano: actas del primer Curso de Historia y Urbanismo Medieval organizado por la Universidad de Castilla-La Mancha. Coord. por Jean Passini, pp 51–70

27. Di Stefano R (1989) Façadisme et speculation. ICOMOS information. Revue Trimestrielle. Edizioni scientifiche italiane, Napoli, pp 1–3
28. Pica V, García-Pulido JL (2013) La vivienda de raigambre andalusí y castellana como ejemplo de arquitectura doméstica sostenible: análisis de algunas experiencias en su rehabilitación. Actas de las Jornadas Internacionales de Investigación en Construcción. Madrid: Eduardo Torroja Institute for Construction Science (Spanish National Research Council), 21st-22nd November of 2013

Ichnography, Orthography and Scenography as Forms of Tracing the Past: A Reconstitution of the Roman *Forum* of *Ebora Liberalitas Julia*

Mariana Martins de Carvalho

Abstract The research presented in this paper is an attempt to understand the central core of a Roman city—the *forum*—from three different forms of representation: *ichnography* (plan), *orthography* (elevation) and *scenography* (perspective). Our laboratory is the monumental centre of Evora, located in the Alentejo, Portugal, where the urban ruins of the Roman temple are located. Here, we are attempting to trace the Roman forum of *Ebora Liberalitas Julia*, in the Roman province of *Lusitania*. In order to comprehend the remaining structure as a whole, a dissection of each fragment was carried out. This was analysed and all the archaeological remains were drawn over the current city plans. This was followed by hypothetical ichnographic, orthographic and scenographic projections which were started rising from this ground. These three forms of expression are the result of thought and invention, such as described by Vitruvius, in his chapter on the fundamental principles of architecture (I, II, 2). In these representations, it is possible to see how each one is increased with a dose of interpretation and analysis of hypotheses. The developed representations are supported not only by archaeological sources, but also by iconographic, literary and historic sources. To study and to draw Roman architecture in an urban context, it is essential to recognize the urban history and the phases through which each building has passed, in order to provide data that will add knowledge to the remaining architecture. Although the purpose of this study is to recognize Roman architecture, it is impossible, in an urban scenario such as this one, not to recognize other periods as well. Considering the importance of the sources and their semantic expression, this essay tries to fill the gap between different types of representation in order to increase the possibilities of a more accurate reconstitution. The ruins, and the archaeological remains that were exhumed, will be our experimental field to try to trace the presence of the past and the different configurations of this monumental centre, as carried out through evidence, knowledge and imagination.

An erratum/a correction to this chapter can be found at
https://doi.org/10.1007/978-3-030-11422-0_23

M. M. de Carvalho (✉)
Faculty of Architecture, University of Porto, Porto, Portugal
e-mail: msmcarvalho@arq.up.pt

G. Amoruso and R. Salerno (eds.), *Cultural Landscape in Practice*,
Lecture Notes in Civil Engineering 26,
https://doi.org/10.1007/978-3-030-11422-0_17

Keywords Urban ruins · Evora *forum* · Representation · Ichnography · Orthography · Scenography

1 Introduction

This essay is part of a research project[1] that studies the drawing of the observation and analysis of the ruin and its history, aiming to provide hypothetical representations of ancient architecture. In this way, drawing is considered fundamental to the process of knowledge, interpretation and perception of these Roman remains. This study finds its roots in the areas of architecture, archaeology and history, and intends to foster the sharing of knowledge between those disciplines using drawing as a research tool.

The ruin is the fragment of the 'monument', its remaining matter associated with the collective memory and the legacy of the past. The ruins "can, in most cases, play their own part, thanks to the *imagination*, which sees in them a sign of past events, granting them specific values. Thus, the ruins become sources for historical knowledge (…)" [1].

However, the remains alone do not bring understanding. They have to embody each significance and cultural memory to become intelligible. This intelligibility is brought about through all the documentation that complements the remains and that has been recorded over the centuries; that is to say its 'documents', its historical knowledge. Then, the 'monument' becomes more than its shape. It is also composed of all its iconographic and written records which carry its historical value. These 'memory materials' are the subject of work for understanding, representing and creating new significances for the archaeological remains. And drawing, in this study, is the privileged method to achieve this.

The remains of the monumental centre of Evora, provide the support for the creation of new 'documents' that contribute to the reading and invention of the 'monument'. Consequently, the available 'documents' of the archaeological remains of the *forum* have been collected, and from them an analysis and a critical selection have been made, thus producing new meanings and new documents for future studies.

Besides collecting the documents, it is also necessary to understand this architectural project using the prevailing language of that time. The *forum*, in Roman architecture, is an enclosed space, a group of buildings for collective use, a place of public utility. A place that is, according to Vitruvius, a "common place" [2] and also a "place of memory" [3], as Pierre Gros puts it. Therefore, religious, administrative, legal, commercial buildings, as well as commemorative monuments and honorific inscriptions were all gathered together in the same square.

[1]This paper forms part of my Ph.D. research, supervised by Prof. Pilar Reis and Prof. Marta Oliveira, at the Faculty of Architecture of the University of Porto (FAUP) and funded by the Portuguese Science and Technology Foundation (FCT).

This space reflected a "specific architectonical image, adapted to its functionality, or rather to all the functions that from part of urban life" [4]; and it reproduced the identity and history of the city, promoted by its citizens using the *Urbs* as a reference, through its architectural and iconographic programme. In the provincial cities this was very visible, as in our case, the *forum* of *Ebora Liberatitas Julia* (Evora), situated in the *Conventus Pacensis* in the *Lusitania* province, and, especially, in the *forum* of its capital, *Emerita Augusta* (Merida), that was built in the image of the capital of the Empire.

The *forum* of *Ebora* must have been designed starting with the choice of the architectural order, followed by the construction technique—considering the local materials available—and by the composition of the buildings and open areas. In this case, those choices reveal knowledge and dexterity related to the classical language and decorative vocabulary, but at the same time they are enriched by other choices, creating a more varied ornamental grammar, a result of the knowledge of the tendencies and the characteristics of the site.

2 Towards a Hypothetical Reconstitution

The action of drawing is indeed interpretative, from the first survey to the presentation of a hypothetical virtual model, and the ways of seeing of each author are crucial. We wish to comprehend, through the example analysed, the tools of analysis applied to the archaeological, historical, architectural and urban interpretation in order to reconstitute an image of this Roman architecture.

Starting from the knowledge learned about the history, the iconography, the monumental architecture of the ancient city and the analysis of excavation reports, undertaken until the present day, as well as identification of the known virtual reconstructions, a hypothesis will be made to reconstitute the spatial configuration of the *forum* complex. In this operation, it was necessary to compare the *forum* and its formation with other architectural complexes, to try to find those that suggest similar characteristics and solutions, in accordance with the scale of the Iberian provinces.

In this case, since it is integrated within a consolidated city and not in an excavated archaeological site, the challenge will be even more speculative, raising possible hypotheses that must later be validated by archaeology. For this essay the archaeological remains of the sacred area of the temple (*temenos*)[2] will be the experimental field for a hypothetical reconstitution.

The first excavation work around the temple was carried out by the scholar Joaquim Cunha Rivara, in 1845 [5]. The area in front of the temple and the surrounding tank was exhumed but soon was left to oblivion. It was only in the 1980s that this monument was studied again, thanks to Theodor Hauschild, of the *Deutsches Archäologisches Institut*. This architect coordinated major excavation works for ten

[2]The temple of the forum is from the imperial age and it was probably built during the time of Claudius-Nero.

years which sought to locate the *forum* deployment area [6]. In these excavations the surrounding tank was partially excavated, together with an ancient cistern and some parts of the cryptoporticus. The works proceeded to the south searching for the Roman square, and reaching the ancient Archbishop's Palace, nowadays Evora Museum [7]. A few minor excavations were made in this area specifically for monitoring infrastructure installation works in the city. An attempt to excavate the whole tank was also undertaken, but this project was not fully implemented given the danger to the temple's stability and the poor conditions of the remains [8].

3 Drawing as a Research Tool

"The basis for all the work the architect has to do, either in the study of an old building or in excavations, is rigorous documentation by surveying plans and elevations" [9].

The focus now passes to the remains from the Roman times. However, remains of other times cannot be neglected in this analysis. The monumental area allows us to zoom in and to study the successive reuse of this space, acknowledging which structures were continuously maintained and which were abandoned.

The large record of archaeological remains from the city's most decisive historical periods (since Islamic, medieval and modern times), provides a more detailed and comprehensive analysis of the fragments. These records do not consist of a plan involving a single layer. They create a succession of layers corresponding to different time periods, that must be studied through sections, starting from the ancient period and continuing up to the present day city levels. The idea of a city made from fragments is something very well recognized in Evora. This approach to the existing fragment and simultaneously to the previously existing buildings allows for a clearer reading of this continuously transformed area.

In this attempt at representation based on a previously stipulated grammar, we will try to verify hypotheses of urban patterns and interpretations of the existing structure. We seek to distinguish what is archaeological evidence and what is conjecture, which may be expressed by a line type, by a hatch, or through a detailed caption/key, depending on the drawing.

It is important to contextualize the interpretation drawing within the so-called 'reconstitution' drawing. It is considered that the reconstitution drawing covers several types of representation. An interpretation drawing may be a reconstitution drawing; however, it is not intended for dissemination. It is an analysis and a research process developed over an archaeological object, a practice of thinking about the space and its materialization. It is a drawing of 'restitution' but never, as sometimes called, 'reconstruction'. This can evolve into more widespread means of communication through becoming a scenographic view, where 'simulation', 'recreation' and 'projection' are worked on.

An interpretation drawing is usually a linear representation. It must always guarantee the accuracy of the structures which have been found and it must be aware of their gaps and the levels of 'imagination' to which they have been subjected. The drawing is expected to be executed in a wise way, conscious of the tectonics of materials and spatial composition. But sometimes these reconstitutions: "are too often cursorily done, with little thought and research given to the archaeological evidence, the nature and date of the building, the functions of different parts of the building, parallels form other sites, and the architectural validity of the reconstruction" [10].

It is necessary to know how to work with different levels of uncertainty because doubt grows as the building rises. When the data for reconstitution are not sufficient, the drawing must create room for interrogation, leaving some parts of the uncompleted built structures. And if the data are not conclusive testing using several reconstitution hypotheses may be considered. The result is an open drawing, able to receive new data that may appear in the future. This contributes to knowledge of the site and to archaeology itself, as it can encourage the carrying out of excavations in certain places to prove (or not) a hypothesis.

Interpretation begins with the recognition of each part of the building and, if there is enough information, this can be extended to speculation about the whole city. For this, it is also necessary to know its topography, in order to understand the urban layout. However, one may identify clues in the existing urban fabric to try to identify a possible Roman pattern.

Thus, besides the recognition of the excavated and visible remains, several other inputs are involved in the drawing, the urban morphology of the existing city being one of the most important.

In the work being presented here, during the recognition of the Roman remains of the city of Evora, we were confronted with an enormous variety of drawings. From these, a subtraction exercise was carried out highlighting the Roman remains. Sometimes, this choice is almost impossible, so room has to be left for the interrogation that will follow all the strands of the drawing. These interrogations will be translated into somewhat rebuttable hypotheses that will enrich the entire process of interpreting the traces.

To assist this analysis, we also used the available surveys of the interiors of the buildings in order to contribute to their interpretation and, in fact, there are some overlapping incidences.

4 Forms of Representation

When Vitruvius describes the fundamental principles of architecture he explains that the forms of representation are: ichnography (plan), orthography (elevation) and scenography (perspective). Given this, we tried to create a possible image of the sacred area following this method.

The ichnographic and orthographic projections were made by assembling different sources, especially iconographic and literary documents as well as archaeological sources. The scenography arises from all these sources, with some added imagination.

4.1 Ichnography

Aiming to create a reconstitution drawing we compiled a reliable work base by collecting all the known archaeological traces and representing them on the plan of the existing monumental centre of the city. To do that, it was used the aerial photogrammetric survey of the city and the survey drawings of the surrounding buildings (namely, the *Lóios* Convent, the Inquisition Courthouse, and the *Cadaval* Palace) to create a base, on which we introduced, one by one, all the archaeological records available in the reports of the excavations carried out from 1987 to 2012 (Fig. 1).

These drawings could not be more unlike, because they were made by different archaeologist at different times. It was therefore necessary to find ways to create a common language between them in order to gain an accurate perception of the

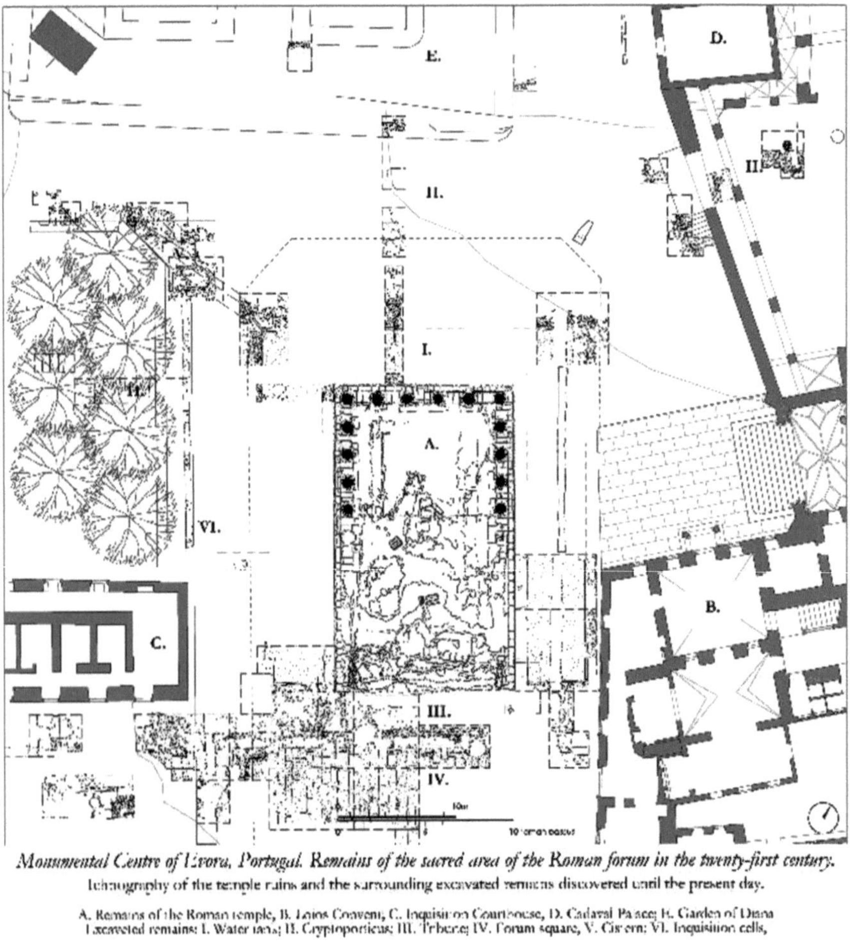

Fig. 1 Ichnography of the temple ruins and the surrounding excavated remains

remaining structures and of the constructions that was probably above them. We worked with diverse layers of chronological information, and each one of these was able to provide important clues for the reconstitution. Sometimes, these layers are compatible, other times they are contradictory. Such confrontations are also important for the interpretation.

After the field data was gathered, it was possible to speculate about the interpretation of the remains. This was the phase of conjecture, of raising hypotheses; a process which refers not only to archaeology, but also to history and architecture. Therefore, the whole operation has to be supported on the theoretical study of classical architecture based on the available treatises, as well as on the observation of the geographical links concerning the Iberian context, to confer scientific accuracy and the argumentation that the interpretation requires (Fig. 2).

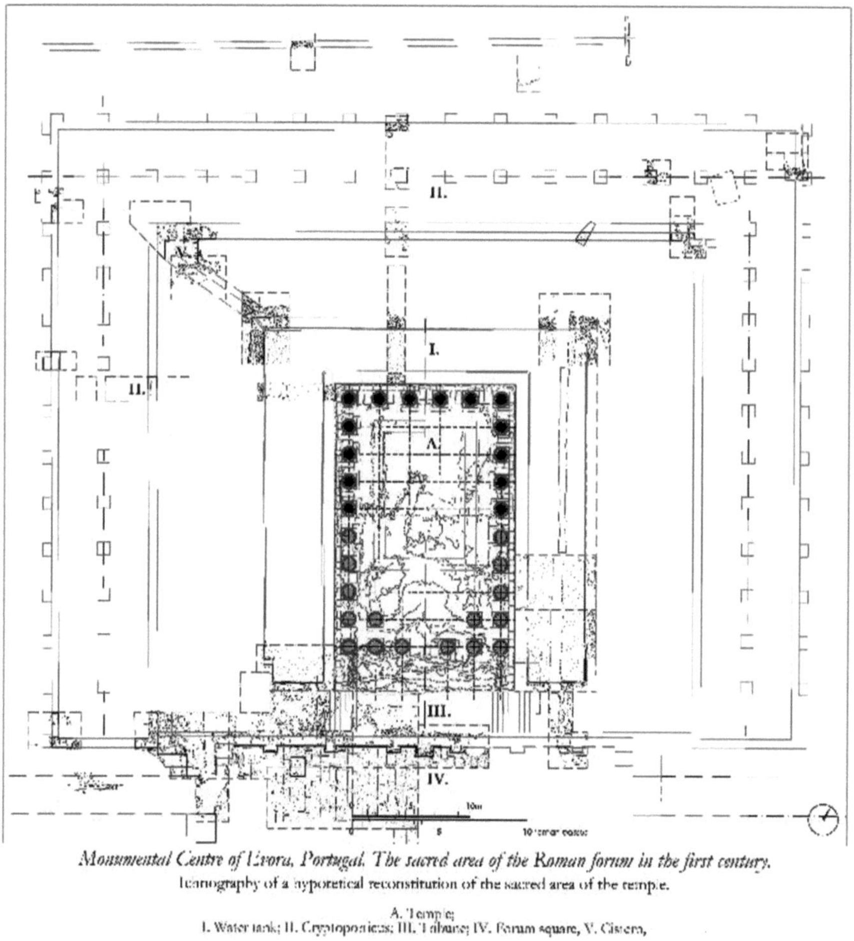

Fig. 2 Ichnography of a hypothetical reconstitution of the sacred area of the *forum*

Thus, the interpretation drawing is placed between the real and the conjectural. It will never be 'true', but only an effort at representation, based on rigorous study and knowledge about the ruin. It presents the known data: the pieces and the fragments. However, here, the records and knowledge blend with intuition and deduction, seeking to recreate the hidden and destroyed information. For example, the number of columns on the side facades of the temple is a mystery. Supposedly, a hexastyle temple should have 11 columns [11] but this was almost impossible to have occurred because of the wall with pilasters that was found in front of the temple. This wall defined the tribune, an elevated platform from the Roman square. Furthermore, the remains of the *opus caementicium* of the *podium* tend to indicate a central staircase starting in the tenth column. A ten columns' façade is not canonical, but it seems to be the only possibility considering the archaeological record. Therefore, we tried a hypothesis of the temple with 6 by 10 columns, with a central staircase and two more staircases on each side. Between the temple and the Garden of Diana, besides the archaeological remains of the tank, the north gallery of the cryptoporticus was also found which involved all the sacred area around the temple. In this excavation, the boundary walls of the cryptoporticus (the floor level was approximately two metres below the current level of the street) were fortuitously found. Between those walls a pillar foundation was also found which could probably be the middle structure of the cryptoporticus. This shows remarkable similarities with the structures in the *forum coloniae* of Merida. What is more, thanks to the excavations undertaken in the courtyards of the Inquisition Courthouse and *Cadaval* Palace, it has been possible to trace the limits of the U-shaped cryptoporticus. Crossing the layers of chronological information, it can be seen that the boundary wall of the courtyard of the Inquisition Courthouse was built over the remains of the cryptoporticus. During the archaeological excavations, many structures were found dating back to the modern era, that belonged to the old Inquisition Palace, some of them coinciding with the Roman walls. In addition to these findings, the remains of an old cistern, deactivated in the middle of the 1st century AD, were also found.[3] On the west side of the temple structures from the modern era, belonging to the cells of the former Inquisition Courthouse, were found.

4.2 Orthography

With the temple of Evora, we had the chance to work with the orthographic projection, due to the partial preservation of the north facade. We will try to disassemble and interpret the remains so as to be able to reconstruct a hypothesis of this

[3]The existence of water was an essential element in the Roman city. Besides the supply necessary for daily use, water would have a connection to worship, and its sacred character was represented in the public and sacred areas of the city.

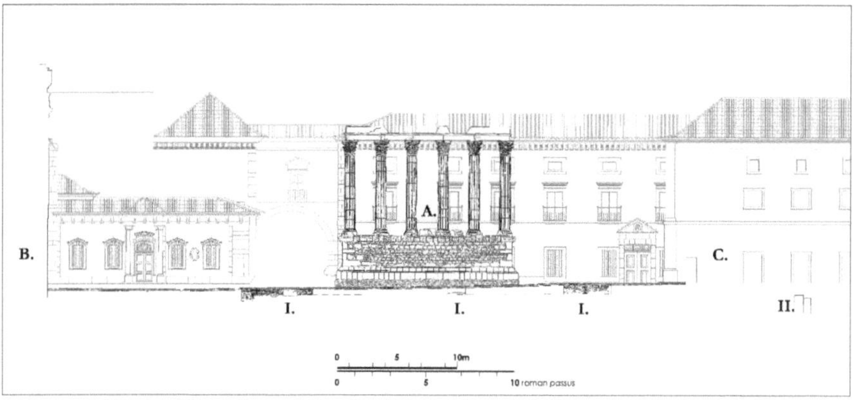

Monumental Centre of Evora, Portugal. Remains of the sacred area of the Roman forum in the twenty-first century.
Orthographic projection of the north elevation of the temple ruins and the surrounding excavated remains.

A. Remains of the Roman temple; B. Loios Convent; C. Inquisition Courthouse;
Excaveted Roman remains: I. Water tank; II. Cryptoporticus;

Fig. 3 Orthographic projection of the north elevation of the sacred area

side of the sacred area. By having at our disposal the photogrammetric survey of the temple, we have drawn the orthographic projection over it. This linear drawing provides a more precise representation and a more reliable information which can be used as a basis for further interpretations. With the available surveys the elevations of the surrounding buildings were also drawn (Fig. 3).

Over the survey drawing, we have superimposed a reconstitution hypothesis, where the thickness of the Roman coating materials were considered, and an attempt to fill the remaining stones with mortar was made: the bases and the capitals were 'restored' and the shafts were created; the regular stereotomic coating of the podium refers to antiquarian representations [12]; in the entablature the remains of the architrave and the frieze were also filled with mortar. The architrave fasciae were drawn and the frieze was left without decorations because there is no evidence of the original decorative elements. The cornice and the pediment were an invention made from similar models and, of course, also from the treatises. After having a hypothesis of the orthographic projection of the temple, we then proceeded to deal with the surrounding area. Based on the archaeological evidence there is the configuration of the bottom of the tank, made with *opus signinum*. We do not know what the level of the *temenos* was so it is guessed that it could have been between the foundation and the base of the podium. The remains found in the Inquisition Courthouse provided the information for a hypothetical reconstruction of the cryptoporticus section. The walls are too destroyed to know if there were some kind of elevated windows like the ones discovered in the *forum coloniae* of Merida. This is also the case with the covering. In the interpretation drawing we have considered that this was made with vaults, as in the well-maintained cryptoporticus of the *forum* of *Aeminium* (Coimbra). Above the half-buried structure there would have been a double portico. If the cryptoporticus structure is extended, it is possible to try

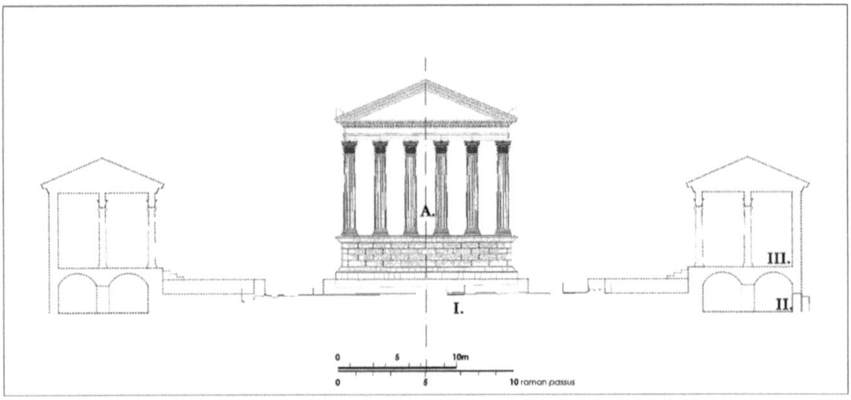

Monumental Centre of Evora, Portugal. The sacred area of the roman forum in the first century.
Hypothetical orthographic projection of the north facade of the temple.

A. Temple;
I. Water tank; II. Cryptoporticus; III. Portico;

Fig. 4 Orthographic hypothetical projection of the north elevation of the sacred area

to configure the portico. The portico was elevated which means there could have been a staircase or a wall coronate with a balustrade. There is no evidence of this portico; this is just an attempt based on similar examples, such as the flavian *forum* of *Conimbriga* (Fig. 4).

4.3 Scenography

The scenographic drawing differs from the iconographic and orthographic drawing because it adds matter that helps its understanding and contextualization; it becomes a privileged means of transmitting the evocative image of the ruin, since it graphically recreates its architecture and also explores the ambiences and the environment of that time.

In this way, these are drawings that, generally, contain plenty of information; in addition to spatial perception, animations, characters, textures, colours, decorative elements, among others, are introduced. This recreation tries to approach an imagined 'real'.

It is a riskier drawing (and it is important to bear this in mind), both for the individual conceiving it and the person observing it, since it always refers to an evocation of something that, however much data and remains there may be, will always be filled with fiction.

However, the impossibility of obtaining a reliable image of the building does not make it less important, since it provides greater knowledge about the observed object and, above all, allows us to question the reconstitution, to discuss it and to

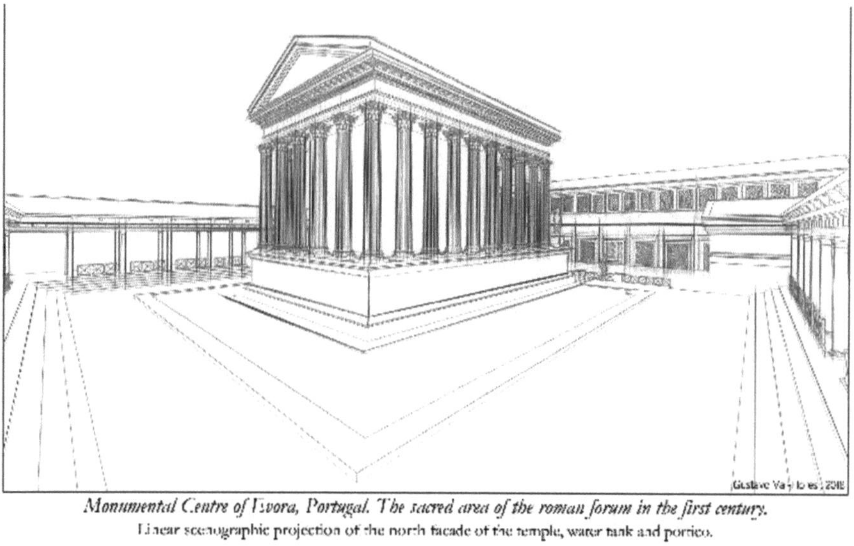

Fig. 5 Linear scenographic projection of the sacred area

develop it, so that the next one can be even more 'real': "[w]hatever the amount of evidence, it is usually worthwhile attempting a reconstruction, since this can lead to new ideas and highlight problems, which in turn will raise standards of excavation and research" [13].

Two scenographic views from the same point of view were conceived especially for this essay, to try to explain the gradation of 'imagination' that these forms of representation may have. The first is a linear image, just like the projections made earlier, which provides a more accurate model of the architectural shape (Fig. 5). However, the great potential of this kind of model is to create an imaginary narrative, and so a second view is also presented that constitutes a challenge to capture the atmosphere of this sacred area (Fig. 6).

5 Drawing the Ruins, Inventing the Past

To study and to represent Roman architecture in an urban context, it is essential to identify all the recognizable remains within each excavation. These overlaid structures provide data that will add knowledge about the remaining architecture.

The *forum* of *Ebora* allows several readings, and this merely forms a contribution that punctuates the information given so far. Nevertheless, this knowledge, grounded in history, archaeology and architecture, may form the basis for other studies. This attempt seeks to contribute in some way to the compilation of this

Fig. 6 Scenographic projection of the sacred area

information. In a domain like this it is impossible to obtain certainties. Here one more hypothesis was tested that tried to include new data. However, this is just one more moment in the narrative of this complex; future research will add more information and more accurate perceptions.

Nonetheless, the main purpose of this exercise was not to produce a 'real' image but to valorise the importance of the sources, and how these determine the interpretation. As we go on developing the four dimensions, more difficulties occur but the more challenging becomes discovery. Thus, the process of interpretation deals with uncertainty. Hypotheses are thrown up, but at the same time doubts remain. What is at stake is to elucidate and enhance the archaeological heritage as a form of knowledge. The field of interpretation is somehow free because it does not want to prove anything, but rather to instigate awareness of the importance of the practice of structured and articulated excavations.

In this case we have considered the *forum*, a dominant place in the Roman city. Here three forms of representation were tested. In the ichnographic representation we developed a trace map, from a collection of records from various sources, enabling the development of a more accurate interpretation. In this case, the representation of the collected data is the most important, and the hypothesis of the configuration of the Roman *forum* is limited to a very thin grey line. The orthographic projection provides the possibility to test the hypotheses raised by the iconography. The first orthographic representation presents the fragment data placed on the current section of the city. Above this, a conjectural elevation is made, with a dashed line to indicate a possible configuration of the building. To close, two scenographic views, supported by the previous representation forms, present a fictional image but with significant support from available knowledge.

The specificity of this drawing shows a method of representation, a process that starts from a rigorous drawing and proceeding towards an increasing level of conjecture. As such, is very important to have a solid foundation in the beginning, a precise ichnography and orthography, in order to create a more sustained scenography. This interpretation process offers new images of the past, new narratives about a specific place. And, in this sense, it contributes to the collective imaginary of that place, to the continuous construction of a cultural landscape.

Acknowledgements The author would like to recognize the valuable contribution of Gustavo Val-Flores, Municipality of Évora, the creator of the virtual model of the *forum*. Assignment co-financed by the European Regional Development Fund (ERDF) through the COMPETE 2020 —Operational Programme Competitiveness and Internationalization (POCI) and Portuguese national funds from the FCT under the POCI-01-0145-FEDER-007744 project. Illustrations from 1 to 4 are by the author, 2018; 5 and 6 are virtual models elaborated by Gustavo Val-Flores, 2018.

References

1. Carena C (1984) Ruina/ Restauro. In: Gil F (ed) (1984) Memória—História, Enciclopédia Einaudi (Vol 1). [Lisbon]: Imprensa Nacional—Casa da Moeda, p 129
2. Vitruvius (I, III, 1), Maciel MJ (2009) Vitrúvio: tratado de arquitectura. IST Press, Lisbon, p 40
3. Gros P (2011). L"Architecture Romaine: début du IIIe siècle av. J.-C. à la fin du Haut-Empire (3ª ed. Vol 1. Les monuments publics). Paris, Picard, p 207
4. Reis MP (2010). Tanques, fontes e espelhos de agua nos fora lusitanos. Studia Lusitana, p 286
5. Rivara JC (1845) O Templo de Diana e o Aqueduto de Sertório em Évora. Biblioteca Pública, Évora
6. Hauschild T (1991) El Templo Romano de Évora. Cuadernos de Arquitectura Romana, 1, pp 107–117; Hauschild T (2001) Évora. Relatório preliminar sobre as escavações junto ao templo romano, 1989–1992. As construções. Lusíada. Arqueologia, História da Arte e Património, Série 1, pp 69–91, Hauschild T (2010) Algumas observações nas construções do foro de Ebora Liberalitas Iulia. Ciudad y foro en Lusitania Romana, Vol IV, pp 27–36; and the recently published report: Hauschild T, Teichner F (2017) Der römische Tempel in Évora (Portugal). Bauaufnahme, Ausgrabung, Wertung (Vol 35). Wiesbaden, Reichert
7. Gonçalves A, Hauschild T, Teichner F (1997) Intervenção Arqueológica no Museu de Évora, Centro Histórico de Évora, 1996. In Actas do 3º Encontro de Arqueologia Urbana, Almada, pp 123–136
8. Hauschild T, Sarantopoulos P (1995–1996) O tanque de água do templo romano de Évora. Notícia preliminar da intervenção arqueológica de 1996. O Arqueologo Português, 4ª serie, pp 429–440
9. Hauschild T (1981) El trabajo del arquitecto en las excavaciones. Caesaraugusta, p 150
10. Adkins L, Adkins RA (2009) Archaeological illustration. Cambridge University Press, Cambridge, p 137
11. Vitruvius (III, II, 6), Maciel MJ (2009) Vitrúvio: tratado de arquitectura. IST Presss, Lisbon, p 113

12. Carvalho MM (2018) From the ruin to the temple, passing by the butchers. A reconstitution of the "Temple of Diana" in Evora, based on literary and iconographic sources. In: Amoruso G (ed), Putting Tradition into Practice: Heritage, Place and Design, Milano, Springer, pp 147–159
13. Adkins L, Adkins RA (2009) Archaeological illustration. Cambridge University Press, Cambridge, p 133

Shortsighted Solutions Versus Long Term Planning

Effects of Rapid Infrastructure Developments in the World Heritage Site of Gjirokastra, Albania

Kreshnik Merxhani and Valmira Bozgo

Abstract It is a case study of an approved infrastructure project for a new road known as the "Bypass", which endangers the Historic Center of Gjirokastra, a World Heritage Site since 2005. The project's initial aim was the diversion of automobile traffic from the city's Old Bazaar, for a more tourist friendly pedestrian area. After few procedural problems, the project revealed a diversion form the original aim and technical solutions that posed a threat to the *Outstanding Universal Values* for which the city enjoys its UNESCO World Heritage status. This generated a wide and largely publicized debate among professionals opposing the development, and the government representatives that put through the proposal. Several technical and legal arguments were presented to responsible national and international authorities. Currently the project is in a "limbo" state, having been suspended by means of several media declarations from national authorities, while still being depicted in the local development plan. The case study will be discussed for its technical and professional problems, disagreements with the priorities of the historical center and the threat on the *Outstanding Universal Values* of the city. The bypass case, will be further utilized to illustrate the effects that shortcut solutions with no sound bases on strategic studies, can have on an urban historical landscape. Possible resolutions will be discussed for breaking the trend of rushed "politically impressive" developmental projects that present negative ramifications on historical landscapes.

Keywords Historical urban landscape · World heritage site · Infrastructure development · Heritage under pressure · Mobility and management plan

---

K. Merxhani (✉)
Polytechnic University of Tirana, Tiranë, Albania
e-mail: ark.kreshnik@gmail.com

V. Bozgo (✉)
Environmental Planner, Tiranë, Albania
e-mail: vbozgo@gmail.com

G. Amoruso and R. Salerno (eds.), *Cultural Landscape in Practice*,
Lecture Notes in Civil Engineering 26,
https://doi.org/10.1007/978-3-030-11422-0_18

1 Introduction

It has been 57 years since Gjirokastra has been declared a Museum City. Since then, changes and challenges in the management of the site have been numerous. After 1990 a very different set of challenges began, taking into consideration the serious political and socio-economical shifts that the whole country experienced. Abruptly Gjirokastra saw the dissolution of the administrative and professional structures that administered the conservation of the Museum City, while just as confusing were the changes in legislation and the role of the state in this process. In 2005 Gjirokastra was listed as a World Heritage Site, being in compliance with UNESCO's criteria (iii) and (iv):

> **Criterion (iii):** Berat and Gjirokastra bear outstanding testimony to the diversity of urban societies in the Balkans, and to longstanding ways of life which have today almost vanished. The town planning and housing of Gjirokastra are those of a citadel town built by notable landowners whose interests were directly linked to those of the central power.
>
> **Criterion (iv):** Together, the two towns of Gjirokastra and Berat bear outstanding testimony to various types of monument and vernacular urban housing during the Classical Ottoman period, in continuity with the various Medieval cultures which preceded it, and in a state of peaceful coexistence with a large Christian minority, particularly at Berat [1].

The world heritage site is valued for the embodiment of the diversity of the urban societies of the Balkans, and the longstanding ways of life, which have today almost vanished. However Gjirokastra fosters a living historic center, thus the preservation of the Outstanding Universal Values deserve constant and careful consideration. The town planning and housing of Gjirokastra are those of a citadel built by distinguished landowners that had high connections to the government of the time. The site contains various types of monuments and vernacular urban housing of the Classical Ottoman period, built in continuity with the various medieval cultures; demonstrating also a state of peaceful coexistence with the Christian minority, creating thus an architectural and urban ensemble that is deemed unique. The historic center is comprised of a total of 615 monuments enlisted as Ist (56 monuments) and IInd category (559 monuments) [2].

2 Development Plans

One of the challenges still present in Albania is rapid, unplanned development. For this reason in 2013 when a new political party took power in the country, a moratorium was imposed on all construction, except for those for the purpose of housing. After this the country went through an administrative reform that merged more than 300 small administrative units called communes into a total of 61 Local Government Units. Then new master plans started to be drafted for the development of these new 61 administrative units. However due to political reasons and pressures very little time was allowed for the drafting of these plans, while in fact they needed much studying and due diligence taking into consideration the diverse types of territories that the new municipalities enjoyed.

3 The Bypass

On November 10, 2015, the Albanian Government through the Ministry of Culture and the National Territory Planning Agency (NTPA)—opened a call for an international competition for the "Design of a Bypass for the UNESCO protected City of Gjirokastra". The objective and rationale of this project was to avoid automobile traffic from the Old Bazaar of the Historic city, turning this into a pedestrian zone as part of the Ensemble of the World Heritage. Thus, the initial aim of the project was to avoid the very activities that would compromise the historical and cultural values of the city, values that have put Gjirokastra in the UNESCO World Heritage List. However, based on the analysis of a group of experts of various backgrounds and fields it has been assessed that the project seriously damages the Outstanding Universal Values (OUV) of the city of Gjirokastra. This group of experts identified 5 main groups of arguments that divulge the crosscutting issues and the negative impact of the project as it was approved on September 13, 2016, by the National Council of Restoration [3]. During the only public consultation session held for the project, these shortcomings were presented however they were not considered further by the responsible authorities. At this stage a detailed technical report was drafted by the "Forum for the Protection of the Values of Gjirokastra", in collaboration with a wide participation of different professionals who studied the multi-level effects of this project. The Forum's report introduces 5 main group arguments and enough supporting evidence to demonstrate the jeopardizing effects that this project will have on the entire cultural heritage ensemble of Gjirokastra, if implemented. Below is an extract of the list of the identified technical and legal arguments, showing the perceived damages of this project to the structural integrity as well as the Outstanding Universal Values of the city of Gjirokastra.

3.1 Procedural and Administrative Problems

The project is in breach of the initial Terms of Reference, which specifically request a solution in the surroundings of the Historic Center and Buffer zone. However, the approved project footprint stands on a protected green zone inside the Historic Center and in a potential Archeological Zone. The winning project announced by the competition, and the approved version, significantly differ in the fact that the latter doubles the estimated cost, and has a different footprint (almost 100 m reduction). Also two Cultural Monuments (traditional houses), have been removed from the List of Monuments due to the fact that they stand on the footprint of this project.

However, the Administrative Court of the First Level of Gjirokastra has repealed the administrative order which removes the traditional houses from the cultural monuments list due to procedural violations.

3.2 Legal Problems

The project goes against (i) the Law on "Cultural Heritage" No. 9048, date 7.4.2003 (amended), specifically Article 29 that prohibits interventions and new constructions in Historic Centers and Protected Zones, (ii) the Albanian Charter on Restoration (D.C.M No. 426, Date 13.7.2007) criteria and principals that request any solutions for transport to be designed outside of the Historic Centers and the Bazaar Zone, (iii) the Regulation on the "Protection, Integrated Conservation and Management of the Historic Center and Protected Areas of the City of Gjirokastra" (D.C.M No. 619, date 7.7.2015), more specifically Article 5, 8 and 9 that prohibit any new and permanent constructions within the Historic Center and their green areas.

3.3 Technical and Professional Problems

There are at least eleven studies starting from the 1980s, showing the planned project area as geologically vulnerable and in need of immediate intervention. These studies also depict a deep geological tectonic fault passing underneath the castle, making this a problem area in terms of geology and seismicity.

Additionally, this project has a very high impact on the landscape, transforming one of the views which currently serves as the Emblem of Gjirokastra Municipality, the Gjirokastra Regional Council and several other organizations in the city. The bearing walls of the new planned road also visually compete with the Castle violating thus its esthetics and that of the surrounding monuments (image below). The project is not in accordance with the UNESCO recommendations, which suggest that a thorough transportation infrastructure study of the Historic Center of Gjirokastra be prepared prior to any decision for the "bypass". This recommendation is expressed in the decision adopted by the World Heritage Council on the 39th session in Bonn, Germany (28 June–8 July 2015). The project is not based and does not contain a detailed study of the infrastructural or mobility situation in the Historic Center and Protected Zone of Gjirokastra. Further potential negative impacts on the environment are also not taken into consideration by the project documents.

3.4 Disagreements with the Priorities of the Protected Historic Center of Gjirokastra

The project is approved notwithstanding the lack of a management plan for the Historic Centre (a repeated request from UNESCO) and the Protected Area. Furthermore the project is not in accordance with the priorities and urgent needs for preservation and conservation of the Historic Centre of Gjirokastra. The latest 2015 study prepared by an international NGO in collaboration with the local office of the Regional Directory of National Culture and Epoka University, shows that in the Historic Center of Gjirokastra there are 169 monuments of the I and II Category in a state of high risk for total collapse (Figs. 1, 2 and 3).

3.5 The Negative Impact of the Outstanding Universal Values (OUV)

This project directly affects both of these criteria by introducing an aggressive intervention that is unharmonious to the urban and natural environment, degrading the cultural values that are an integral part of the urban planning of the Historic

Fig. 1 Original proposal of the competition

Fig. 2 Current and approved proposal by the National Council of Restoration

Fig. 3 The view of the approved "Bypass" and how it competes with the castle by completely transforming the characteristic landscape of this area

Center of Gjirokastra, but also by demolishing two vernacular houses, while transforming the urban typology of the historic zone.

The severity of this project is intensified due also to the fact that the approved the Master Plan of Gjirokastra, includes the footprint of this Bypass, giving this project thus the highest level of approval.

Gjirokastra has been a World Heritage Site since 2005, and now, almost 12 years later, a new professional movement formed as a direct response to this damaging project, and is now asking for the repeal of the "Bypass" which is putting the heritage values of Gjirokstra under critical pressure. As a result, responsible national and international authorities have been called into action by means of public forums, technical meetings and official requests, in order to abolish the project. This measure of abolishment was promised in meetings, in the presence of the national and local authorities as well as in the media, however, we remain pending on the follow-up with the appropriate legal acts.

The of the newly created movement, is a more effective use of public money, to be put in service of the preservation of the unique and unrepeated cultural heritage ensemble of Gjirokastra. The so called "bypass", if planned right, can not only fulfill the original aim of diverting traffic from the Bazaar, but also to aid public services such as—waste collection and management, improvement of emergency and risk response, as well as the renewal of the urban transportation of Gjirokastra. A better allocation of this investment can be envisioned for the improvement of the infrastructural network, offering alternative solutions for mobility, but only following the two prerequisite studies; (1) the management plan for the Historical Center and the Protected Area of Gjiroastra, and the (2) the mobility/infrastructural study.

4 A History of Neglect

For many years before and after inscription in the World Heritage List, Gjirokastra has suffered from neglect and denial of the problems facing its cultural heritage, lack of policies, and misappropriation of both public and private funds. As a result, the city is now in imminent danger of losing its Outstanding Universal Value. Understanding Gjirokastra's critical state, in 2011 the UNESCO World Heritage Committee indicated that Gjirokastra would be placed on the List of World Heritage in Danger if it did not meet certain requests, including the development of a management plan: (http://whc.unesco.org/en/decisions/4490) [4].

However, none of these requests have been met (with the exception of a moratorium on new constructions in 2014). In particular there is, until today, no Management Plan, no Tourism Strategy, and no Conservation Plan that would meet essential criteria, which must be applied to such documents. Most recently, more than 200 vernacular buildings listed as cultural monuments of 2nd Category located in the core and buffer zone of the WH Property, have been deleted from the list without proper study, consultation with experts or transparent decision-making process, strapping the historic center of about one-third of its substance in cultural monuments and opening the way for further loss of authenticity and integrity [5].

5 Monument List Reductions of 2016

In March 2016 the Albanian authorities after a site assessment, reduced the list of monuments of the IInd category from 559 to 323 objects. This reduction according to the authorities was based on the lack of specific listing and requests from UNESCO through Reactive Mission of 2012. The civil society engaged in preserving the Outstanding Universal Values of Gjirokastra have repeatedly asked for the specific criteria used in this assessment, however no further information or clarification has been issued by the authorities on this issue.

From a careful consideration of the 2016 monument listing while taking into account the state of the monuments, we identify the following discrepancies:

1. The reduction of the number of monuments by 43%, leaving out object that deserve the monument status.
2. Discrepancies in object names when compared to the UNESCO or previous lists of the National Institute of Monuments. In all other lists monuments are known by the family name of the property owner which in most cases is a patriarch or the grandfather of the current residents.
3. New monuments appear in the list for the first time. Some of these are illegal construction, built during the1990-ies, containing none of the characteristics of the monuments of the Historic Center and the Protected Zone. While monument embodying OUV, have been delisted.
4. The listing names 35 monuments with very vague descriptions such as “2–3 floors stone building with gray walls and stone roofs, concrete slab”. This causes much confusion and a total disconnection with the archive documents.

The procedure followed for the new listing is unclear and does not fall in line with previous procedures (study, monument passport, other documentation, elevations, and photographical material, and a technical report).

Below is an example of a new listing and a listing that has been redacted, showing a complete lack of understanding for the OVU (Fig. 4).

5.1 Lack of a Management Plan and the Torresi Study [6]

During the 41st session of the WHC, the state party declared that the list is not final. In their decision WHC/17/41.COM/18, p. 117, WHC urges the state party to finalize proceedings, however no developments have taken place since, while the list remains an approved administrative decision by the Institute of Monuments of Culture. Many of the recent problems with the historic center of Gjirokastra, come from the lack of an approved and functional management plan for the site, such as it is repeatedly requested by UNESCO. However the Municipality of Gjirokastra, in collaboration with a group of Italian experts (Studio Torresi), finalized a study that was almost eight years in the making. The study evaluates geological and seismic

Fig. 4 New listing Example (Dec. No. 60—Dt. 11.03.2016) in Palorto Neighborhood No. 60/8) of inappropriate new building in Historic Center damaging the integrity of the entire ensemble and the OUVs. Photo taken from the 2014 State Report described as New Building within the Historic Center of Gjirokastra

risks, as one of the most important challenges of the city and further proposes a restoration plan. For this reason it is also quoted on the official site of UNESCO [7]. However the plan was not approved and Gjirokastra has now entered into a period that the local NGOs refer to as the History of Neglect. The problems of this History of Neglect in our view are as follows:

1. Lack of capacities and unclear role of local and national institutions for the preservation of the OUVs of Gjirokastra.
2. New development projects that could do more harm than good, not being based on a proper management plan for historical center, and neglecting the priorities of the historical center and the protected zone of Gjirokastra.
3. Degrading environmental conditions and loss of green space in the historical center which goes against the principles of landscape and geological conservation.
4. Institutional and social drain; only recently Gjirokastra's archive was moved to Fier. Being the district with a 5% annual decline in population the dislocation of institutions and loss of people seems to correspond to the degradation of monuments due to in-use, and lack of targeted investments.

5. Legalizations [7] of illegal constructions, a process that introduces abusive interventions into the protected zone and historical center damaging the site's integrity and reducing its OUVs (Figs. 5 and 6).

With regards to the illegal construction the government of Albania has recently made real progress with stabilizing the situation with a moratorium on construction in 2014 that put an end to the practice. As a result of that in 2015 by means of Decision of the Council of Ministers procedures were laid out for legalizing some of this construction when deemed necessary, and where the criteria of the object built was not in disconformities with general development plans. In the case of historical and protected zones this DCM was clear. No legalization process can be undertaken in such and area. A year later this decision was amended by DCM no. 756 dt. 26.10.2016, which adds that legalization can be undertaken in protected zones for objects that have lost their protected status. This, combined with the discrepancies of the new listing in Gjirokastra creates a problem that gives way to legalizing external extensions or even floor additions to objects that are not on the list, or where the current listing is unclear, seriously endangering the OUVs of the architectural and urban ensemble of Gjirokastra.

In the 41st session of the World Heritage Committee held in Krakow, Poland we drafted a Resolution to stop the History of Neglect in Gjirokastra. The problem of

Fig. 5 Traditional House in Hazmurat neighborhood that is now delisted. As can be seen the house, features, traditional roof, kiosk, windows and stone walls and has only some minor interventions like the metallic eaves in ground floor that can easily be removed. Photo by Arch. Kreshnik Merxhani—June 2014

Fig. 6 Illegal intervention within the Dunavat quarter ensemble. Photo by Arch. Kreshnik Merxhani, June 2014

the delisting of monuments was also stated in this resolution. We are still facing the fact that the original monument list is risking serious drainage. In the conditions of the lack of a management plan we fear more monuments will fall through the cracks of forgetfulness, while illegal interventions are being legalized and the state of the Outstanding Universal Values of Gjirokastra will decline.

6 Conclusions

The Gjirokastra bypass is studied here as an example of rushed planning for political impact, or under political pressure, but with the potential of having very negative effects on the real value of a city that is unique and in dire need of careful integrated planning that is fit for a city of world heritage. A very large role to play has also the national government with the enactment of national policies such as construction permitting procedures and legalization acts that in the case of urban centers with specificities need certain adjustments, or in such cases a detailed local plan can be taken into consideration. However when this plan is lacking or not approved the situation becomes delicate, and the city will show it with the loss of its authenticity. Short term solutions can only be overthrown by careful planning and strategic documents drafted and enacted at the appropriate time. The case of the bypass, as a project that is now approved, notwithstanding the opposition it received for destroying the values of the urban landscape of the city and having important technical deficiencies, while risking the fate of the city castle, may have been easily avoided with an approved management plan. A plan that has been in the making ever since Gjirokastra entered that prestigious list of world heritage sites. With the lack of an integrated management plan it is no wonder that monuments will also start disappearing not only from neglect and lack of maintenance but also due to a string of policies that even though are very good nationally, could have grave consequences in a specific locality.

References

1. UNESCO (2018) Historic centres of Berat and Gjirokastra. UNESCO World Heritage List, World Heritage Convention, 05 Mar 2018
2. Commettee, World Heritage (2005) Decision of the world heritage committee. World Heritage Scanned Nomination, UNESCO, 15 July 2005, https://whc.unesco.org/uploads/nominations/569rev.pdf
3. Merxhani K (2018) Heritage under pressure: report on the new bypass road construction project that seriously damages the historic center of Gjirokastra—UNESCO world heritage site. Academia.edu, 07 Feb 2017, 05 Feb 2018
4. UNESCO (2011) Decision: 35 COM 7B.82 historic centres of Berat and Gjirokastra (Albania) (C 569 bis). Committee Decisions, World Heritage Convention, 05 Mar 2018
5. Riza E (2017) Kumbaro te gjeje 236 monumentet kulturore, interpretim diletantesk i ligjit, nuk e shpeton nga pergjegjesia Julia, V. Gazeta Sot, 02 June 2017, p 2
6. Torresi T (2006) Piano di Ricupero per Centro Storico—Gjirokaster. Gjirokaster Municipality
7. DCM no. 280, dt. 1 Apr 2015, amended by DCM no. 756, dt. 26 Oct 2016

Functional Heritage. Reconnecting with the Iron Web

William Howse and Renata Jadresin-Milic

Abstract Historic buildings in New Zealand have previously been redundant in the face of contemporary development, although the potential for adaptive reuse is increasingly in popularity. This project developed from a personal passion for New Zealand's architectural heritage, and the relevance it has in contemporary cities. The fundamental aim of this project was to explore and develop a design strategy for the adaptive reuse of historic buildings, with the intention for the chosen building to become functionally relevant in the 21st century. The selected site is located in Mornington, Dunedin, which is the currently neglected historic former Mornington Cable-Car Depot. As a general methodology for analysing the site condition a systematic study of: architectural, historical, and urban analysis was conducted to provide insight into which approach for the adaptive reuse was appropriate. Additional applied research methods in this project include: field trips, examination of relevant literature and precedents, and studying different designs concepts through drawings and 3D model making. The final design proposes *Alternative Representation* as a new and experimental approach to heritage. The final approach was used as a mechanism for the historic building to maintain its integrity, while enhancing the functionality of the building and revealing selected portions of the architectural cultural heritage. Therefore, the architecture creates a reminder to the contemporary city of their heritage, while the building fulfils its original function as a cable-car depot; with the end objective to stimulate change in the treatment of heritage fabric in the urban landscape of New Zealand.

Keywords Heritage · Historic knowledge · Adaptation · Preservation · Alternative representation · Authenticity · Identity

---

W. Howse · R. Jadresin-Milic (✉)
School of Architecture, Unitec Institute of Technology, Auckland, New Zealand
e-mail: rjadresinmilic@unitec.ac.nz

G. Amoruso and R. Salerno (eds.), *Cultural Landscape in Practice*,
Lecture Notes in Civil Engineering 26,
https://doi.org/10.1007/978-3-030-11422-0_19

1 Introduction

It is generally accepted that there is no remaining native Maori architecture in New Zealand from pre-European settlement still in existence. Resultantly, the majority of New Zealand's architectural heritage originates from the arrival of European settlers throughout the 19th century and is, therefore, comparatively young. For this reason, the origins of New Zealand's colonial architecture are primarily based on the transferal of architectural design ideas and beliefs from the United Kingdom. The former Mornington Cable-Car Depot, the case study for this project, is an excellent example of this transferal of architectural ideas from Europe to New Zealand, as the depot building is designed in industrial Victorian classical style, commonly seen throughout Europe.

Contemporary New Zealand is redefining its relationship with the past through the recognition that its historic buildings are integral components of the built city fabric [1, p. 8]. Heritage New Zealand is the government agency, which fundamentally promotes the importance and relevance of cultural heritage. Heritage New Zealand lists historic buildings and sites based on historical and cultural significance, with each building and site appropriately categorised. Historic buildings or sites which are deemed to have 'special' or 'outstanding' historical significance are listed as Category One. Category Two listed buildings are deemed to have historical significance. Numerous other categories are used to list heritage areas, as opposed to specific buildings, as well as Maori heritage [2]. However, listing by Heritage New Zealand does not guarantee protection from demolition or safeguard historic buildings from becoming redundant or deteriorating to a condition where they are beyond repair. New Zealand's local councils ensure the protection of historic buildings from demolition through their own individual council listings, and have obligations under legislation for the management and protection of cultural heritage. Legislation includes: The Resource Management Act 1991, Heritage New Zealand Act 2014, Local Government Act 2002 and Building Act 2004 [3]. Therefore, the listing of the former Mornington Cable-Car Depot for protection is the responsibility of the local Dunedin City Council. However, as of January 2018, the building is unprotected.

The Dunedin City Council is currently in the process of finalizing its Second Generation District Plan (2GP) which will include the listing of additional historic buildings, that have not been listed under the current district plan [4]. In other large New Zealand cities, especially Auckland and Wellington, the rate of city redevelopment due to the significant population increase in the latter half of the 20th century led to significant demolition of heritage buildings. However, Dunedin did not experience such a boom in population, and resultantly the city contains the largest portion of its original built-heritage comparatively to all other major New Zealand cities. Dunedin is currently benefiting from this wealth of architectural heritage, and is going through a period of revitalisation where many historic buildings are being adapted for reuse, following decades of neglect. The tourist potential of historic buildings is being recognised by the Dunedin City Council and property developers, which has led to the development of numerous heritage precincts in the city. As a result, Dunedin is currently marketing itself as the 'Heritage Capital of New Zealand'. The adaptive reuse of historic buildings is not only playing an important economic role by attracting tourists to the city, but is also encouraging new businesses to open in previously underutilised and largely abandoned parts of town.

As an alternative to the loss of unlisted architectural heritage in New Zealand, this project examines the potential for an historic building and its function to develop with the city; ensuring the long-term preservation of built-heritage. An agreed strategy has not yet been developed for the adaptive reuse of historic buildings in New Zealand, as both conservation and legislative guidelines are usually subjective.

An example of the controversy around the topic is the Dunedin Lightrail Heritage Trust proposal. The proposal includes constructing a new cable car building in the Mornington Park across the road from the original Mornington Cable-Car depot building (c1882–1925). The proposal additionally seeks to build a new line along the original cable-car route that operated from 1883 to 1957, as a heritage tourist attraction. The original Dunedin cable-cars, on permanent loan from Christchurch's Ferrymead Heritage Park, are planned to be used on the new line [5]. Nevertheless, while a new line is required, a new building does not necessarily have to be built.

This project proposes an alternative in which the original historic cable-car building could be reused, with the intention to demonstrate that the adaptive reuse of the original former Mornington Cable-Car building, is a viable option. The possibility of constructing additional cable-car lines was explored. Two cable-car lines that originally connected with the Mornington Cable-Car Depot were identified: the Eglin Road and Maryhill lines. Both lines were evaluated for their potential to enhance the cultural heritage value of the site and improve the feasibility of the project. The Maryhill cable-car line (1885–1955) was identified as having the greater potential to improve the financial viability of the programme as it was the steepest cable-car line in the Southern Hemisphere. Re-establishing the Mornington cable-car line would not necessarily result in the adaptive reuse of the depot building becoming financially viable, these projects frequently require more than one programme to fund the adaptive reuse. As part of the alternative proposal the function of the historic building can be redeveloped by providing space for other related functions. In this project these functions are a cable-car maintenance facility, a museum, and an art gallery.

Dunedin is the only main centre in New Zealand without a facility for the conservation and maintenance of historic cable-cars and trams. This presents an opportunity to further expand the programme for the Mornington Cable-Car Depot, as the New Zealand Rolling Stock Register [6] identifies that there are currently eight Dunedin cable-cars and trailers in the process of conservation work. The adaptive reuse of the former Mornington Cable-Car Depot would allow these cable-cars once they have been repaired ongoing maintenance and use. Including a contemporary cable-car maintenance facility would further expand the cultural experience of tourists and the general public.

2 Methodology of the Project

The proposal of a historic building to be adapted to a contemporary function requires an understanding of current local conditions, current and past theoretical debates, a precedent study, and site-specific research. During the research process and project development, the following was undertaken:

- The potential market of Dunedin was considered and evaluated; it was concluded that: there is a real need for the proposed reuse.
- The physical analysis of the building was undertaken; the requirements of the local building code were checked and it was concluded that the structural stability and zoning allowed the proposed reuse.
- An architectural and historical evaluation was carried out; it was concluded that the building meets the criteria of the National Register, the historic fabric exists, the authentic materials and workmanship that give the building its character and integrity are preserved; and it is feasible to preserve them.
- The context of the site and urban situation were examined to discovered constraints and opportunities; a thorough inventory and analysis of the external opportunities and constraints which have the potential to influence the project were considered.

2.1 Theories

Theoretical debate and controversy about the adaptation of historic buildings for reuse reached its peak in France and England during the nineteenth century. The controversy originated from authenticity and questions such as: should it be allowed for an historic building to be restored to its original condition when much of the building may no longer exist? If restoration is allowed does this result in a fraudulent imitation of the original? Since this old dilemma continues as an on-going argument in heritage conservation, obtaining general information about the conservation and reuse of historic buildings included both current and past debates. Literature which compares old and new theories/approaches was studied. This was followed by further literature which was explored to gain an understanding about the political, and social implications of adaptive reuse. This method concluded with evaluations about the various theories, which aided in developing design guidelines for the project.

The initial key publications used through the research process were; Glendinning: *The Conservation Movement* [7], Jokilehto: *A History of Architectural*

Conservation [8], Plevoets: "Adaptive Reuse as a Strategy towards Conservation of Cultural Heritage: A Survey of 19th and 20th Century Theories" [9], Scott: *On Altering Architecture* [10], and Hill: *The Double Dimension: Heritage and Innovation* [11]. Glendinning and Jokilehto provided an understanding of the origin of debates in heritage conservation, in conjunction with detailed information on specific debates such as Scrape (Restoration) and Anti-Scrape Philosophies. Jokilehto's publication was further used to understand policy implications that derived from these debates, including the formation of the Society for the Protection of Ancient Buildings in 1877.

New Zealand historian and expert on cable-cars and trams Stewart, identifies in his publications many of the technical aspects as well as the social significance that cable-cars played in people's lives prior to the 1960s [12]. Stewart was used as a source for understanding which design features and programme requirements are essential to operate cable-cars. His publications were further used to investigate how a rebuilt cable-car network could be integrated back into the community, whilst being historically sympathetic. Furthermore, policy literature and legislation which is regarded as being essential in a modern society to maintain responsible actions toward historic buildings [8] were used: SPAB, England [13], ICOMOS New Zealand Charter [14], and ICOMOS Australia Burra Charter [15]. From these discussions, specific design implications and opportunities are identified that can be used for the adaptive reuse of the former Mornington Cable-Car Depot.

According to ICOMOS New Zealand Charter the building is '*tangible*' evidence of continuity between past, present, and future due to it being one of the oldest buildings in Mornington Village. The building also has strong intangible values because, for many decades, it was the social hub of the community as the centre of the Mornington public transport system. These '*tangible*' and '*intangible*' values could be acknowledged if the building is again used as a cable-car depot; through the use of an appropriate design intervention and development of a programme. In addition, there could be further implications for the Mornington Cable-Car Depot when taking the local district plan into consideration. Although not currently listed, the building has one of the largest, longest and prominent individual street façades in Mornington village which, under the district plan, is a requirement for it to possess significant town-scape or heritage value. The building can be seen on the ridge line from other Dunedin suburbs and, according to the requirement of Schedule 25.1 in the district plan [16], '*the building's appearance can be observed from a specific area*'. The building further meets the requirement in the district plan for having '*multiple street frontages*' as two façades are visible. For these reasons exceptional care should be taken with the design for adaptive reuse.

For comprehensive understanding of how to develop a coherent strategy for determining the heritage value of a building, the following sources were used; Austin: *Adaptive Reuse: Issues and Case Studies in Building Preservation* [17], and Clark: *Adaptive Reuse of Industrial Heritage: Opportunities & Challenges* [18]. Literature that provided information about economic valuation and viability of cultural heritage projects, especially adaptive reuse, includes; Fitch: *Historic Preservation: Curatorial Management of the Built World* [19], Bedate: *Economic valuation of the cultural heritage* [20], Murtagh: *Keeping Time* [21], and Rabun: *Building Evaluation for Adaptive Reuse and Preservation* [22]. Research about the economic benefits of heritage buildings, conservation and adaptive reuse in nearby Australia that were examined included Bullen and Love [23, 24]. These references provided a valuable insight about the decisions made in regards to adapting or demolishing existing buildings to meet changing economic, social and sustainability requirements.

Recent scientific references were studied which focus on current situations, and on the topic of decision making in reuse of historic buildings: Wang, "A multi-objective decision-making process for reuse selection of historic buildings" [25], which provided a comprehensive design methodology and the criteria for the selection of the best reuse of historic buildings, that governments, developers, owners or architects and architectural historians may apply to their projects; Lucchi, "Multidisciplinary risk-based analysis for supporting the decision making process on conservation, energy efficiency, and human comfort in museum buildings" [26] informed this research about new cross-disciplinary approaches that have been developed, starting from existing standards, best practices, policies, guidelines, techniques, procedures, and tools at international level; Rodrigues, "Adaptive reuse of buildings: Eco-efficiency assessment of retrofit strategies for alternative uses of an historic building" [27] informed the research about the main challenge of adaptive reuse to reconcile historic preservation and sustainable design: to perform an eco-efficiency assessment for a historic building to assess alternative retrofit strategies and uses; Tadeu, "Energy retrofit of historic buildings: Environmental assessment of cost-optimal solutions" [28] and Ascionea, "Design the refurbishment of historic buildings with the cost-optimal methodology: The case study of a XV century Italian building" [29] were valuable case studies of how to combine the building protection requirements and the application of energy efficiency measures; Roberti, "Energy retrofit and conservation of a historic building using multi-objective optimization and an analytic hierarchy process" [30] informed the research about a methodology for heritage preservation to be compatible with demanding energy saving and comfort requirements. International operational guidelines that informed the research about world heritage properties that may support a variety of ongoing and proposed uses, ecologically and culturally sustainable are: Intergovernmental Committee for the Protection of the World Cultural and Natural Heritage [31]; and Pedersen [32] explaining how historic buildings can become important assets to develop a local tourism industry and how to use tourism as a tool to benefit conservation.

2.2 Precedent Studies

The application of the theoretical approaches and their associated implications was further investigated through a study of precedents that provided insight into different design strategies that can be used for the adaptive reuse of neglected historic buildings. To provide an understanding of a broad range of design approaches to adaptive reuse, and their implications for the envisaged adaptive reuse of the former Mornington Cable-Car Depot, specific precedents were chosen according to their degree of intervention, and defined as: minimum intervention, medium intervention, and extensive intervention. Such a comprehensive analysis enabled the adaptive reuse of historic buildings to be compared and evaluated based on the degree of intervention and number of changes made to the heritage fabric, and were then evaluated in regard to the overall effect it would have on the Mornington Cable-Car building.

Minimum Intervention: The former Administrative Centre of Pantin in Paris completed in 1972 by architect Jacques Kalisz, and adapted to function as the National Dance Centre in 2004. This project was specifically selected for use as a precedent for the following reasons: the minimal impact of the intervention on the integrity of the building, the integration of new circulation patterns, and the use of both texture and color to distinguish new design decisions from those of the original architect. Adopting the use of a contextual design intervention of the Administrative Centre of Pantin, at the Mornington Cable-Car Depot, would likewise minimize change to the existing fabric. In a similar way, the emphasis of new intervention through the application of color and texture could be employed, which would allow the different historical layers of the building to be clearly distinguished.

Medium Intervention: The former medieval Castelvecchio in Verona, adapted by Carlo Scarpa from 1956 to 1964, was specifically selected for use as a precedent for two reasons: juxtaposition, or the contrast between new and former, takes account of the fact that parts of the existing building are from different eras; unique design techniques that are utilized to exhibit historic artefacts. To clearly identify the different parts of the former Mornington Cable-Car Depot that were built at different times, similar design strategies could be used. Using Scarpa's idea where artefacts were highlighted in the Castelvecchio Museum, heritage features were identified and showcased through the use of juxtaposition. Such items, that acknowledge the original functions of spaces within the Mornington Cable-Car Depot, also have the potential to enrich a tourist-based museum programme.

Extensive Intervention: The former General Electric Power Station in Madrid, adapted to function as a museum and cultural centre known as Caixa Forum, by Herzog and du Meuron. This type of strategy can be controversial as it often results in the demolition of existing heritage fabric, and was selected as a precedent for the following reasons: its relationship with the surrounding environment, the construction of intervention both above and below the original building, and the techniques used for the new openings made in the historic facades. The Caixa Forum is an example of demolition, which permitted complete reconfiguration of the available space. Additions were constructed which stand out through the use of contemporary design and materials so that the addition is clearly acknowledged. However, the Caixa forum approach has the potential to severely compromise the cultural heritage values of the Mornington Cable-Car Depot, due to the extensive demolition and removal of heritage fabric. Alteration would minimise the level of demolition as more of the building would be kept intact allowing retention of the majority of heritage fabric.

In conclusion, although all three design strategies of minimum, medium and extensive interventions appear to have been successful in revitalising each historic building and enabling it to be adapted for reuse, there were no attempts in the projects to 'reconstruct' or 'restore' lost architectural elements or features. This is likely due to issues around authenticity. However, this provides a unique opportunity for this project to explore the 'reconstructing' of lost/previously demolished architectural elements, in a manner which retains the integrity and authenticity of the former Mornington Cable-Car Depot. The reconstruction of lost architectural elements could also be explored to determine how these reconstructed forms could not only augment understanding of the cultural heritage but to also enhance the functionality of the adapted building.

2.3 Mornington Cable-Car Depot Building and Site Research

Historical Analysis: Identifying the cultural heritage values of the Mornington Cable-Car site from when the historic building was initially constructed in 1882 was of crucial importance for the project. An analysis of the building and the cable-car system itself was undertaken to enable understanding of the building's original function as a cable-car depot. This required a historical analysis and an architectural analysis. The historic and architectural analysis included several research stages and data collection from various sources and institutions that were accessed to obtain this information.

The historical analysis was conducted by visiting the city archives at the Council Buildings in the Octagon, Dunedin. Copies of floor plans from 1960 to 1995 were obtained to be referred to throughout the design process. Likewise, the historical photos of the building were obtained which helped any parts of the building that had been demolished or moved, over time, be identified. A visit was made to the Hocken Library in Dunedin to source additional photos and plans, and The Alexander Turnbull Library [33] was used to access aerial photos taken by Whites Aviation of Mornington suburb. To supplement the gathered sources and to provide an additional primary source of information, several field trips - site visits were made. The Mornington Cable-Car Building was inspected and information on any remaining physical evidence of historical areas of interest was later used to inform design decisions (Fig. 1). This evidence included: blocked up original window openings (Fig. 1a); original 1903 roof trusses (Fig. 1b); original double-hung sash windows (Fig. 1c); remains of original cable-car track (Fig. 1d); remains of cable-car machinery ironwork on flooring (Fig. 1e); original High Street Cable-Car rope (Fig. 1f); remains of staircase on basement wall (Fig. 1g); evidence of original cable-car cutting in nearby suburb of Roslyn (Fig. 1h). During the field trips an

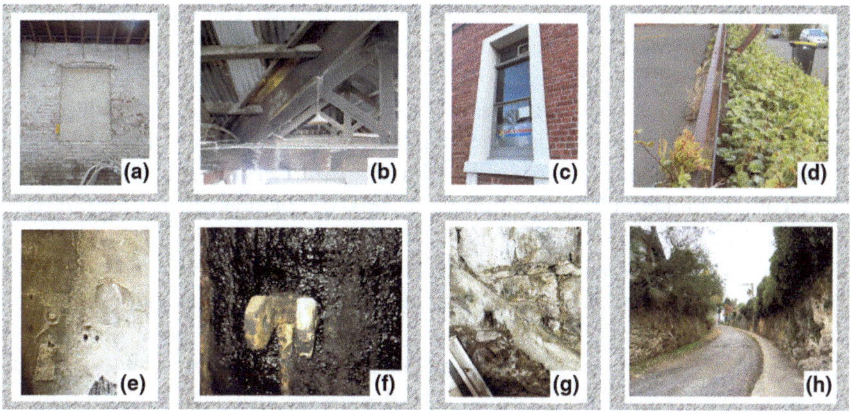

Fig. 1 Areas of particular historical interest discovered on site visits. These features are proposed to be preserved and/or embellished as a part of the adaptive reuse. *Photos* by Author

approach was made to the Senior Heritage Planner of Dunedin City Council, the chairman of the Dunedin Light Rail Heritage Trust, and the current owner of the building. The owner of the building provided access into the interior of the building what was supplemented by the videos and photographs taken at the time of the visit.

Architectural Analysis: The former Mornington Cable-Car Depot is constructed of double brick and masonry load bearing exterior walls with a timber joinery and truss roof structure. The exterior walls of the front 1903 part of the building are exposed red brickwork while the exterior of the 1925 rear addition has a cement stucco finish that has been painted over in recent years. Both the interiors of the 1903 and 1925 parts of the building are industrial in nature with exposed brickwork. The interior of the 1882 masonry basement walls is finished in a lime wash render. The roofing materials for all parts of the building are corrugated iron, while the basement ceiling/ground floor is constructed out of timber joists.

Architectural analysis highlights the horizontal language of the 1903 front facade, rectangular form of the 1882 basement, and distinctive triangular roof form of the 1925 era (Fig. 2). Existing architectural language was used to inform the height, proportions and scale of new design intervention. At the rear of the building, the circular addition was proportioned off the 1925 engine house. This allows what is an unmistakably contemporary circular form to take visual cues from the language of the historic building. Further design decisions were informed through architectural analysis of the existing building; the 1903 part of the building was raised to a height that maintains the horizontal language of the front façade. The materiality of new additions was also informed through a systematised study of the materials used on the different eras of the depot building; the *Alternative Representation* of the chimney is proposed in rusticated steel to mimic the brick material of the original and now demolished chimney. Opposite the depot building in Mornington Park, the height of the entry to the underpass was centered on the height of the cornice of the depot building front façade. The main findings summarized here are: Minor architectural changes were made to the building from 1925 until the closure of the cable-car system in 1957 (Fig. 3). After the cable-cars had ceased operation, the depot was used to store the bodies of the new trolley-buses that had arrived from Britain. The building was subsequently leased to local businesses before being sold to a private buyer in 1961. The first post cable-car era modifications to the building took place in 1961–62; the machinery and rails were removed, and the maintenance pits filled in, which facilitated the building to function as warehouse and retail space. After the 1950s, the original ornamental cornice on the front façade and the small chimney in the original staff dining room were removed.

Through archival research, including historical newspaper articles and photographs, coupled with site visits, it was discovered that the 1882 Mornington

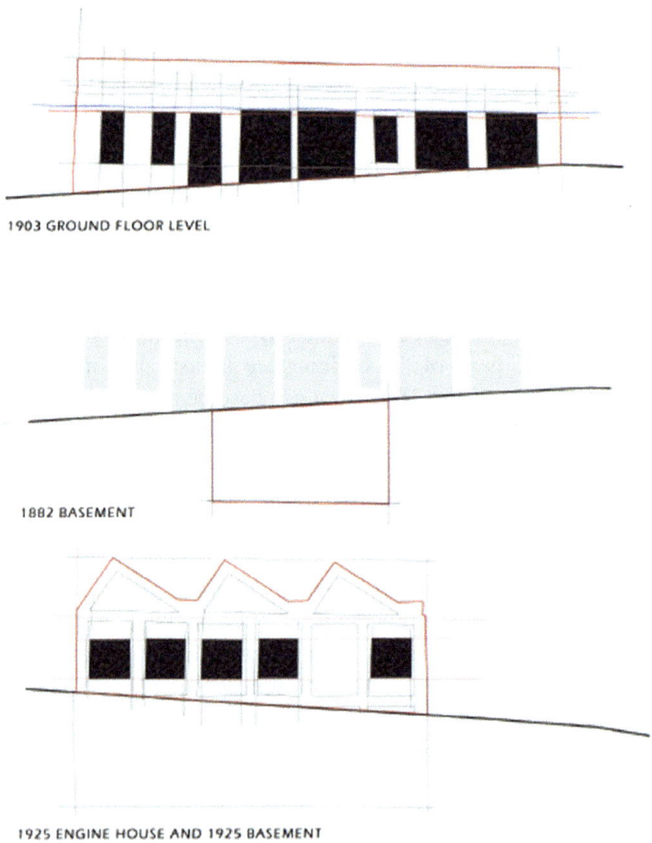

Fig. 2 Architectural analysis of the proportions of the 1903, 1882, and 1925 parts of the depot building. Original drawings from the Dunedin City Council Archives adapted and redrawn by Author

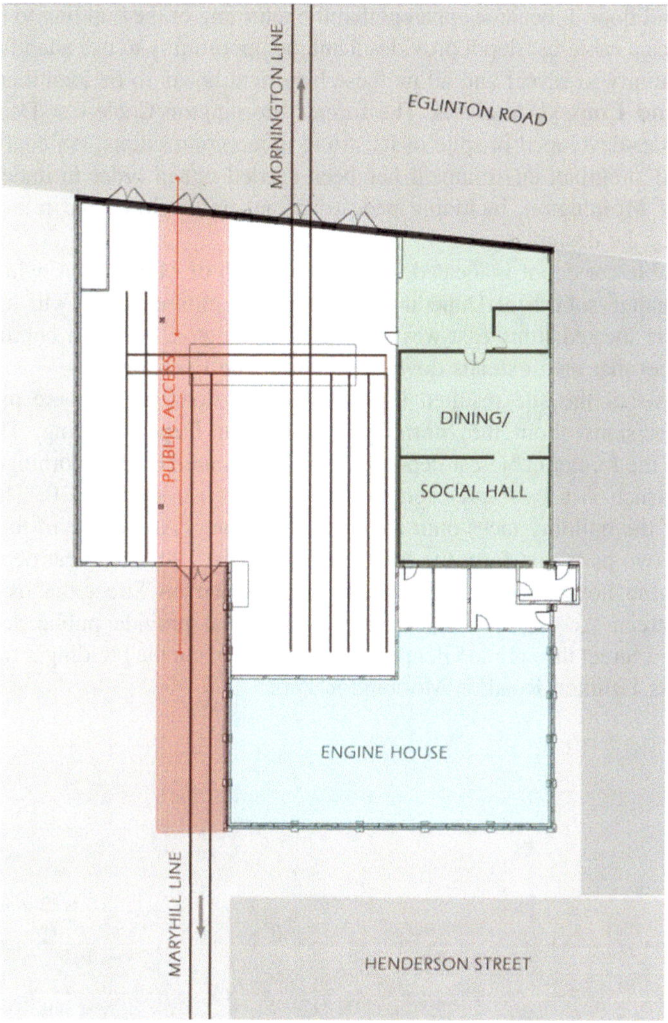

Fig. 3 Mornington Cable-Car Depot floor plan—ground floor interior layout 1957. Original drawings from the Dunedin City Council Archives redrawn by Author

Cable-car Depot building had been expanded and rebuilt between 1903 and 1925. The greatest event occurred in early 1903 when the entire cable-car building burnt down apart from the basement. The building was rebuilt in late 1903 incorporating the original basement and in addition, a new substantial brick chimney was built for burning coal to power the cable-car machinery. In 1925, the chimney, which had been monumental on the Dunedin skyline, was demolished and replaced with an engine-house for electric machinery (Fig. 4). The building, therefore, has three historical layers: the 1882 basement, the 1903 level of the depot, 1925 basement and ground floor. It became apparent that the returning of the building to its original function as a cable-car depot provides a unique opportunity to use adaptive reuse as an opportunity to reveal and allow these historical layers to be identified.

Site and Context Analysis: The former Mornington Cable-Car Depot is currently under-developed in spite of its strong urban connections. An analysis of the urban and suburban environment has been carried out in order to understand the suburb of Mornington, including pedestrian and traffic flows and relevant implications for the design.

The cable-car depot is located within the suburb of Mornington, which adjoins The Exchange suburb of Dunedin City (Fig. 5). Eglinton Road, which the depot building is located, runs east-west through Mornington, and is a continuation of High Street that then extends down through to The Exchange.

Analysis of the site resulted in several conclusions which were relevant for design decisions about the Mornington Cable-Car Depot building. The eastern facade of the former cable-car depot building is obscured by the Mornington Health Centre, which is a two level commercial space constructed c1970. The western facade of the building faces onto a shared access-way. At the rear of the building, there are two parts that form the southern façade: the 1903 cable-car depot and the 1925 engine house, which faces south onto Henderson Street that is a no exit service street. Across the road there are steps that provide public access from Henderson Street through to Glenpark Avenue. Opposite the building's front façade and across Eglinton Road, is Mornington Park.

Fig. 4 The rear of the Mornington Cable-Car Depot showing the chimney c1910 prior to demolition and replacement with the 1925 engine house. *Photos* from Alexander Turnbull Library [33] and by Author

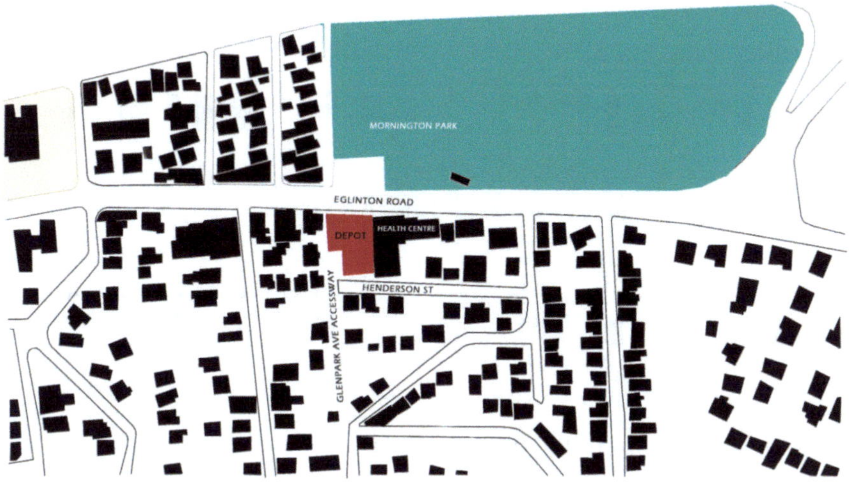

Fig. 5 Urban analysis of the site and context. Drawn by Author

2.4 *Design Trials*

In the next phase of the project, physical and digital design concepts were produced using information gathered from the various methodologies. Multiple concepts based on different design propositions, were used as a design tool to investigate their effectiveness when making the building functionally relevant. The effectiveness of each concept was then evaluated based on how much of the building would be retained, which parts of the building could be retained based on the degree of historical significance, and how well the concept would function to meet the needs of, and satisfy, the proposed programme. Therefore, the evaluation considered the functionality and the degree of retention of the existing Mornington Cable-Car building while making it functionally relevant in the 21st century. Each concept was developed, analysed and tested. Design concepts were experimented through the use of physical and digital models, as well as through drawings, to explore their potential effectiveness. As the concepts progressed, more developed design models were digitally produced. The primary software used was a combination of Autodesk's AutoCAD and Autodesk's Revit, which allowed a combination of 3D modelling and 2D drawing. As the concepts progressed, more developed design models were digitally produced through Autodesk's Revit.

Theoretical Design Trials I (Fig. 6): A series of initial design trials were undertaken to evaluate possible approaches for the adaptive reuse of the former Mornington Cable-Car Depot. Conclusions were subsequently made on an appropriate strategy for the adaptive reuse. The focus of the trials was to develop a design response that not only adapted the building for reuse, but also preserved and revealed the cultural heritage of the site.

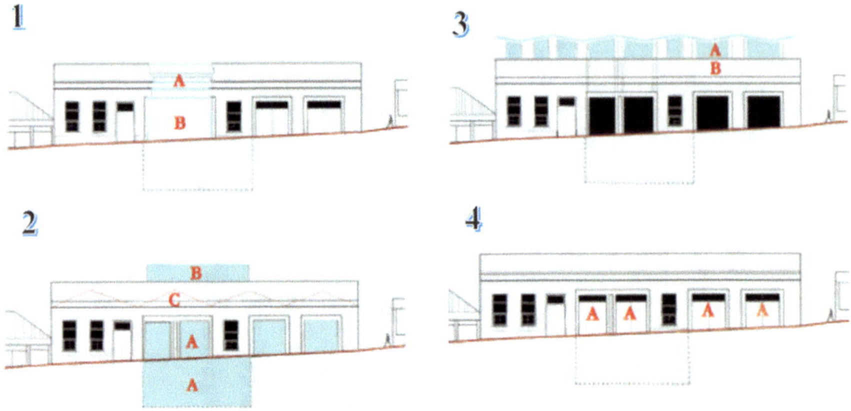

Fig. 6 Theoretical Design Trials I. Drawn by Author

Design Trial One: Minor demolition can be used to reveal layers of construction from different eras. Minor demolition allows any junctions from the different eras to be readily identified (refer A) and also provides the opportunity to uncover original circulation routes through the building. This can stimulate ideas about new design intervention within areas that have been demolished (refer B). However, minor demolition requires the demolition of heritage fabric, it could compromise the cultural heritage values of the building and site [34, p. 11]. Nevertheless, the uncovering of the historical aspects of the building would contribute to visitors' heritage experience.

Design Trial Two: The insertion of new design intervention within the upper and basement levels (refer A) and in the roof space above the existing building (refer B) can be used to enlarge and adapt the current usable space. However, this would result in historic aspects, such as the 1903 roof trusses (refer C), would have to be demolished and perhaps discarded. An advantage of using this design would be that any increase and/or adaptation of existing space would permit the introduction of a new programme(s) for the building.

Design Trial Three: An additional floor level could be added above the building (refer A). The addition could be partially concealed behind the parapet (refer B) so that change in the building's appearance would be minimal. As an alternative, the changes could be made obvious so that it is clear that an addition has been made to the existing form. The benefit of this strategy is that the cultural heritage value of the building and site would be preserved.

Design Trial Four: The building could be returned to its original configuration as a cable-car depot. However, to return the building to its original configuration would require the restoration of lost architectural features (refer A) which are evident in the few photographs that are available. A more complete restoration would pose problems due to the lack of early plans of the building and other relevant photographs. Restoration of any features could be considered 'dishonest' as it would compromise the authenticity of the building.

Conclusion from Theoretical Design Trials I: Design Trials Two and Three do not compromise the authenticity of the existing depot building because new interventions would be clearly distinguishable due to the use of contemporary materials. Both designs would provide the additional space that is required by the programme. However, Design Trial Two involves discarding heritage fabric whereas Design Trial Three achieves the requirement of providing additional space without compromising the integrity of the building. Although providing additional space for new functions, Design Trial Three does not permit acknowledgement of the various eras of the current building. For this reason, a second series of theoretical design trials was undertaken. The focus of the next design trials was not to place greater emphasis on one part of the building than another, as though one part of the building has greater cultural heritage values than the other parts, but rather that each part has its own unique cultural heritage values which may benefit from a unique design approach.

Theoretical Design Trials II (Fig. 7)

Existing: In the first diagram colour is used to show the different historic periods/eras of the depot building. However, in the building there is an overlap that obscures the different eras of construction. Outside, the basement is covered by the walls of the 1903 build whereas previously it was clearly visible as being the basement of the building. Consequently, it is difficult to clearly identify each of the historical layers and their cultural heritage value.

Design Trial One: This initial design focuses on separating the building into two parts to highlight the 1925 build from the two earlier building eras. The issue with this design is that the 1882 and 1903 eras are still blended and not clearly identifiable from each other.

Design Trial Two: This design involves lifting the upper level, which would accentuate the different parts of the building to draw attention to the basement and the later parts of 1903 and 1925 build. However, this would involve separating the 1925 upper level and its 1925 basement, which creates an issue as they are of the same era.

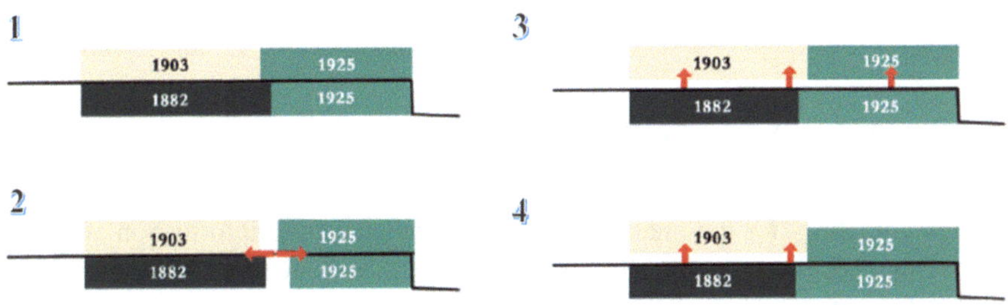

Fig. 7 Theoretical Design Trials II. Drawn by Author

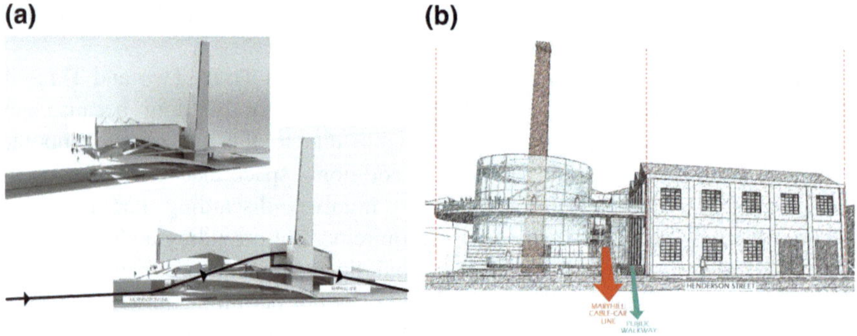

Fig. 8 **a** Physical concept model of raising the 1903 part of the depot building. **b** Digitally produced sketch of proposed rear addition to depot building. Model and sketch by Author

Design Trial Three: This design involves separating the 1903 upper level from the 1882 basement which would permit the three eras of the building to not only be identified clearly but the basement to be valued as the only part that survived the 1903 fire. By leaving the 1925 basement and 1925 upper level of the building connected, and remaining as one, would minimise the extent of the changes. However, separating the basement and upper level of half of the building might make the building appear 'disjointed'. Despite this, it could provide a creative opportunity for augmenting the disjointedness so the junction itself is more pronounced between the different eras.

Conclusion from Theoretical Design Trials II: Using the third design, where only part of the building is separated, would mean that demolition can be avoided. Keeping the 1925 basement and its upper level intact, allows each era of the historic building to be clearly identified as the three eras will be distinct from one another. Preserving and clarifying each era would add emphasis about the history of the depot building and permit it to retain its authenticity. Acknowledging the different eras permits understanding of the tangible and intangible values of the Mornington Cable-Car Depot building as part of the cultural heritage of Dunedin.

The successful components of Design Trial Three were incorporated into the further design concepts, using both physical (Fig. 8a) and digital modelling (Fig. 8b), until the final design, when it was deemed that the project's design aims and objectives had been achieved.

3 Results and Discussion

Design guidelines for the final design were formulated from information and ideas gained through the research of relevant precedents and literature in the field. This further gave direction to the design concepts according to the principles of intervention deemed suited for the different parts of the depot building.

3.1 (Three) Design Approaches—(Three) Heritage Protection Methods

Three design approaches became evident from the research, and they were used in combination to inform the design proposal for the adaptive reuse of the former Mornington Cable-Car Depot. The first two approaches of *Adaptation* and *Preservation* have been generally based on the ICOMOS New Zealand Charter definitions and have been viable, generally accepted approaches for the adaptive reuse of historic buildings. The third design approach *Alternative Representation*, was developed for the purposes of the project and have been based on the ICOMOS definition of 'reconstruction' but has not been limited to it. The design approach for the Mornington Cable-Car Depot uses a combination of *Adaptation*, *Preservation*, and *Alternative Representation* in such a way that demolition is minimised, the building is adapted for reuse, the heritage fabric is preserved, authenticity is maintained, and the cultural heritage is both exposed and highlighted.

Adaptation: *Adaptation means the process(s) of modifying a place for a compatible use while retaining its cultural heritage value. Adaptation processes include alteration and addition* [14, p. 8].

As adaptation is the alteration made to the existing building fabric it was decided that the alterations would, in this instance, be kept to a minimum if demolition of heritage fabric is required. Alterations would be made where the functioning of the Mornington Cable-Car Depot building was compromised by its current form, and to reveal aspects of cultural heritage that are currently obscured.

Preservation: *Preservation of a place involves as little intervention as possible, to ensure its long-term survival and the continuation of its cultural heritage value. Preservation processes should not obscure or remove the patina of age, particularly where it contributes to the authenticity and integrity of the place, or where it contributes to the structural stability of materials* [14, p. 6].

Preservation will be used for those areas that, in their current state, contribute to the building's authenticity and integrity. This implies that only maintenance will be used in particular areas, so minimising removal of heritage fabric.

Alternative Representation: *Alternative Representation* is a method defined by the author and is applied when the original form is reproduced using different material(s) to the original. Therefore, the new form is clearly identified as 'new' despite being in the form of the lost element. *Alternative Representation* is therefore recognised as different to Reconstruction as defined by ICOMOS New Zealand Charter. *Reconstruction is when a form is reconstructed as closely as possible to the original, which involves using similar, although new, materials* [14, p. 7]. *Alternative Representation* might also involve the newly reproduced form being used for a function different to that of the original form.

Alternative Representation is used in this project when documentary evidence exists in order to reproduce the form that has been lost. This would involve judgement as to their degree of importance. The intention is to augment understanding of the building's cultural heritage, while contributing to the functionality of the adapted building.

Combining these approaches in the design process enabled the historic building to be adapted for reuse, whilst still retaining its authenticity, as well as to highlight the cultural heritage values of the building and site. It was decided that an attempt to restore the existing building would not be used as a strategy. As a result, there are no ethical issues as the authenticity of the building will be not be corrupted. Instead, the design focuses on adapting the building for reuse so that the original function of the building can be retained and brought back to life.

3.2 Final Design

For the final design it was important to acknowledge the three different eras of the building, since it permits an understanding of the tangible and intangible values of the Mornington Cable-Car Depot building as part of the cultural heritage of Dunedin. It meant firstly, keeping the 1925 basement and its upper level intact, which allows each era of the historic building to be clearly identified. In the design process it was thought that preserving and clarifying each era would add emphasis about the history of the depot building and permit it to retain its authenticity. However, it included a strong and even provocative design decision; raising of the 1903 level of the depot from the 1882 basement (Fig. 9b). The raising of one part of the building was considered seemingly radical in the first moment and as an option that could compromise the integrity of the building. However, from the further investigation undertaken to establish how the design could be developed appropriately, it was concluded that the separation of the 1903 part of the depot building from the 1882 basement would create a gap between the two levels. This would create a view for passersby of the basement, cable-cars and exposed machinery. This opening would further enhance the experience of cable-car passengers, as they would have a view up to street level from the basement which currently lacks natural light.

Lifting part of the building proved to be important to separate the 1882 era from the 1903 era of the building, and to create a distinctive threshold when entering the 1925 part of the building. However, equally significant was to maintain the horizontal look of the façade at the front of the building. This meant that the roof-line must be considered when the gap is created between the basement and the upper level of the building. It was decided that any change must be subtle so that the alteration in the roof-line is barely perceivable. The original lettering of *Mornington Municipal Buildings. 1904* was proposed to be reproduced on the parapet using *Alternative Representation*, including the missing cornice on the parapet, so that the form of the building is kept, but the changes can be easily identified.

The adapted depot building has a cable-car and pedestrian entry from the underpass (Fig. 9a) a platform, machinery for operating the cable-cars, and the cable-car maintenance facility. These are all located in the existing basement as well

Fig. 9 **a** Final design showing north-south sectional perspective/underpass. **b** Final design showing raised 1903 part of the Mornington Cable-Car Depot building above the original 1882 basement, with gap between. Renderings by Author

Fig. 10 Final design showing rear addition with the *Alternative Representation* of the chimney. Render by Author

as the eastern and western additions either side. The western addition to the basement was designed to be underneath the existing ground floor level and an asphalt car park area, that is not currently used. An extension to the current building was decided to go above this new part of the basement (Fig. 10) as it would not require demolition of any part of the existing building. The new addition's appearance makes it distinct from the original through the use of contemporary materials. It further concluded in the design process that the addition of another era will contribute to the contemporary relevance of the Mornington Cable-Car depot building. The new addition will further include the *Alternative Representation* of the original chimney, now constructed from perforated panels.

In the 1903 upper level, the social hall and dining room walls were designed to be reproduced using *Alternative Representation* in the place where the walls originally were (Fig. 11a). The larger space formerly the social hall, was designed to display artworks for visitors (Fig. 11b). It was concluded that leasing this space

Fig. 11 **a** Final design showing the *Alternative Representation* of original interior walls to function as art gallery space. **b** Final design showing the junction between the 1925 and 1903 parts of the building showing restaurant (below) and gallery spaces above. Renders by Author

would contribute to funding the changes for the adaptive reuse of the building. Both the museum and the maintenance area are located in the enlarged basement. The maintenance area could be leased to Dunedin Light-Rail Heritage Trust to be used for cable-car repair and refurbishment. For this reason; theory, practical investigation, and design were complementary within the project, promoting an integrated approach in order to find the most effective solution to the historic building.

4 Conclusions

Overall this project provokes discussion on four main aspects relating to the adaptive reuse of historic buildings.

Firstly, the returning of a historic building to its original function brings to light multiple unique opportunities and dilemmas for the design. Adaptive reuse is often defined as the changing of a function, which differs greatly from that of the original. Reverting a historic building to a version of its original use requires careful consideration about authenticity. In this project, the integrity of the building was retained using architectural strategies, including *Alternative Representation*. Such strategies allowed the building to preserve its integrity, 'reveal' its historical layers of construction whilst also increasing the functionality of the building.

Secondly, the former Mornington Cable-Car Depot is not currently listed for heritage protection. Therefore, this project aims to raise awareness of the existence of the building and more importantly, draws attention to the New Zealand heritage dilemma regarding the need to protect cultural heritage that lies outside of 'heritage status'. The project further illustrates that there is a difference between the New Zealand heritage context and that of 'older' countries, especially European. A 21st century New Zealand context has alternative issues to consider, and this is shown through the preservation of the c1925 addition to the cable-car depot, as part of the design. The 1925 part of the building may not be considered 'architectural heritage' in countries with a longer history pre-dating the Victorian period.

Thirdly, this project suggests a combination of *Adaptation*, *Preservation*, and *Alternative Representation*. The latter is a specific design position and decision to take. The defining of *Alternative Representation*, as the project developed, allowed the different historical periods and demolished architectural features of the building to be understood and 'reconstructed' to enhance the functionality of the building. *Alternative Representation* is a new approach, which differs from the guidelines of the ICOMOS New Zealand Charter that promotes *reconstruction* for the rebuilding of demolished architectural features or elements. However, *reconstruction* involves using the same kind of material as the original, although its contemporary version, to make a clearly defined distinction between periods. Therefore, *Alternative Representation* is a variation on this approach, using different materials, and the same form, to clearly differentiate between new and old.

Finally, this design project suggests a 'bold' design move; the raising of one historical layer of the building. Although primarily for the purpose of revealing the history and sequence of the historic building's construction, the design move seeks to enhance the functionality of the building. When the concept of lifting the 1903 part of the building was first considered, it was thought that this decision could provoke reaction as it might be considered that separation of the levels could prejudice the integrity of the building. The final design seeks to ameliorate these potential concerns through the use of strategies in keeping with the architectural language of the existing building. Architectural and historical analysis, and the precedents, provided techniques that were extrapolated so that the building would not only reveal its cultural heritage, but retain and protect it.

The project for the Mornington Cable-Car Depot building raises another important question about what are possible approaches to education of future architects today. Throughout the research and design process the world standards in the field of education in conservation were considered. The experiences of methodologies applied in numerous schools around the world, in the recent years were examined. This included methodologies that had been carried out in order to investigate possibilities of protection, preservation and adaptive reuse of historic buildings and sites [35]. The intention was to broaden awareness of different ways of solving complex issues while investigating, evaluating and renewing historic areas. Becoming acquainted with the contemporary methodological approaches to the historic areas and buildings allowed the old cable-car building to be presented in its both immediate and wider context, both in its built and natural environment, being a quite specific and valuable cultural landscape. Problems and methods of defining its potential for the development; comparative analysis of advantages and disadvantages of different options and approaches to their protection and preservation, contributed to this strong and decisive attitude towards revitalization of this particular cultural and historic area of value.

Future studies in this area of this research, especially in the New Zealand context, should focus on the implications of earthquake strengthening on the preservation and adaptive reuse of architectural heritage. This is particularly relevant in New Zealand cities such as Christchurch and Napier, which have lost significant

Fig. 12 Aerial view of final design. Render by Author

historic buildings in earthquakes. Additional studies could also focus on the financial feasibility and costings of adaptive reuse and earthquake strengthening of the chosen historic building. The final design proposed by this project could further be assessed against sustainable design and eco-efficiency guidelines to determine appropriate retrofit strategies.

The final design for the project strives to fulfil what was intended, that the cultural heritage is retained and valued but that the functionality of the building makes it suitable for reuse, while still maintaining its integrity. Each era is acknowledged for its contribution not only to the building itself but as a record of what took place in the history of Dunedin and could stimulate a momentum for change in the Dunedin urban landscape (Fig. 12).

Acknowledgements The project resulted from one-year research that is final part of the Master of Architecture (Professional) programme at the Architecture Department, Unitec Institute of Technology in Auckland, New Zealand. The project was nominated to represent Unitec Architecture Department at the NZIA Central Innovation Student Design Awards—SDA for Master of Architecture Professional students within New Zealand in 2017 (https://www.nzia.co.nz/awards/student-design-awards). William Howse is the author of the project and the explanatory document "Functional Heritage. Reconnecting with the Iron Web". Dr. Renata Jadresin Milic supervised the research work on the project and writing of the document throughout the year. William Howse and Renata Jadresin Milic wrote this paper together.

References

1. Bollack FA, Frampton K (2013) Old buildings, new forms: new directions in architectural transformations. Monacelli Press, New York. ISBN 9781580933698
2. Heritage New Zealand. Pouhere Taonga. Available online: http://www.heritage.org.nz/the-list/about-the-list. Accessed 5 Jan 2018
3. Heritage New Zealand. Pouhere Taonga. Available online: http://www.heritage.org.nz/protecting-heritage/local-government. Accessed 25 Dec 2017
4. The Dunedin City Council, Second Generation District Plan. Available online: https://2gp.dunedin.govt.nz/2gp/index.html. Accessed 5 Jan 2018
5. Proposal to Reintroduce the Dunedin High Street Cable Car Line. Dunedin Cable Car Trust Publication, Booklet handed out 2017
6. New Zealand Rolling Stock Register. Available online: http://www.nzrsr.co.nz/. Accessed 5 Jan 2018
7. Glendinning M (2013) The conservation movement: a history of architectural preservation: antiquity to modernity. Routledge, New York. ISBN 10:0415543223; ISBN 13:978-0415543224
8. Jokilehto J (2017) A history of architectural conservation. Routledge, New York. ISBN 13:978-0750655118, ISBN 10:0750655119
9. Plevoets B Adaptive reuse as a strategy towards conservation of cultural heritage: a survey of 19th and 20th century theories. WIT Transactions on The Built Environment. Available online: http://www.academia.edu/1473330/Adaptive_Reuse_as_a_Strategy_towards_Conservation_of_Cultural_Heritage_a_Survey_of_19th_and_20th_Century_Theories. Accessed 25 June 2017
10. Scott F (2008) On altering architecture. Routledge, New York. ISBN 10:0415317517; ISBN 13:978-0415317511

11. Hill J (2004) The double dimension: heritage and innovation. ACT: Royal Australian Institute of Architects, Canberra. ISBN 1447-1167
12. Stewart G, Olphert L (1997) Fares please! The horse, steam & cable trams of New Zealand. Grantham House. Wellington, New Zealand. ISBN 1 869340647
13. The Society for the Protection of Ancient Buildings. SPAB: SPAB's purpose—regular maintenance of old buildings, restoration, repair, authenticity, preserve. Available online: https://www.spab.org.uk/what-is-spab-/spab-s-purpose/. Accessed 15 June 2017
14. ICOMOS New Zealand Charter for the Conservation of Places of Cultural Heritage. ed. ICOMOS (New Zealand Revised 2010). Available online: http://icomos.org.nz/charters/. Accessed 20 Nov 2017
15. Burra Charter. (Australia 1999), 6. Available online: http://icomos.org.nz/charters/. Accessed 20 Nov 2017
16. Dunedin City District Plan. Available online: http://www.dunedin.govt.nz/__data/assets/pdf_file/0006/147327/Schedule-25.1-Townscape-and-Heritage-January-2015.pdf. Accessed 5 Jan 2018
17. Austin RL, Woodcock DG, Steward WC, Forrester RA (1988) Adaptive reuse: issues and case studies in building preservation. Van Nostrand Reinhold, New York. ISBN 0442208154
18. Clark J, Wolkenberg T et al (2013) Adaptive reuse of industrial heritage: opportunities & challenges. Heritage Council of Victoria, Melbourne. Available online: http://heritagecouncil.vic.gov.au/wp-content/uploads/2014/08/HV_IPAWsinglepgs.pdf
19. Fitch JM (1990) Historic preservation: curatorial management of the built world. University of Virginia Press. ISBN 978-0813912721
20. Bedate A, Herrero LC, Sanz JA (2004) Economic valuation of the cultural heritage: application to four case studies in Spain. J Cult Heritage 5:101–111. https://doi.org/10.1016/j.culher.2003.04.002
21. Murtagh WJ (2006) Keeping time. The history and theory of preservation in America. Wiley, Hoboken. ISBN 0-4714-7377-4
22. Rabun JS, Kelso R (2009) Building evaluation for adaptive reuse and preservation. Wiley, New York. ISBN 978-0-470-10879-6
23. Bullen PA, Love PED (2010) The rhetoric of adaptive reuse or reality of demolition: views from the field. Cities 27:215–224. https://doi.org/10.1016/j.cities.2009.12.005
24. Bullen PA, Love PED (2011) Adaptive reuse of heritage buildings. Struct Surv 29:411–421. https://doi.org/10.1108/02630801111118243
25. Wang HJ, Zeng ZT (2010) A multi-objective decision-making process for reuse selection of historic buildings. Expert Syst Appl 37:1241–1249. https://doi.org/10.1016/j.eswa.2009.06.034
26. Lucchi E (2016) Multidisciplinary risk-based analysis for supporting the decision making process on conservation, energy efficiency, and human comfort in museum buildings. J Cult Heritage 22:1079–1089. http://dx.doi.org/10.1016/j.culher.2016.06.001
27. Rodrigues C, Freire F (2017) Adaptive reuse of buildings: eco-efficiency assessment of retrofit strategies for alternative uses of an historic building. J Cleaner Prod 157:94–105. http://dx.doi.org/10.1016/j.jclepro.2017.04.104
28. Tadeu S, Rodrigues C, Tadeu A, Freire F, Simões N (2015) Energy retrofit of historic buildings: environmental assessment of cost-optimal solutions. J Build Eng 4:167–176. http://dx.doi.org/10.1016/j.jobe.2015.09.009
29. Ascionea F, Chechea N, De Masia RF, Minichielloa F, Vanolib GP (2015) Design the refurbishment of historic buildings with the cost-optimal methodology: the case study of a XV century Italian building. Energy Build 99:162–176. http://dx.doi.org/10.1016/j.enbuild.2015.04.027
30. Roberti F, Oberegger UF, Lucchi E, Troi A (2017) Energy retrofit and conservation of a historic building using multi-objective optimization and an analytic hierarchy process. Energy Build 138:1–10. http://dx.doi.org/10.1016/j.enbuild.2016.12.028

31. Intergovernmental Committee for the Protection of the World Cultural and Natural Heritage (2008) Operational guidelines for the implementation of the world heritage convention. United Nations Educational, Scientific and Cultural Organisation, World Heritage Center. Available online: http://whc.unesco.org/archive/opguide08-en.pdf
32. Pedersen A (2002) Managing tourism at world heritage sites: a practical manual for world heritage site managers. World Heritage Manuals. UNESCO World Heritage Centre. Available online: http://whc.unesco.org/uploads/activities/documents/activity-113-2.pdf
33. Alexander Turnbull. Available online: http://natlib.govt.nz/. Accessed 20 Nov 2017
34. Rausch C (2017) Global heritage assemblages: development and modern architecture in Africa. Routledge, New York. ISBN 13:978-1138219472, ISBN 10:1138219479
35. Roter Blagojević M, Milošević G, Jadresin Milic R, Nikolić M (2013) A new life of the Ottoman Fortress Ram. J Cult Heritage 14–35:S20–S24. ISSN 1296-2074. http://dx.doi.org/10.1016/j.culher.2012.11.021

The Ottoman Landscape of Büyükçekmece. A Case of Oversight or Misinterpretation of the Past?

Luca Orlandi

Abstract The focus of this paper is the Ottoman site at Büyükçekmece Lake in Thrace, incorporated today in the outskirts of Istanbul. The Ottoman site analyzed as case-study was designed in the second half of the 16th Century by the master-builder Sinan for the Sultan Suleiman and Grand Vizier Sokollu Mehmet Pasha. The Büyükçekmece complex, a post station on the road, included a monumental stone bridge, a caravanserai, a small mosque and a fountain, and for centuries was an important center along the Imperial road connecting the capital with the West. As a result of uncontrolled expansion of the outskirts of Istanbul, starting in the 50s and 60s of the 20th Century, the site has been radically compromised, with the complete transformation of the rural landscape that was surrounding the complex, transforming it into an industrial area interspersed with residential zones. A recent project consisting of tourist-sports center and an entertainment park done by the Great Municipality of Istanbul in order to revitalize the area and attract more visitors, had compromised the historical buildings and the original site, making impossible to read the traces of the Ottoman past in that region. The impact on the historical buildings has been completely ignored and the redevelopment process did not take into account the cultural values of the architectural elements, their legacy and their relation with the landscape. This paper describes how regenerating the Ottoman heritage in Büyükçekmece, had compromised the whole site, giving no more possibilities to read the landscape that once characterized the area.

Keywords Sinan · Ottoman heritage · Ottoman landscape · Thrace

---

L. Orlandi (✉)
Department of Architecture, Istanbul Technical University, Istanbul, Turkey
e-mail: orlandiluca@itu.edu.tr

G. Amoruso and R. Salerno (eds.), *Cultural Landscape in Practice*,
Lecture Notes in Civil Engineering 26,
https://doi.org/10.1007/978-3-030-11422-0_20

1 The Architect and the Patronage: "Building" the Landscape at Büyükçekmece Lake

The post station, *menzil külliye* in Turkish, and the bridge designed by the great Ottoman architect Sinan between 1563 and 1567 were built close to the shores of the Marmara Sea and the Lake Büyükçekmece in Thrace and they represented one of the most important achievement concerning a monumental work along the Imperial road connecting the capital of the Ottoman Empire, Istanbul, with Edirne, the ancient Adrianopolis [1].[1] Situated approximately 35 km from the city center, this post station in Büyükçekmece[2] was the first stop for those leaving the city and the last one for those travelling from West towards the Ottoman capital. According to several scholars, the imposing bridge built nearby is one of the best accomplishments made by Sinan as engineer among its vast production. It is also the only building mentioned in the epitaph carved in marble above Sinan's tomb in the Süleymaniye complex, written posthumously by his poet friend Sâi Mustafa Çelebi. In the 16th century, the area of Büyükçekmece Lake used to be covered with vast swamps and subjected to sudden floods after heavy rains. Sultan Suleiman wanted a long and solid stone bridge to cross the waters, for the safety of travelers and to move faster the army during the military campaigns towards West. In the complex built on the road by Suleiman the Magnificent—completed by his successor Selim II—assisted by the Grand Vizier Sokollu Mehmet Pasha, is possible to see the small but important post station also designed by Sinan. The complex consisting of a small mosque, or *mescid*, with a peculiar free standing minaret, a caravanserai opposite it, as well as a public fountain in an open space, all gathered around the caravan route at the head of the bridge towards Istanbul, provided a place to pray, lodging and refreshment for the travelers (Fig. 1).

In order to understand the milieu created by the Ottomans in their provincial areas of their vast empire, we should emphasize how the territories were changed in a more suitable environment for the new Muslim conquerors. The annexation and transformation of those territories, once in the hands of the Byzantines, especially in the Balkans areas but in Anatolia as well, was a constant practice very well analyzed in the studies of Cerasi [2].[3] Büyükçekmece is not an exception and confirms this attitude of the Ottomans to absorb and renovate the lands according to their needs, like the establishment of pious foundations, *külliye* or *imaret*, with mosques as fulcrum, the caravanserais for the travelers, the public baths and fountains,

---

[1]There is a vast literature on the works of Sinan, but specifically concerning the territorial scale, the landscape and the routes to the Balkans it is possible to find many update references in Orlandi [1].

[2]The inhabited center of Büyükçekmece, called Athyras in ancient times, is located on the eastern shore of the homonymous lake, at the point of confluence with the Marmara Sea, and constitutes an autonomous municipality that lies within the administrative boundaries of the extreme western suburbs of the Great Istanbul Municipality.

[3]About the fundamental studies conducted by Maurice Cerasi in the past years on the Ottoman towns [2].

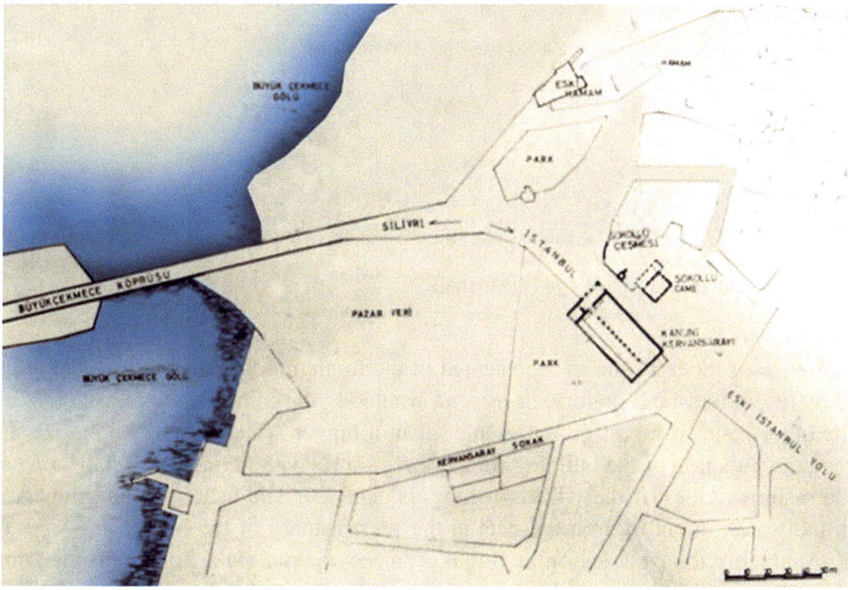

Fig. 1 Plan of the post station at Büyükçekmece (re-elaborated from Küçükkaya (1990), Mimar Sinan Dönemi İstanbul-Belgrad arası menzil yapıları hakkında bir deneme, Essay on the station post buildings between Istanbul and Belgrad in Sinan "s age), *Vakıflar Dergisi*, XXI

the formation of new neighborhoods (*mahalle*), the bazaar areas and the infrastructures like roads, aqueducts and bridges. All those elements, with specific and define function combine with the typical vernacular houses were essential part in the formation of the Ottoman town, as asserted by many scholars in their researches in the past and in more recent times [3–5].[4]

Today it is very difficult in Turkey and particularly in cities like Istanbul, mainly due to the fast modernization process, to recognize how urban environment was especially in the 16th century, since many material traces have been lost forever. Other sources, like travelogues, commentaries, memoirs and drawings or sketches, can be helpful to understand that environment that once existed. For instance, in the *Şehnâme-i Selim Han*, (Book of Kings of [Sultan] Selim Khan) ordered by the Sultan Selim II to commemorate his life and his military actions, a miniature made by Nakka Osman well depicts the area of Büyükçekmece towards the end of the 16th century. In this miniature, we can clearly recognize the stone bridge with the structure composed of four large humpback bays; each of them consisted of arches and piles; on the right side of the bridge the village of Büyükçekmece is drawn including the fountain, the caravanserai with a pitched roof and a large adjoining courtyard. Behind this wall is a mosque. It has not been possible to ascertain the

[4]Among the scholars and their researches we can briefly mention the following ones: Pinon [3], Lowry [4], Akin [5].

Fig. 2 The Büyükçekmece bridge (© Orlandi)

existence of this mosque as represented in the miniature, because it does not seem to represent the small oratory, or *mescid*, wanted by Sokollu Mehmet Pasha but, on the other hand, it might be represented it in a bigger scale only to emphasize the religious function of the building. In the foreground, with richness of details, one of the main activities of the village, fishing, is depicted. The fishermen are immersed in the water, intent to drop the nets in the great estuary of the lake.

Analyzing the post station at Büyükçekmece we can make some consideration about the past, the state of the art of those structures and their present conditions. Undoubtedly, the Büyükçekmece Bridge, also known as Suleiman Han Bridge, is the most interesting structure of the whole complex, with its total length of about 700 m and divided into four large humpback bays. The elegance and at the same time the practicality of this system of bridges connected to each other is immediately perceptible as a thin mass of gray stone, inserted in the surrounding landscape, dominated by slight hills and mainly horizontal lines: the lake and the sea (Fig. 2).

The strength of the whole stone structure from the solid pillars is underlined by the sharp cusps that form the humps of the four bridges—certainly an inspiration from Seljuk's art—and is flanked by the linearity and rhythmic repetitiveness of the 28 arches that allow the water to flow through, echoing the design of ancient Roman bridges.[5]

The second important structure is the caravanserai, known as *Sultan Süleyman Han Kervansarayı*, built by Sinan in 1566; it is also known as Sokollu Hanı, referring to the Grand Vizier Sokollu Mehmet Pasha, who supervised the work after the death of the sultan. The building consists of a large rectangular hall whose measures are 48 m long by 22.30 m wide. The short sides are raised in rectangular shape with a triangular tympanum on which rests the wooden pitched roof, covered externally by tiles. The perimeter walls are built in alternating courses of stone and brick and in the interior, there are some raised platforms in stone that run along the two major walls along the entire length of the building; they were used by travelers

[5]In 1566, the Sultan Suleiman the Magnificent died during a military campaign against the Austrian Empire while he was besieging the city of Szigetvár in Hungary, not being able to see the work completed. But the buildings at Büyükçekmece were carried out by his son and heir Selim II, who maintained Sinan as the site architect, under the supervision of Sokollu Mehmet Pasha, the new Grand Vizier of the Empire, appointed by Suleiman just one year before his death.

to rest and eat, not being in direct contact with the damp ground of the hall. Along the walls there are 12 fireplaces for cooking or warming up and 24 niches for storing personal belongings. Three rows of pillars (33 in total) run between the two raised platforms to support the truss beams of the large roof; while the two sides are made of wood, the central one was made up of eleven stone pillars. On the short side facing south-east there are several openings of different shapes, alternately round and elongated with ogive, with geometric patterns that let the internal lighting of the room; those visible todays are no longer the originals and the effect from outside is more the result of an overlapping of layers which occurred after the recent restorations of the building in 1988, rather than an integration into the wall surface. The loopholes have similarities with the openings that were used during the period of the Seljuks (between the 11th and 14th centuries) in the *ribat*, a type of fortification along the caravan routes (Fig. 3).

On the short side facing north-west there is a portico covered by a sloping roof that precedes the real entrance of the caravanserai which includes a closed space on the left, used as accommodation for guests and travelers. Also, this part is built in alternating courses of stones and bricks and two large arched windows on the south-west side allow natural indoor lighting. There were no traces left of the portico, as evidenced by the survey made between the 50s and 60s, and all this part of the caravanserai was completely rebuilt during the restoration works. During World War Second, despite the neutrality of Turkey to the conflict, the area was heavily militarized, used as strategic point near the sea and close to Istanbul. A series of reinforced concrete bunkers, some of which can still be seen in the area, were built all around the complex and soldiers used the caravanserai as accommodation place. After the war period, the ruined building was used by farmers to

Fig. 3 The Sultan Süleyman Han caravanserai and the new access to the area (© Orlandi)

store the seed oil obtained from the sunflower fields in the neighboring agricultural areas. Subsequently, in the years 1965–66 the area was subjected to a heavy restoration under the control of the General Direction of the Pious Foundations, the *Vakıflar Genel Müdürlüğü*. During these renovations to repair the large roof, which in the meantime had collapsed inside the structure, destroying its internal supports, the wooden pillars and joints were replaced by new ones in reinforced concrete, as well as the stone capitals.

The other important structure of the station post is the small mosque, the Sokollu Mehmet Pasha oratory, or *Sokollu Mehmet Paşa Mescid*, that was part of the same complex and was ordered by Sokollu Mehmet in 1566 to be used for prayer by travelers and by those who stayed in the station post. It is a building of modest proportions, located on the north side of the caravan route, in front of the caravanserai (Fig. 4).

The measures of the prayer room are 9.50 × 7 m and the building is built in alternating courses of stones and bricks. Internally the structure is no longer readable and it is difficult to find the original details due to the repeated bad restorations where both layers of plaster on the original wall surfaces and wooden beads along the walls have been added. Even the decorations on the walls and ceiling are of low quality and poorly done without following any criteria. From the outside, the four-pitched roof is covered with leaden sheets stretched over the external portico, supported by square-section wooden pillars. These were completely replaced during the restoration in 1962–63 and subsequently, during the last restorations, they were again replaced by a wooden structure and with glass windows, closing completely the portico and transforming it into yet another room to add more space for the worshippers. The new room thus has three wooden walls

Fig. 4 The Sokollu Mehmet Pasha oratory, seen from South-East (© Orlandi)

Fig. 5 The façade of the Sokollu Mehmet Pasha oratory (© Orlandi)

with large openings with white PVC frames and the fourth side, the one that constituted the stone façade of the mosque, was rudely plastered, except for the entrance portal alone[6] (Fig. 5).

Fortunately, the other three external sides of the mosque have been preserved in their entirety and there are no obvious alterations on the masonry surfaces. In the courtyard, near the portal insert in the stone precinct that connects the mosque with the caravan road, there is an odd minaret, completely detached from the building; a combination of a pulpit and a minaret, *minber-minare*, a not very usual structure in Sinan's design. This structure is a fine stone work, inserted between the courtyard portal and the perimeter wall of the enclosure: a staircase composed of 12 steps leads to the octagonal loggia, where geometric decorations in the stone panels are enclosed the ogive windows. These panels, thanks to eight slender pillars, form the base for the bulb hat that encloses the turret-loggia (Fig. 6).

As a completion of the *menzil* at the head of the bridge, stands the Suleiman fountain, *Sultan Süleyman Han Çeşmesi*. It was ordered by Suleiman in 1566 when he was still alive, to provide a refreshing place for travelers; it is likely that the sultan saw it under construction—as well as the bridge—as he left to Hungary for the last military campaign of his life. This fountain is tripartite, consisting of three panels, each of them with a large arch carved into its walls; the length is about 6 m while the height reaches 4 m[7] (Fig. 7).

[6]The marble arch above the entrance to the prayer hall is left intact and an Ottoman inscription carved on its surface is still readable: «Kelime-i tevhid» (The real Word).

[7]In the central panel, from which water spurts, an inscription, dated 1566, celebrate the Sultan and its endowment: "This fountain was built by the sultan Suleiman khan victorious in 974 H. bring new fresh water to the world".

Fig. 6 The *mimber-minare* structure in the Sokollu Mehmet Pasha oratory (© Orlandi)

Fig. 7 The Suleiman fountain (© Orlandi)

2 The Regeneration Project of "Cultural Park" and the Ottoman Heritage Today

The urban-scale project in the Büyükçekmece area, built in the late 90s for a total of nearly 200,000 m^2, bore the signature of the Municipality of Büyükçekmece Technical Office and the aim of the project was to develop a new plan for the urban redevelopment of the area.[8] The main concept was to open the entire area to the public, providing it with various services: a large outdoor theater, a new public bath, a sports center with tennis and soccer fields, a new building used by the Municipality itself for cultural events and administrative activities, as well as small buildings used for refreshment services such as cafes, bars and restaurants. It should not be surprising that in Turkey this kind of operations and projects are very often designed and implemented without any local or international competition and almost always without the involvement of local authorities or experts, citizens" associations, Chamber of the Architects or other participants at district level, such as local and public committees etc. Another attitude of the local municipalities is to reserve absolutely discretionary choice when selecting consultants or academics for their preliminary studies. This well-established practice, even looking at the most recent developments in the country, worsened exponentially. The decisions, as well as the projects, are super-imposed in a way solely linked to the immediate interests of the municipalities or its directors at that specific moment, without a vision to foresee the possible developments and without taking preventively in account the objective difficulties that could be found in the area; doing so often, the projects turn out to be the design of something absolutely artificial, created as it is "a flat table", where the problems that could arise are simply discarded for the simple reason that the spam between the design process and the executive phase must be done in least possible time (Fig. 8).

The crucial problem of the Büyükçekmece area project was obviously that of reviving the *menzil külliye* built by Sinan and thus enhancing the historical area, or rather the little that had survived of this important caravan center, giving it a new look for attracting tourist and giving it a cultural meaning at the same time; the Kültur Park's project was subjected to a real makeover and was born from what looks like a "Disneyland-style" idea to glorify Sinan, through an interpretation—perhaps a bit "too simplistic" and naïve—of the great architect. An enormous statue of Sinan that "should" portray the Ottoman architect of the great project commissioned by Suleiman the Magnificent and continued by the great vizier Sokollu Mehmet Pasha, was thus placed halfway between the area occupied by the fountain, the caravanserai and from the oratory and the bridge head (Fig. 9).

[8]All the information about the restoration process and the project done by the Municipality of Büyükçekmece is presented in their website: retrieved from: http://www.bcekmece.bel.tr/en-us/Buyukcekmece/Ilcemizde-Yasam/TarihiYerler/Pages/Büyükçekmece-Menzil-Külliyesi.aspx.

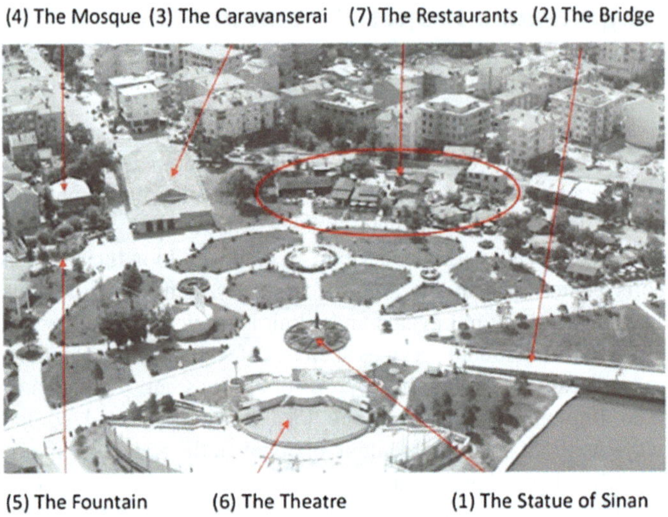

Fig. 8 The new "Culture Park" project

Fig. 9 The Statue of Sinan in the center of the park (© Orlandi)

All the functions, services and gardens that form part of the "cultural park" are developed concentrically around this statue, including—ironically—even the real "historical monuments". The boundaries of the new park are very limited: to the west by the lake and by the access to the Sinan's bridge, while in the opposite direction we reach the land belonging to the caravanserai; in the remaining two directions, respectively north and south, the park is bounded by the large theater behind which there is a sports center, and by the catering services located in small structures. Within this large circle, of which the great statue of Sinan is the fulcrum, green areas to rest and walk and refreshment areas that have been designed for the public, open only to pedestrian traffic and not to vehicular traffic. During the realization of the project, great attention was paid to the distribution of these green spaces, with fountains, pools and paved pedestrian paths completed by urban furnishings; they are also found in the middle of the flowerbeds and the greenery—apparently more didactic than aesthetic—statues of historical figures both Ottomans and those from the republic period, in which everything must be put on display, as if the "Sinan theme park "was not enough (Fig. 10).

The most important and imposing architectural work carried out in the park is certainly the large open-air theater, with 5000 seats, which hosts important cultural events during the summer, such as dance shows, concerts, theatrical performances and other shows. The sports complex behind the theater occupies an area outside the park in a northerly direction, very close to the lake; here there are different playgrounds and indoor and outdoor areas—allowing both summer and winter use—for different games like basketball, soccer, tennis and volleyball. Near the sports fields, used above all by young people who go to play or play sports, there is also a reinforced concrete structure that houses a contemporary Turkish bath, probably replacing an existing old one, lost during the last centuries. The town hall building in the immediate vicinity of the urban park and the Sinan caravanserai are periodically used by local folk groups, dance or guitar courses and various artistic events. During the religious month of Ramadan, the caravanserai is also used for the promotion of local food products from the Büyükçekmece area, with stands and gastronomic tastings, and also serves for performances by small theatrical companies or for the traditional Ottoman shadow theater: Karagöz-Hacivat. From the data relating to some activities, for example, it appears that for the I International

Fig. 10 Statues, landscape and new buildings in the park (© Orlandi)

Fig. 11 A trash bin in the park; in the background the open-air theatre and other structures (© Orlandi)

Summer Dance Festival, held in 2002, more than 100,000 visitors were welcomed. The lake and the gulf are still used for fishing and as a reserve of water, even if there are also small tourist boats. There are also many restaurants scattered around the park, especially along the banks of the lagoon lake, in the direction of the new highway bridge, which runs parallel to that of Sinan, about half a kilometer away from the Sea of Marmara. The area has been landscaped to give the opportunity to have lunch or dinner outdoors with a panoramic view of the Sinan Bridge. Then there are the *Yöre Evleri*, the traditional houses, which house bars, kiosks and rooms of all kinds and shapes, incredibly small in size, generally made of wood and reminiscent of rustic village houses or mountain huts, almost a miniature reproduction, built more for what could be an audience of "dwarfs and elves" than for ordinary customers. Trying to look like a delicious corner of the past, perhaps alluding to a healthy and genuine cuisine like that of the past, the "mini buildings" are named with toponymies of Turkish cities like "The Mansion of Trabzon", *Trabzon Konağı*, "The House of Cyprus", *Kıbrıs Evi*, the Inn of Urfa, *Urfa Lokantası*, and so on. The fairy tale atmosphere that you breathe while walking in the park is also underlined and marked by the colorful flower beds, paths and cobbled streets, fountains surrounded by round pools and drinking fountains, and white statues plastic or fiberglass depicting smiling children or forest animals, carrying fake wicker baskets used as (real) trash bins (Fig. 11).

3 A "Post Facto" Evaluation of the New "Cultural Landscape" at Büyükçekmece

Although the effort of the Municipality of Büyükçekmece was to launch cultural activities in the area affected by the historical presences of Sinan artifacts, and inspired by them as intention, it is difficult to acknowledge whether the goal was achieved or not. In fact, faced with an evident overall improvement in the area and more than positive economic balance coming from the influx of the park's cultural initiatives, it does not seem that the objective of safeguarding the area from the point of architectural-urban view of the existing artifacts is achieved. It seems that the designers of the area have not understood the importance of the ancient complex and its appropriate environmental integration in a project of a cultural urban park. The entire area around the lake has, in fact, been damaged in the last thirty years by illegal building activities, by a non-discipline in organizing the building fabric and by the proliferation of establishments without any urban planning control behind it. The Ottoman road, with its layout, the *menzil külliye* and the great bridge of Sinan, should have been the center of the project, the axis that—in an ideal sense—would have allowed finding a balance between the already compromised surrounding environment and the architectural works. Instead, these elements, no longer part of any historical and environmental milieu, are set in the park's project, in a literal sense, like big "Disneyland" silhouettes from a Hollywood film scene about the Ottoman Empire. The statue of Sinan, disproportionate and clumsy, has been placed in the center of a series of concentric circles halfway between the lake—on the side of the bridge head—and the caravanserai, unmistakably erasing any sign of any original route and marginalizing the authentic buildings made by him. Moreover, the road that once crossed the bridge doesn't connect anything today. It is just a pedestrian path to go back and forth from one side to the other and in a way; it is good that at least is not use by vehicles, since it was close to traffic. But the problem is that the bridge has become a self-standing monumental object, just a tourist attraction, real this time, not artificially reconstructed, trapped between the expanding town of the Municipality of Büyükçekmece and the industrial areas on the opposite bank. Within this approach the *genius loci* was not respected at all, and despite the conditions in which it are located, akin surrounded by factories, highways and industrial plants, the buildings of the complex could at least be safeguarded for what was left of them in a proper manner; instead they have been treated individually, forever erasing any material connection and leaving them alone, in their bare monumentality, thinking only of their uniqueness, without any physical context to rejoin them ideally with the past. The best example of this cultural incomprehension is nevertheless given by the great open-air theater reechoing in shapes the Ottoman fortresses; (Fig. 12) the construction was carried out following the harmonic lines of the concentric circles that depart from the statue of Sinan—always that one. In the minds of the designers the fire generated by the center of the square, the circles, the Sinan's statue, would have contained

Fig. 12 The "Ottoman" theatre (© Orlandi)

perfectly—Euclidean it could be said—the circles of the step and stage stands, with a small detail of which they do not seem the designers have noticed.

The ideal background of the theater, the "natural" scene that could—and should—be the great lake, for the joy of the spectators, with the spectacular bridge and open space, in the dullness of "wrong" points of view, is been completely neglected. And so, the great theater gives its back to the lake, and there is no possibility from the tribune to enjoy the sight of a minimal panorama, if not the one offered by the pathetic statue. In conclusion of what has been said so far and to confirm that general interest of projects, when motivated by priority urgencies, can move in the right direction or, as in the case for the area of Büyükçekmece, in wrong, when local interests of such administrations not attentive to the meaning of protection of a historical environment are involved. Through his work, Sinan has represented a very important milestone for all the Mediterranean culture and thanks to his architecture; Sinan has elevated the Ottoman civilization, too often forgotten at other latitudes, to the highest ranks. The case of Büyükçekmece has been explored in details in this paper, trying to understand what went wrong in the regeneration process, according to the common idea that an historical heritage should be well preserved and put under protection but also be serviceable to the users. Regarding such heritage, the lack of adequate preliminary and interpretative investigations and the absolute lack of coordination of the whole project gave rise to a previous "danger" for the conservation and protection of the already existing heritage. In conclusion, this case-study can be seen as a negative model of "cultural operation" totally detached from any real will to intervene with the protection of the assets, considering it as part of a fragile architectural-landscape system. The historical stratifications that through the centuries well defined a milieu—that in other

Fig. 13 The bridge and its environment today (© Orlandi)

works I identify as "Ottoman landscape", [6][9] should have been the keystone in understanding the complexity of the problem of assessing and implementing the historical landscape heritage into today's architectural urban interventions (Fig. 13).

References

1. Orlandi L (2017) Il Paesaggio delle Architetture di Sinan—L'esempio della Tracia, Istanbul: Ege Yayınları, Istanbul
2. Cerasi M (1988) La Città del Levante. Civiltà urbana e architettura sotto gli Ottomani nei secoli XVIII-XIX. Milano: Jaca Book and Cerasi M (2005) La Città dalle Molte Culture. Milano: Libri Scheiwiller
3. Pinon P (2008) The Ottoman Cities of the Balkans. The City in the Islamic World, Volume 2. Brill
4. Lowry HW (2008) The shaping of the Ottoman Balkans, 1350–1550: the conquest, settlement & infrastructural development of Northern Greece. Bahçeşehir University Publications, Istanbul
5. Akın N (2001) Balkanlarda Osmanlı dönemi konutları. İstanbul: Literatür
6. Orlandi L (2015) In search of an "Ottoman Landscape": Sinan's works in thrace as expression of tangible heritage. A/Z ITU J Fac Architect 12(2)

[9]This subject was already presented in a wide range in other publications done by the author [6].

Ottoman Heritage in Southern Balkans: The Multicultural Port Town of Kavala

Velika Ivkovska

Abstract Multiculturalism is one of the keystones in understanding the complexity of the Ottoman civilization. Especially in the provincial areas of the vast empire established by the Ottomans in almost six hundred years, it is possible to observe how multiculturalism and multi-ethnical components were a fundamental tool to glue together such diverse of populations and religions, in which each community had the possibility to maintain their distinctive collective identities, culture, rituals and practices. In the case of the town of Kavala in the Southern Balkans, that passed from Byzantine into the hands of the Ottoman rule after its conquest in 1391, it is possible to track how the town—through the centuries—became more and more a multicultural center, due to the economic and social dynamics in the society itself, up till the dramatic events of the Balkan Wars in 1912. The Ottoman town developed rapidly around its harbor and the trade activities and this paper wants to outline the urban environment of Kavala and its growth in the Ottoman time, from the conquest until its annexation to Greece. In a contemporary perspective, today it seems significant to widen the perception of how a vast tangible heritage left by the Ottoman civilization, crossing the borders of different nations, religions, customs and cultures, especially in the Balkan context, could be analyzed and incorporated in a bigger system to redefine the importance of such multicultural asset in a global manner.

Keywords Ottoman heritage · Vernacular architecture · Provincial architecture · Balkans

V. Ivkovska (✉)
Department of Architecture, Bahçeşehir University, Istanbul, Turkey
e-mail: velika.ivkovska@arc.bau.edu.tr

G. Amoruso and R. Salerno (eds.), *Cultural Landscape in Practice*,
Lecture Notes in Civil Engineering 26,
https://doi.org/10.1007/978-3-030-11422-0_21

1 The Lands of Rumelia and the Ottoman Penetration Towards West

This study's' fundamental component is the knowledge concerning the vast architectural and urban production occurred in that part of Southern East Europe, the Balkans. In those lands, somehow still in search of a new identity after the collapse of the Yugoslavian regime in 1992 and after the consequent tremendous civil wars and land divisions, an astonishing and substantial existence of monuments and infrastructures, dated back to the Ottoman time, is more than visible. In these lands that have always been defined by strong multi-ethnic components and multi-confessional religious system, the traces left by the Ottoman domination, during their escalation of conquests towards West, were mainly characterized by architectural artifacts, inhabited centers, infrastructures and transformed landscapes, stratified over time. Those material traces represent unequivocal strong and direct link with the European history in general and with the important role played by the Ottomans and Turkish civilizations in those border countries for long time. In fact, for almost six hundred years, the Balkan peninsula was called *Rumelia*, the 'land of the Rum' [1][1] and included within its boards several countries historically different from each other, such as: European Turkey, Slovenia, Hungary, Romania, Moldova, Bosnia and Herzegovina, Croatia, Serbia, Albania, Macedonia, Bulgaria, Montenegro, Kosovo and last but not least, Greece.

The rapid growth of the Ottomans in these lands was dictated by an agenda in which military and strategic purposes on one hand and increasing commercial and diplomatic relations on the other, were implemented and improved by moving populations from other countries to repopulate the lands, by the creation of new centers around existing ones, passing through the renovation of ancient, and almost in ruins infrastructures and route networks, planning of new towns and staging posts, in other words, by the Ottomanisation of the conquered territories.

In every historical period, these lands have always been used by the most diverse populations and civilizations as an obliged passage between the West and the East. In the Ottoman era, three large main roads to West connected the capital Istanbul with the Balkans, with the regions north of the Black Sea and with central Europe. These roads therefore played a fundamental role in all relations with the Western powers, both in terms of commercial exchanges or military and strategic purposes. The Ottomans were able to establish and develop along those roads new cities and villages, commercial centers or simply multi-functional complexes. Therefore, they built infrastructures like bridges to cross lands and aqueducts to bring water to the towns; designed new roads to increase relationships, trades and commerce between

[1]The term Rumelia, or Rumeli in Turkish, was mainly used by the Ottomans to indicate their possessions to the West; it derives from the Arab word ar-Rūm, used by both the Arabs and other Muslim peoples and by the Turkish tribes from Central Asia. It is a recurrent term in the Qur'an, to indicate the Eastern Roman Empire. Coming from the Greek word Ρωμαῖοι, or Roman, through which the subjects of the Byzantine Empire were designated. Cfr. entry: Rum, in Püsküllüoğlu [1].

people and countries, as well as to move armies, to control or to enlarge the empire's borders, following a policy of territorial expansion, and they transformed the landscape, both in architectural and urban scale, by designing functional elements, urban features and setting landmarks on it [2].

The so-called 'Left Road' or *Sol Kol* in Turkish, was the western highway of the Rumelia road system, connecting the capital of the Ottoman Empire to Greece and to Italy. Starting from Istanbul, after crossing Büyükçekmece and Silivri, it bended along the coast of the Sea of Marmara, crossing centers like Tekirdağ, Malkara, Ferecik, and reaching into the higher lands of Dimetoka, Komotini (Gümülcine), Kavala, Pravişte, Lanzaka, and Larissa then dividing into two other branches, one towards south, direct to Athens and the Peloponnesus, while the other was continuing until the coast of Ionian Sea.

The road, which in ancient times was named Via Egnatia (Fig. 1), was mainly a military road built and used by the Romans, and was for a long time the commercial road that directly connected—through the lands of today's Greece and

Fig. 1 Via Egnatia and the settlements on its route (re-elaborated from Cerasi M., 1988. La Città del Levante. Civiltà urbana e architettura sotto gli Ottomani nei secoli XVIII–XIX. Milano: Jaca Book)

Albania—the two capitals of the Roman Empire, Rome in Italy and the new "Rome", Constantinople. To reach Rome from Constantinople the land road roughly covered the above-mentioned centers except that at the point of Kavala it was coming down by the coast of the Aegean Sea due to the high hills of the Rodopi mountains who were coming down to the shores of the sea. This land route was extending towards west reaching up to the Adriatic Sea in Durres, in present-day Albania, and then continued through the Ionian Sea and reached the town of Egnatia—from which the road took its name—a small port on the Apulian coast between Brindisi and Bari, and from there the road finally was continuing to the capital [3].

To summarize, by penetrating the Balkans along those routes and spreading this road network, the Ottomans were also able to integrate in their possessions the conquered territories preserving the local cultures and at the same time they were able to assimilate and even get stimulated by those cultures they entered in contact with. Far in fact, from some old and misleading interpretations about the generic concept of the existence of a stereotype "Islamic city", the researches done on the Eastern Mediterranean and the Balkan peninsula showed clearly that those areas have always been effected by a sort of multiculturalism, related to the multiplicities of several population living in those areas and as well as the coexistence of different religions [4].

2 Ottoman Kavala Throughout the Centuries (1391–1912)

The first Ottoman conquest of Macedonia was accomplished by Evrenos Bey between 1383 and 1387, the date of Thessaloniki's surrender. [5] The conquest of the town of Christoupolis, later renamed Kavala, occurred in 1387, but the fate of the area had been decided sixteen years earlier after the battle of Tsirmen, when the Ottomans commence the rapid conquest of cities in Macedonia [6]. The conquest was completed in the last decade of the 14th Century, when the town of Christoupolis town was completely destroyed and leveled to its foundations, leaving nothing behind for almost one hundred years. According to a codex from the monastery of Panteleimon on Mount Athos [7]:

> In this year, the city of Christ, the city of Christoupolis, was destroyed by the incredible Mohammedans, and destroyed by the foundations, with speed, and the inhabitants were distributed to various districts and places.

In 1478 a tax register appeared in the Ottoman State Archives related to the population of the town of Kavala. From the conquest of Christoupolis in ca. 1391 up until the appearance of this tax register from 1478,[2] the site was vacant of

---

[2]This tax register, which is housed in the Department of Ottoman Archives (Başbakanlık Osmanlı Arşivi) in Istanbul, (under TT. d 7, s. 88–89) states, under nefs-i Kavala (inhabitants of Kavala), that in 1478 Ottoman Kavala was a settlement with total population of 467 inhabitants.

any settlement and Kavala, that was built almost a century later, was in fact a newly named and established Ottoman village/town, which name's origins are still unknown [8].

Kavala lay at the point where Via Egnatia was coming down to the shore because of the high mountains to the north. The caravans that were commuting along this route were at constant attacks by the corsairs lurking along the coast [9]. Piri Reis, a famous admiral of the Ottoman fleet in that period, in his *Kitab-i Bahriyye* (Book of The Seas) from 1521, gives a most detailed depiction of the Mediterranean world, and this book is the most accurate portolan of the 16th Century (Fig. 2). It is from him that we learn that Sultan Selim I built the fortress in Kavala, on the top of the peninsula: "West of the river Karasou (Nestos), at the top of a mountain, there is the strait they call Kavala. There, the late Sultan Selim Han (1512–1520) had built a castle" [10].

Fig. 2 Map of the Island of Thasos (from Piri Reis, *Kitab-i Bahriye*, 1521, Istanbul University Rare Books Library)

2.1 *The Rise of Kavala Under Sultan Selim I, Suleiman the Magnificent and Grand Vizier Ibrahim Pasha*

The aftermath of the conquest is visible in the fortress that Sultan Selim I (1512–1520) built on the top of the hilly peninsula. From the urban plan point of view, a great improvement of the town was done in the period of Suleiman the Magnificent (1520–1566). It was the Grand Vizier Pargalı Ibrahim Pasha who put the built program high in his agenda with first priority given to the city's safety. He, thus, reinforced and extended the old walls [11]. Inside of this new, supplementary wall, new squares were fashioned and buildings erected; three mosques are mentioned to have existed during this era, the coastal one was the central (Fig. 3), while the city also included of baths and caravanserai [12]. The charitable Muslim institutions in the city included a *medrese*, *hamam* two *kervansarays*, dervish lodge, *mekteb*, *han* and a soup kitchen [13].

The French traveler Pierre Belon's account of his visit to the northern Aegean port town of Kavala in the early spring of 1547, [14] discusses the role played by Sultan Süleyman's Grand Vizier, Ibrahim Pasha, in endowing a number of charitable works on behalf of the inhabitants of Kavala, and makes it clear that their services were in no way restricted to Muslims. Belon states that the *kervansaray-*

Fig. 3 Ibrahim Pasha Mosque, today converted into the church of St. Nicholas (© Ivkovska, 2016)

imaret, or inn-soup kitchen which Ibrahim Pasha built as part of his *vakıf* (religious foundation), was open to all regardless of their religious affiliation [15].

> Taking into account that there are hardly any hostelries in Turkey let us speak about the great building which İbrâhim Paşa erected in Kavala, which the Turks call a Carbasharra (sic. Kervansaray). He also built a mosque next to the hostel, where all who pass by are lodged and fed. Our group was only three in number, with our horses, and we were given food for three days in succession without paying anything and without any trouble… Nobody, be he Christian, Jew, Muslim or idolater is refused here [15, p.107].

Belon also provides us the information that Kavala had already been settled by Jews, brought over, by the Ottomans, from Hungary [16].[3] This must probably be dated to 1527/28 after the Ottomans forces, that were co-commanded by Sultan Süleyman and his Grand Vizier Ibrahim Pasha, took Buda and Pest in 1526 [17]. Why the Jews were brought and offered shelter to Kavala by Ibrahim Pasha is unknown. Probably it was due to his concern with all the aspects of the physical infrastructure of the town he spent so much time creating, so he took direct action to increase the size of its population [17]. The Jews established in Kavala an extremely developed and progressive community with a continuous presence spanning over four centuries [18]. The phase of prosperity for the city essentially commenced with the exploitation of its commercial port. During the 17th and 18th century, the establishment of the French consulate in 1701, the Venetians inaugurated their consulate in 1746, and the French Commercial House established in 1771 were the moving force for the people of the city, motivating them to engage in trade activities [18, p. 44]. In 1789 goods from France, and specifically from the port of Marseilles, were imported via Kavala to cater for the needs of the Macedonian and Thracian inland. Correspondingly, other products, such as wool or rice, which were produced in the nearby region were accumulated in Kavala and exported to France [18, p.189].

3 The Development of Kavala as Important Center in the Mediterranean Sea for the Tobacco Industry

The development of Kavala in the early 19th Century was principally effected by its favorable geographic location, and being an important port in the north coast of Eastern Macedonia, there were the appropriate conditions to establish new commercial activities; the port was being used for shipment of tobacco, that which harvest and production in the area started growing and increase rapidly during the century. By the end of the 18th and throughout all the 19th Century, Kavala was marked by a period of great development. The most important figure that brought up this development was Mehmed Ali Pasha. Mehmet Ali Paşa (Kavalalı Mehmet

[3]For further information, see Merle [16].

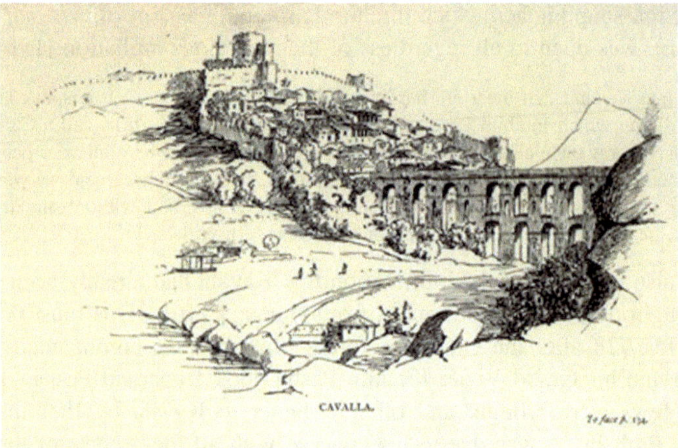

Fig. 4 Drawing of Kavala from the book by Mary Walker (Old Tracks and New Landmarks, 1897)

Ali Pasha in Turkish) was born in this small Aegean port town in Ottoman Rumelia, the European part of the Ottoman Empire, in the year 1184 Hicri (August 27, 1771–April 17, 1771) [19].

> The period of Mehmed Ali's reign, which started in 1805, when he was appointed by the Ottoman Sultan as *wali* of Egypt and ended in 1848 with his deposition as a result of mental illness, offers one of the most interesting epochs of modern Egyptian history. During this period Egypt, while still forming a part of the Ottoman Empire, assumed an increasingly independent stance, and was finally granted as a hereditary domain to Mehmed Ali by the Sultan 'Abd al-Majid in 1841 [20].

The evidences of Mehmed Ali's affection of his birthplace were to be seen in the town. He had rebuilt the aqueduct, built three centuries earlier by Süleyman the Magnificent and his Grand Vizier Ibrahim Pasha (Fig. 4), and brought the water inside the walled town, where the Muslims lived.[4]

It was known that Kavala was a Turkish town and that there were not any Christian churches within the gates [21]. The town was double walled and occupied by Turks. The entrance to the walled town was through not over-wide gate. The town's urban layout was consisted of bends and corners and shady passages with mosques and minarets [21] (Fig. 5).

After the Greeks started moving into Kavala, from various parts of the country because of the need of workers due to the expansion of the tobacco industry in the region and Kavala being the export port for this stock, the Christians, after getting a permission from the Ottoman authorities, built their quarter in the plane located behind the neck of the peninsula (Fig. 6).

[4]Fraser in pictures from the Balkans from 1906 states about Kavala that: "It is just the place for a Turk. It makes even an Infidel feel like a Turk".

Fig. 5 Postcard from Kavala showing the old walled town, the new structures built in the plains by the harbor and the many minarets (HLAK, beginning of 20th century)

Fig. 6 Edward Lear's drawing of Kavala, 1856 (HLAK)

The profession of Toutountzi (tobacco) was mainly conducted by Greeks, but also by Armenians, Jews and Turks [22]. The town of Kavala in the late nineteenth century, became one of the most important tobacco centers in the Balkans with merchants arriving both from the Ottoman lands as well as the Western empires, especially Austro-Hungary. As a result of this many tobacco warehouses were built outside the town walls. The construction of these warehouses significantly reshaped the city, shifted its center, changed its character, and its borders. The architecture of those warehouses shaped the identity not only of the town but the society too. These structures in a way erased the borders between the city and the production facility reshaping the life of the tobacco workers [23]. The outlook of Kavala changed

Fig. 7 The tobacco depots of the Commercial Company of Salonica LTD (Tobacco Museum, Kavala, 2015)

significantly by the middle of 19th century [24]. The tobacco trade businesses belonged to Ottoman, Armenian, Jewish and Greek tobacco merchants, members of the diverse communities, millet that coexisted in the Ottoman Empire. The composition of the tobacco workforce itself was also multicultural. The non-discrimination between the workers of the diverse nationalities clearly stated in the charter of the Tobacco Workers International Union that was based in Kavala [23]. The first big company that settled in Kavala in the middle of the 19th century was the House of Allatini, a well-known Italian family of Thessaloniki. Later, the House was renamed into "Commercial Company of Salonica Ltd" (Fig. 7), and relocated its headquarters in London.

The Abbott Brothers moved from Thessaloniki to Kavala in 1858 and in 1884 followed the French Monopoly of Regie, which was the Governments monopoly, and the only cigarettes that were supposed to be smoked in Turkey were Regies [21]. The Austro-Hungarian "The Oriental Tobacco Trading Company Ltd" (Fig. 8) (M.L. Herzog et Cie) based in Budapest settled in Kavala in 1890.

At the same period, many foreign companies started opening their branches in Kavala. In 1901, the "American Tobacco Company" (ATC) moved in Kavala together with "Alston", "Gary", "M. Melachrino", as well as the Jewish "Schinasi Bros" which had its headquarters in New York. More or less at the same period, the Cairo tobacco industries of A. Chelmis, K. Doulgaridis, N. Tsinaklis, M. Melachroinos, Demetrios and the Armenian O. Matossian founded some great tobacco Trade Houses in Kavala. Actually, in 1910, the great companies (Commercial, Herzog and ATC) were employing 6000 workers in their various warehouses (www.ikkik.gr).

Fig. 8 The first offices of M.L. Herzog et Cie, built after 1891 (© Ivkovska, 2015)

4 A Matter of Ottoman Heritage and Revaluation of Kavala Today

Even today, scattered throughout the lands on which the sultans ruled for over six hundred years, the surviving Ottoman architectures mark in an indelible way the territories and the landscapes of the Balkans. In local historical memories, their presence and the common matrix of cultural identity, which from time to time has adapted to local needs, sometimes imposing itself other times accepting the suggestions given by the conditions.

Crossing Balkan countries such as Bosnia and Herzegovina, Albania, Kosovo or Macedonia, and stopping in cities and towns such as Sarajevo, Niš, Skopje, Plovdiv or Thessaloniki, it is still possible to come across—along the main routes or in big and minor centers—mosques, public baths, hans, caravanserais, covered markets and bazaar, vernacular architecture built in wood and stone, as well as bridges, aqueducts and other infrastructures, which testify the incredible achievement done by the Ottoman civilization in different and difficult contexts in which they operated during their long term domination in those lands.

It is worth to mention that numerous studies in the last decades have been multiplied in order to fill the gaps for an overall revaluation of the Ottoman architectural heritage located in the Balkans. More and more scholars are nowadays aware of the richness that the Ottoman civilization brought in those geographical areas and we believe that more in depth research in this sense must be conducted to

Fig. 9 Panoramic view of Kavala today (© Ivkovska 2014)

re-create a sort of "Ottoman landscape" in which all these fragmentary pieces of the puzzle can be placed in [25].[5]

Coming to the case-study presented, the tangible presence of the Ottoman past in today's Kavala is more than visible in this small town on the shores of the Aegean Sea (Fig. 9). In the inevitable and tragic process of reconstruction followed the Balkan Wars, World War First and the Greco-Turkish War, overall between 1912 and 1922, and following the other big urban changes and transformations carried on by the equally inevitable modernization, nevertheless a fragmentary process of reconstruction and revitalization is implement and ongoing in the small town.

Despite the fact that most of the town's Ottoman heritage, considering both the prominent monuments like mosques, aqueduct, hamams, madrasas as well as the civic architecture were neglected for decades, a new wave of treatment and behavior towards this heritage and its protection is ongoing.

The architectural or urban artifacts still existing in Kavala should not only be perceived and experienced as simple monuments or museums or exhibition areas, but they should be integrated as a whole within the environment that contains them. The main strategy should help to identify the asset as a heritage belonging to the civilization that generated it, and therefore to consider it as part of a common historical memory and as an active player of a precise urban context with specific functions. Also, what should be done in order to preserve and protect those goods?

This should be a matter of interest—and a specific duty -from other institutions, foundations or entities that will put appropriate effort, with specific targets and precise goals, and deciding how to intervene. Again, the main problem that lies underneath this awareness is more complex, because it is linked to knowledge and acceptance that this heritage exists. Another issue is to contextualize it so it should be possible to identify the relationships between the territory itself and the architecture or urban fabric built on it. Preserving and revitalizing cultural and

[5]A recent publication on Ottoman architecture, based mostly on the studies in the Balkan region of Thrace and considering the works of the 16th Century great architect Sinan, presents extensively an interesting analysis on the concept of 'Ottoman Landscape'. See: Orlandi [25].

environmental heritage, like in the case study of Kavala, should be a crucial aim in order to promote the sustainability of local communities and increase public accessibility of those areas.

Moreover, the recommendation for the assessment of the architectural heritage of those communities is as important as the duty to conserve the significant built heritage and its values or traditions of previous eras.

Acknowledgements The author would like to recognize the important contributions to the understanding and appreciation of the Ottoman presence and heritage, particularly remarkable in the researchers conducted in the past years by scholars like Professor Heath Lowry whose work has covered most of the territory of Northern Greece from historical aspect, including here the built public and religious structures, documenting the Ottoman monuments in Kavala as well as providing us with important archival documents. Maurice Cerasi who worked mainly with the formation of the Ottoman town in the Levant and in the Balkans was able to understand and interrelated the multi-faced cultural expressions of such varieties of populations in those lands studying their architecture and the urban history. The extensive work of the Greek scholars like Stedanidou, Renentzi, Bakirtzis, Papazoglou, Lichounas, Ageloudi etc. among which many natives of Kavala have given an invaluable contribution to the knowledge, preservation, revitalization and protection of the architectural heritage of the town of Kavala.

References

1. Püsküllüoğlu A (1999) Doğan Kitapçılık A.Ş. Türkçe Sözlük, Istanbul, p 1285
2. Orlandi L (2015) In search of an "Ottoman Landscape": Sinan's works in Thrace as expression of tangible heritage. A/Z ITU J Fac Archit 12(2):61
3. Orlandi L (2017) Il Paesaggio delle Architetture di Sinan—L'esempio della Tracia. Ege Yayınları, Istanbul, p 49
4. Cerasi M (2005) La Città dalle Molte Culture. Libri Scheiwiller, Milano, p 15
5. Smyrlis K (2008) The first Ottoman occupation of Macedonia (CA. 1383–CA. 1403) some remarks on land ownership, property transactions and justice. Diplomatics in the Eastern Mediterranean 1000–1500: aspects of cross-cultural communication. Brill, Boston, p 327
6. Papazoglou KN (2008) Eastern Macedonia: Kavala and Drama—time and space (eds) Region of Eastern Macedonia and Thrace-Eastern Macedonia. "Anglohellenic"—"Pelti" Publications S.A., Komotini
7. Schreiner P (1977) Die Byzantinischen Kleinchroniken. 2 Teil Historischer kommentar. Verlag Der Österreichischen Akademie Der Wissenschaften, Wien, p 343
8. Lowry HW (2008) The shaping of the Ottoman Balkans, 1350–1550: the conquest, settlement & infrastructural development of Northern Greece. Bahçeşehir University Publications, Istanbul
9. Zachariadou et al (1996) The via Egnatia under Ottoman rule, 1380–1699. Crete University Press, pp 9–16
10. Lowry HW (2010) Piri reis revisited: the Kitab-ı Bahriyye as a source for Ottoman history—the Aegean Port of Kavala & the Island of Limnos (İlimli) as Described by Piri Reis. Osmanlı Araştırmaları 35:15
11. Lowry HW (2008) The shaping of the Ottoman Balkans, 1350–1550: the conquest, settlement & infrastructural development of Northern Greece. Bahçeşehir University Publications, Istanbul, p 235 (Chapter VI)

12. Papazoglou KN (2008) Eastern Macedonia: Kavala and Drama—time and space (eds) Region of Eastern Macedonia and Thrace-Eastern Macedonia. “Anglohellenic”—“Pelti” Publications S.A., Komotini, p 169
13. Lowry HW (2008) The shaping of the Ottoman Balkans, 1350–1550: the conquest, settlement & infrastructural development of Northern Greece. Bahçeşehir University Publications, Istanbul, p 234 (Chapter VI)
14. Yerasimos S (1991) Les Voyageurs dans l’Empire ottoman (XIVe–XVIe siècles), vol 117. Imprimerie de la Societé Turque d’histoire, Ankara
15. Lowry HW (2010) The ‘Soup Muslims’ of the Ottoman Balkans: was there a ‘Western’ & ‘Eastern’ Ottoman Empire? Istanbul J Ottoman Stud 36:105
16. For further information, see Merle A (2001) Voyage au Levant-Les observations de Pierre Belon du Mans (eds) Paris
17. Lowry HW (2008) The shaping of the Ottoman Balkans, 1350–1550: the conquest, settlement & infrastructural development of Northern Greece. Bahçeşehir University Publications, Istanbul, p 238 (Chapter VI)
18. Papazoglou KN (2008) Eastern Macedonia: Kavala and Drama—time and space (eds) Region of Eastern Macedonia and Thrace-Eastern Macedonia. “Anglohellenic”—“Pelti” Publications S.A., Komotini, p 185
19. Fahmy K (2009) Mehmed Ali: from Ottoman governor to ruler of Egypt. Oneworld Publications, Oxford, pp 2–3
20. Fahmy K (1998) The era of Muhammad ‘Ali Pasha, 1805–1848. The Cambridge history of Egypt, vol 2. Cambridge University Press, p 139
21. Fraser FJ (1906) Pictures from the Balkans. Cassell and Company, Limited, London, p 172
22. Stefanidou A (2000) The Tobacco processing areas in Kavala. Aristhotelus University of Thessaloniki, School of Polytechnics, Thessaloniki
23. Rentetzi M (2008) Configuring identities through industrial architecture and urban planning: greek tobacco warehouses late nineteenth and early twentieth century. Science studies vol 21 (1). Athens
24. Ziogas CP (1990) Traditional buildings of modern Kavala. Altinitzis Athanasios, Thessaloniki
25. Orlandi L (2017) Il Paesaggio delle Architetture di Sinan—L’esempio della Tracia, Ege Yayınları, Istanbul

A Cultural Reading of the Territory. Practices and Interdisciplinary Approach for the Making of Beauty in Apulia

Valentina Battista

Abstract Beauty is the only idea that revealing itself to the real world. It is the moment when the perceivable is handed over to men. According to the ancient Greek culture, which was inherently visual, beauty represented a well-suited harmony between finite and infinite, an overarching force to grasp those preternatural laws shaping human life, our earthly life. Consequently, the need to protect and preserve cultural heritage is a human rights issue and that cultural heritage is relevant not only in itself but also in relation to its human dimension, in particular in its meaning for individuals and community as well as in their identification and development processes. Since any cultural reading of the territory and its own beauty should start from the sources of law/the legal instruments available to ensure its preservation, we ought to consider the pivotal example of Apulia and the regional law on beauty, an ambitious project of interdisciplinary bottom-up co-drafting/programming, which ties together seemingly antithetic criteria like quality and well-being, focusing on citizen, on his needs and dignity. This law provides tools, methods and actions to experience the beauty of our territory and to appreciate the vast diversity of Apulian identities' mosaic.

Keywords Beauty · Cultural heritage · Development process · Regional law · Apulia

1 A Cultural Reading of the Territory

> Everything beautiful offers us something to understand, not only in itself but in our destiny. (Simon Weil).

Beauty is the only idea that revealing itself to the real world. It is the moment when the perceivable is handed over to men. According to the ancient Greek culture, which was inherently visual, beauty represented a well-suited harmony

---

V. Battista (✉)
Materias s.r.l, Università degli Studi di Napoli "Federico II", Naples, Italy
e-mail: valentina.battista@materias.it

G. Amoruso and R. Salerno (eds.), *Cultural Landscape in Practice*,
Lecture Notes in Civil Engineering 26,
https://doi.org/10.1007/978-3-030-11422-0_22

between finite and infinite, an overarching force to grasp those preternatural laws shaping human life, our earthly life.

A measure, which if observed carefully, in architecture as well as in the other arts, is able to produce *eudaimonia*, happiness. In addition, here is how "a fragment of a temple, wrote Simon Weil, is still beautiful because we recognize the symbol of the temple in its entirety, as everything in the universe is beautiful because it symbolizes the universe".

Beauty, therefore, is an archipelago of different islands of customs, ideas, uses, something real and imaginary, made partly of laws in part of remote uses and partly of necessity and fear. It has a saving function that brings within itself the crisis from which it is born and develops giving life to a cultural reading of territory, where recognize a lot of our heritage and discover richer than any imagination.

Beauty is entrenched to territory, whose intrinsic union has the function of "cultural mediator", of what men create, receive as inheritance and as Paul Claval emphasizes, of the ensemble of representations that enclose the transmission of sensibilities, of ideas and the rules passed down from one generation to the next. If doing beauty produces culture, culture refers us to beauty, highlighting its strong human and creative dimension, the horizon of which man gives testimony [1].

Men do not build their territory exclusively for utilitarian purposes, but for deep reasons of belonging, giving the territory itself a cultural connotation, in which to live and become aware of its cultural identity. At the source of this creativity stands cultural heritage necessary to understand processes, methods and actions of territories cultural construction according to the article 27 of the *Universal Declaration of Human Rights*.

Consequently, the need to protect and preserve cultural heritage is a human rights issue and that cultural heritage is relevant not only in itself but in relation to its human dimension, in particular in its meaning for individuals and community as well as in their identification and development processes.

As stated in art. 7 of Unesco *Universal Declaration on Cultural Diversity* adopted unanimously in Paris during the 31st session of the Unesco General Conference in 2001:

> Creation draws on the roots of cultural tradition, but flourishes in contact with other cultures. For this reason, heritage in all its forms must be preserved, enhanced and handed on to future generations as a record of human experience and aspirations, so as to foster creativity in all its diversity and to inspire genuine dialogue among cultures. [6]

It is confirmed by the definition of cultural heritage according to the *Unesco Convention Concerning the Protection of the World Cultural and Natural Heritage* (1972), which, in describing the categories of monuments, agglomerations and sites to be included in this definition, recognizes their outstanding universal value as an essential identification criterion [3].

Examples of this are, among others, the definitions of cultural heritage provided by the Faro Convention which states that "cultural heritage is a group of resources inherited from the past which people identify, independently of ownership, as a reflection and expression of their constantly evolving values, beliefs, knowledge and traditions. It includes all aspects of the environment resulting from the interaction between people and places through time; it includes all aspects of the environment that are the result of interaction over time between populations and places".

A concrete example is represented by the famous Unesco site of Alberobello with its typical limestone dwellings called *trulli* as an outstanding example of corbelled dry-stone construction, a prehistoric building technique still in use today.

According to the ECOSOC, the concept of culture must not be seen as a series of isolated manifestations or watertight compartments, but as an interactive process in which individuals and communities, preserving their specificities and goals, give expression to the culture of humanity. More specifically, on the question of the human dimension of cultural heritage, it is argued that the definition of cultural heritage is not limited to what is considered to be an exceptional value for the whole of humanity, but rather this includes what is meaningful to individuals and communities.

In the Preamble of the *Unesco Recommendation on Participation by the People at Large in Cultural Life and their Contribution to it* (1976), culture is also defined as a social phenomenon, which results from the interaction of individuals through creative activities and it is not limited to the use of works of art and humanistic content. At the same time, culture is the acquisition of knowledge, the choice of a lifestyle and the need to communicate [4].

Citizens are searching for a sense of belonging. The cultural territory is defined as a form of rooting and attachment to places, according to a cultural principle of identification. Territorial construction is no longer just a matter of material actions, but also of the discourses, values and myths it contains. Men do not only build their environment for a material purpose, but to offer themselves an image of themselves, to become aware of what they share and to experience the relationship to the other.

Whether in the field of perception of the territory, its representations, or its symbolic appropriation, territorial construction results from culture. This link has two meanings: if the territory produces culture (it is only necessary to think of "putting into myth of the landscape"), the cultural product in return of the territory: by the use of emblems, symbols the cultural makes it possible to appropriate a space, to transmit a territorial belonging constitutive of the collective and/or individual identity.

Furthermore, the *Unesco Convention for the Safeguarding of the Intangible Cultural Heritage* (2003) adds a further intangible connotation, stating, in Article 2, that:

> The intangible cultural heritage means the practices, representations, expressions, knowledge, skills – as well as the instruments, objects, artefacts and cultural spaces associated therewith – that communities, groups and, in some cases, individuals recognize as part of their cultural heritage. This intangible cultural heritage, transmitted from generation to generation, is constantly recreated by communities and groups in response to their environment, their interaction with nature and their history, and provides them with a sense of identity and continuity, thus promoting respect for cultural diversity and human creativity.

2 The Making of Beauty: The Best Practice of Apulia Region

Since any cultural reading of the territory and its own beauty should start from the sources of law/the legal instruments available to ensure its preservation, we ought to consider the pivotal example of Apulia and the regional law on beauty, an ambitious project of interdisciplinary bottom-up co-drafting/programming, which ties together seemingly antithetic criteria like quality and well-being, focusing on the citizen, on his needs and dignity.

This law provides tools, methods and actions to experience the beauty of our territory and to appreciate the vast diversity of Apulian identities' mosaic.

From this innovative law, two themes come to light: that of continuity, that is the living evolution of that complex of goods, values, beliefs, knowledge and traditions that form the cultural heritage of a community, and that of self-identification, that is, of the process recognition by an individual or a community of their representative cultural references.

On the one hand, the continuity in the evolution of what constitutes the cultural heritage (mainly in the intangible cultural heritage, but also in the material one if you think about how, for example, the use made of squares, streets, sites or monuments can affect the "way of life" of individuals and their customs social) and, on the other hand, the research by individuals of their identity, through a process of cultural self-identification that involves both the knowledge of the testimonies of the past and the cultural participation in the present.

The regional law on beauty promoted by the *Regional Councilor for urban planning*, Alfonso Pisicchio, represents a relevant best practice. In particular, using an integrated interdisciplinary bottom-up approach promotes methods and tools, useful for managing complex processes, characterized by the participation of actors with different backgrounds and cultures, in which the final recipients often have an active role of co-design and co-creation. In order to pursue its objective, the *Regional Councilor for the Urban Planning* set up a technical-scientific table with the task of drawing up a draft law that indicates the best route to reach the final goal: "beauty" where culture, identity, become transversal elements to safeguard and encourage the sustainable development of the territory. It proposes a true "Law on Beauty" that preserves, protects and enhances the beauty of the Apulian territory, based on the need to put the local community and its rights at the centre of an enlarged cross-disciplinary concept of cultural heritage in a constantly evolving complex society. The draft introduces new skills for the design of services that are more effective, efficient and centred on the needs of citizens, through experimentation and training initiatives. Among the objectives:—orientate the processes of organizational change necessary for the implementation of the law—improve the processes of building policies, in order to make them more dynamic and connected to empirical evidence.

According to Pisicchio, to build a full citizenship is essential to raise awareness of all citizens to the Apulian cultural heritage, artistic and landscape.

The goal is to educate them to its protection, transmitting the value it has for the community and fully enhancing the dimension of the common good and the potential that it can generate for the democratic development of the Region.

For this reason, in accordance with the need to involve the citizens in the ongoing process of defining and managing the Apulian cultural heritage, the scientific Committee has defined necessary steps to put into practice in order to create a greater synergy of competencies among all the public institutional and private actors concerned. After the first phase of the needs analysis, the draft will be open to public consultations with the civil society through a participative co-design process.

This second phase represents a democratic exercise, as it directly encourage behaviors which have to be more respectable of the common heritage, and take appropriate measures to reach the aim of the law, stimulating concrete and factual policies. It legally reshapes the category of "community" "that is in fact indicated as "landscape community", in which what is "beautiful" is decided within participatory processes, in negotiation with the communities, in line with tradition and innovation. This implies new models of management in which even the change must contain elements of recognition, in which the individual always perceives himself as active actor and citizen. The main goal is to create a "heritage community" of people with ideals, principles and values derived from the experience and reinforce a social cohesion by fostering a sense of shared responsibility towards the different places of the territory in which Apulian people leave.

Therefore, the Committee in collaboration with all the local research bodies, such as schools and universities, intends to initiate the actions necessary to develop the reconstruction and enhancement of the different identities of the Apulian mosaic.

The law is aimed primarily at citizens, with meetings and debates, and everyone can make a contribution to enrich the text of the contents.

The diversity of the Apulian identities' mosaic cultures is a common heritage of humanity and must be preserved for the benefit of all. According to the Councilor Alfonso Pisicchio, beauty is also this: "decide together the public thing, discuss it in what the Greeks called *agorà* because every law must necessarily pass through listening to citizens' real needs, not only the theoretical ones".

The Apulian Regional law will be ratified only after an intensive period of dissemination in all the different mosaic territories with all the actors involved.

Cultural heritage is conceived as the "wellspring of creativity". The law on beauty expresses in the most successful form, the correspondence between law and citizens: it is informing citizen, that placed before the mirror of the law, can recognize much of their heritage, and discover richer than any imagination.

This example of a cultural reading of the territory focuses more precisely on the role played by heritage in territorial construction. It is clear that the law on heritage and beauty is useful for territorial legitimation. This process tags the territory, gives it a reality and builds meaning. Beyond this innovative approach, which is concrete and factual, heritage can be a lever for territorial development. Finally, through the heritage-based projects and actions implemented on and for the territories promoted by the law on beauty, it is possible to identify a number of typical configurations that define the preferred spatial organizations.

In addition to the past-present relationship, the law on beauty also allows the Apulian community to project itself into the future through the valorization projects that it underpins.

In this case, of cultural reading of the territory, heritage is not limited to the heritage but it is associated with the present dimension. Even if it can be a matter of past practices, of monuments having lost their original functions ("historical monuments"), heritage has a present reality, just because it often corresponds to a concrete reality.

Heritage would then have a real strength of social cohesion and spatial cohesion. Not only does it give an identity to the territory, but it also legitimizes it as a territory [2].

This image must ensure the cohesion of the members of the group over time and in space.

Giving new life to the "old stones", means also giving hope in the possibilities of a future. Participatory methodologies promoted by the law will play the card of interactivity in order to involve the population to take action and to improve the ability to transform the "natural landscape" in "cultural landscapes" linked to the balance of natural resources, to well-being, to psychic and symbolic health, in other words to the "beauty".

The parameter of the "cultural identity" of the populations requires, to be postulated, a broad discussion and negotiation, as well as a work of symbolic reconstruction and anthropological excavation and collective memories.

The ambition of the law, in fact, aims to safeguard the existing, to protect everything that improves not only urban contexts but also above all human dignity. According to Pisicchio life is not worth living if surrounded by degradation and "ugliness".

3 Conclusions

In the *Unesco Declaration of Mexico City on cultural policies* (1982), it is stated in this regard:

> [...] It is the culture that specifically makes us human, rational beings, endowed with critical judgment and moral commitment. It is through culture that we choose the values to which we appeal and make choices. It is through culture that man expresses himself, becomes aware of his humanity, recognizes his incompleteness, questions his achievements, researches tirelessly new meanings and creates works through which he transcends his limitations. [5]

It emerged, therefore, that the ultimate aim of protecting cultural heritage, intended for the benefit of all humanity, as well as its human dimension, corresponds to the protection of the formation of the cultural identity of individuals, briefly guaranteed by their access, participation and contribution to the cultural heritage itself.

The outstanding example of the “law on beauty” aims to create a heritage community of people, with ideals, principles and values derived from the experience gained through progress and past conflicts, which foster the development of a peaceful and stable society founded on respect for human rights and democracy.

Everyone has the “right to beauty”, to benefit from the cultural heritage and to contributes to its increase and enrichment. However, it is not only a benefit but also a social responsibility to respect the identities of Apulian territories, its heritage and consequently the European common heritage.

Acknowledgements The Author would like to thank the *Apulia Region* and the *Councilor for urban planning*, Prof. Alfonso Pisicchio, for the opportunity to be part of the scientific committee that is promoting and releasing the law on beauty.

References

1. Claval P (2011) Landscapes, identities and development. Zoran Roca Editore
2. Micoud A, Marchenay P, Bérard L, Rautenberg M (1999) Campagnes de tous nos désirs. - Patrimoines et nouveaux usages sociaux, Maison des sciences de l’homme Editore, Parigi
3. UNESCO (1972) Convention concerning the protection of the world cultural and natural heritage
4. UNESCO (1976) Recommendation on participation by the people at large in cultural life and their contribution to it
5. UNESCO (1982) Declaration of Mexico City on cultural policies
6. UNESCO (2001) Universal declaration on cultural diversity

Correction to: Ichnography, Orthography and Scenography as Forms of Tracing the Past: A Reconstitution of the Roman *Forum* of *Ebora Liberalitas Julia*

Mariana Martins de Carvalho

Correction to:
Chapter "Ichnography, Orthography and Scenography as Forms of Tracing the Past: A Reconstitution of the Roman *Forum* of *Ebora Liberalitas Julia*" in: G. Amoruso and R. Salerno (eds.), *Cultural Landscape in Practice*, Lecture Notes in Civil Engineering 26, https://doi.org/10.1007/978-3-030-11422-0_17

The original version of this chapter was inadvertently published with incorrect figures. The correct version of figures is given below.

The original version of this chapter can be found at https://doi.org/10.1007/978-3-030-11422-0_17

G. Amoruso and R. Salerno (eds.), *Cultural Landscape in Practice*,
Lecture Notes in Civil Engineering 26,
https://doi.org/10.1007/978-3-030-11422-0_23

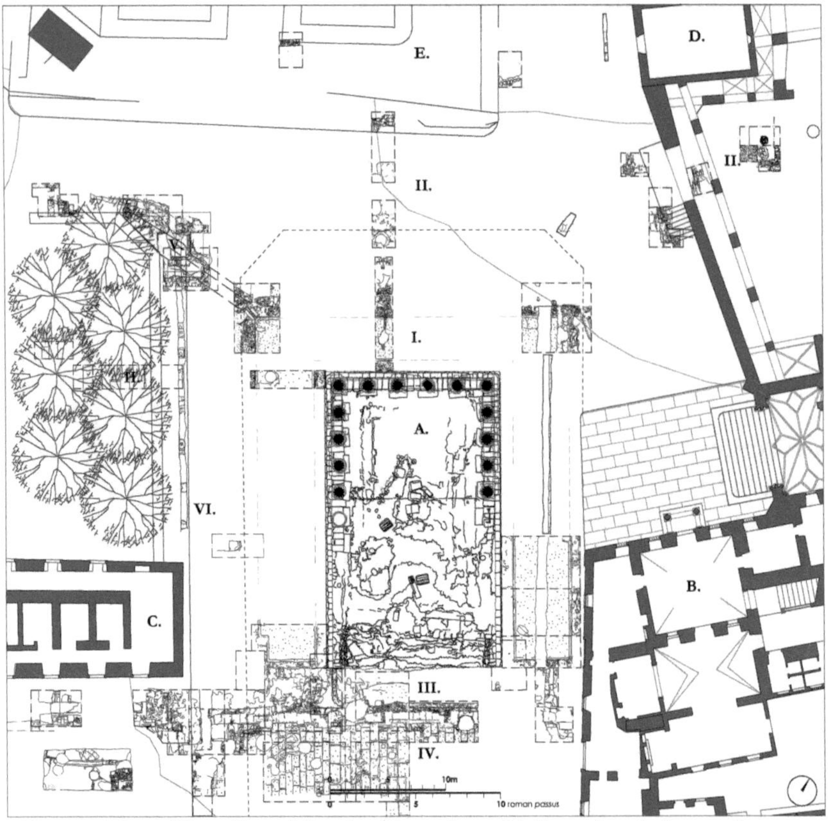

Monumental Centre of Evora, Portugal. Remains of the sacred area of the Roman forum in the twenty-first century.
Ichnography of the temple ruins and the surrounding excavated remains discovered until the present day.

A. Remains of the Roman temple; B. Loios Convent; C. Inquisition Courthouse; D. Cadaval Palace; E. Garden of Diana
Excaveted remains: I. Water tank; II. Cryptoporticus; III. Tribune; IV. Forum square; V. Cistern; VI. Inquisition cells;

Fig. 1 Ichnography of the temple ruins and the surrounding excavated remains

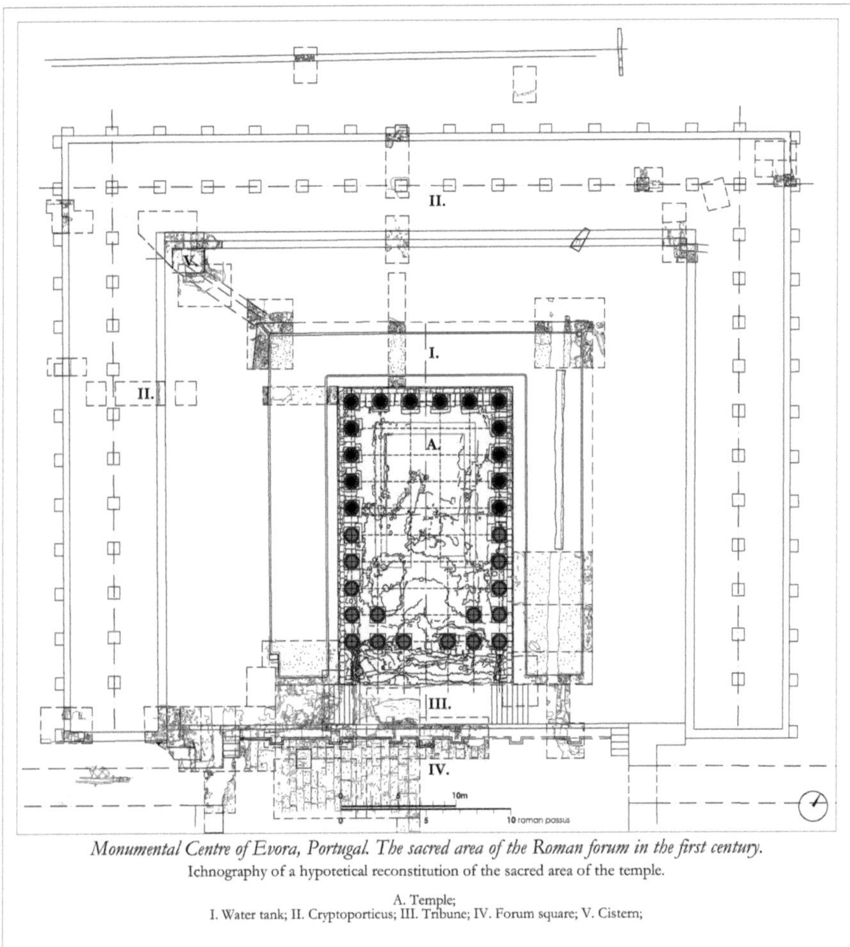

Fig. 2 Ichnography of a hypothetical reconstitution of the sacred area of the *forum*

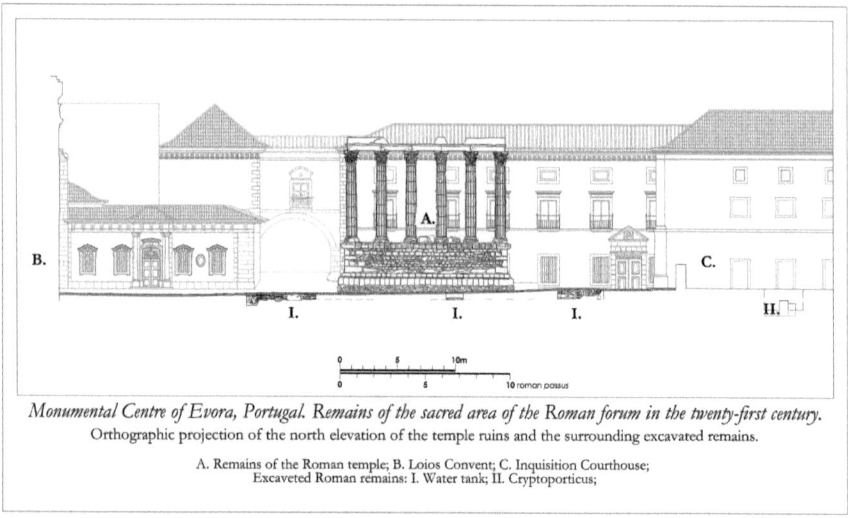

Fig. 3 Orthographic projection of the north elevation of the sacred area

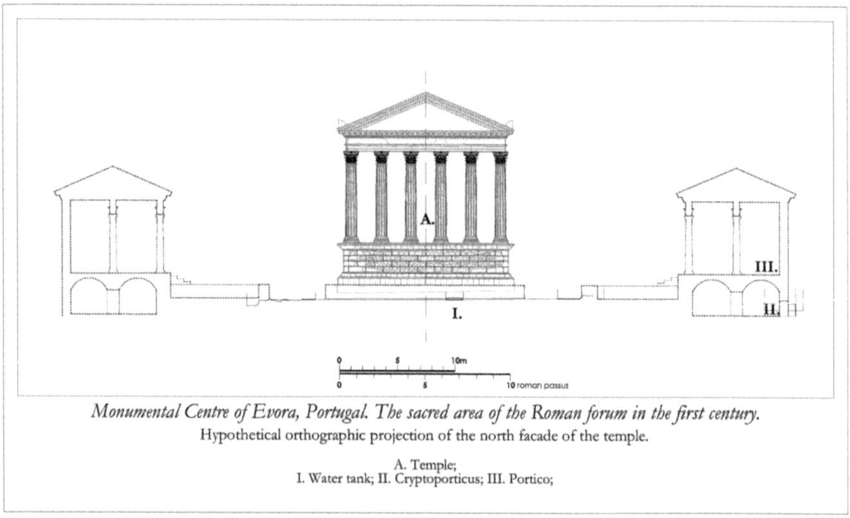

Fig. 4 Orthographic hypothetical projection of the north elevation of the sacred area

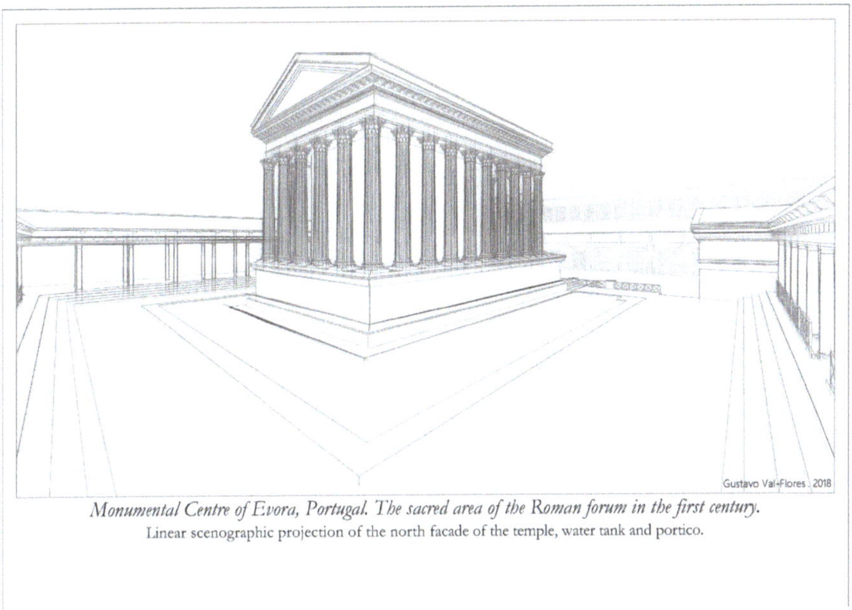

Fig. 5 Linear scenographic projection of the sacred area

Fig. 6 Scenographic projection of the sacred area

MIX
Papier aus verantwortungsvollen Quellen
Paper from responsible sources
FSC® C105338

If you have any concerns about our products,
you can contact us on
ProductSafety@springernature.com

In case Publisher is established outside the EU,
the EU authorized representative is:
Springer Nature Customer Service Center GmbH
Europaplatz 3, 69115 Heidelberg, Germany

Printed by Libri Plureos GmbH
in Hamburg, Germany